Writing and Unwriting (Media) Art History

Leonardo

Roger F. Malina, Executive Editor

Sean Cubitt, Editor-in-Chief

From Technological to Virtual Art, Frank Popper, 2007

META/DATA: A Digital Poetics, Mark Amerika, 2007

Signs of Life: Bio Art and Beyond, Eduardo Kac, 2007

The Hidden Sense: Synesthesia in Art and Science, Cretien van Campen, 2007

Closer: Performance, Technologies, Phenomenology, Susan Kozel, 2007

Video: The Reflexive Medium, Yvonne Spielmann, 2007

Software Studies: A Lexicon, Matthew Fuller, 2008

Tactical Biopolitics: Art, Activism, and Technoscience, edited by Beatriz da Costa and Kavita Philip, 2008

White Heat and Cold Logic: British Computer Art 1960–1980, edited by Paul Brown, Charlie Gere, Nicholas Lambert, and Catherine Mason, 2008

Rethinking Curating: Art after New Media, Beryl Graham and Sarah Cook, 2010

Green Light: Notes Toward an Art of Evolution, George Gessert, 2010

Enfoldment and Infinity: An Islamic Genealogy of New Media Art, Laura U. Marks, 2010

Synthetics: Aspects of Art & Technology in Australia, 1956–1975, Stephen Jones, 2011

Hybrid Cultures: Japanese Media Arts in Dialogue with the West, Yvonne Spielmann, 2012

Walking and Mapping: Artists as Cartographers, Karen O'Rourke, 2013

The Fourth Dimension and Non-Euclidean Geometry in Modern Art, revised edition, Linda Dalrymple Henderson, 2013

Illusions in Motion: Media Archaeology of the Moving Panorama and Related Spectacles, Erkki Huhtamo, 2013

Relive: Media Art Histories, edited by Sean Cubitt and Paul Thomas, 2013

Re-collection: Art, New Media, and Social Memory, Richard Rinehart and Jon Ippolito, 2014

Biopolitical Screens: Image, Power, and the Neoliberal Brain, Pasi Väliaho, 2014

The Practice of Light: A Genealogy of Visual Technologies from Prints to Pixels, Sean Cubitt, 2014

The Tone of Our Times: Sound, Sense, Economy, and Ecology, Frances Dyson, 2014

The Experience Machine: Stan VanDerBeek's Movie-Drome and Expanded Cinema, Gloria Sutton, 2014

Hanan al-Cinema: Affections for the Moving Image, Laura U. Marks, 2015

Writing and Unwriting (Media) Art History: Erkki Kurenniemi in 2048, edited by Joasia Krysa and Jussi Parikka, 2015

See http://mitpress.mit.edu for a complete list of titles in this series.

Writing and Unwriting (Media) Art History

Erkki Kurenniemi in 2048

edited by Joasia Krysa and Jussi Parikka

The MIT Press
Cambridge, Massachusetts
London, England

This book was set in Stone Serif and Stone Sans by Toppan Best-set Premedia Limited. Printed and bound in the United States of America.

Library of Congress Cataloging-in-Publication Data

Writing and unwriting (media) art history : Erkki Kurenniemi in 2048 / edited by Joasia Krysa & Jussi Parikka.
  pages cm.—(Leonardo)
Includes bibliographical references and index.
ISBN 978-0-262-02958-2 (hardcover : alk. paper)
1. Kurenniemi, Erkki, 1941—themes, motives. 2. New media art—Finland. I. Krysa, Joasia, editor. II. Parikka, Jussi, 1976– editor. III. Kurenniemi, Erkki, 1941– Works. Selections. English. IV. Title: Erkki Kurenniemi 2048.
N7255.F53K8739 2015
709.4897—dc23
2015009375

10 9 8 7 6 5 4 3 2 1

Contents

Leonardo/International Society for the Arts, Sciences, and Technology (ISAST)

Leonardo, the International Society for the Arts, Sciences, and Technology, and the affiliated French organization Association Leonardo have some very simple goals:

1. To advocate, document, and make known the work of artists, researchers, and scholars developing the new ways that the contemporary arts interact with science and technology and society.
2. To create a forum and meeting places where artists, scientists, and engineers can meet, exchange ideas, and, where appropriate, collaborate.
3. To contribute, through the interaction of the arts and sciences, to the creation of the new culture that will be needed to transition to a sustainable planetary society.

When the journal *Leonardo* was started some forty-five years ago, these creative disciplines existed in segregated institutional and social networks, a situation dramatized at that time by the "Two Cultures" debates initiated by C. P. Snow. Today we live in a different time of cross-disciplinary ferment, collaboration, and intellectual confrontation enabled by new hybrid organizations, new funding sponsors, and the shared tools of computers and the Internet. Above all, new generations of artist-researchers and researcher-artists are now at work individually and in collaborative teams bridging the art, science, and technology disciplines. For some of the hard problems in our society, we have no choice but to find new ways to couple the arts and sciences. Perhaps in our lifetime we will see the emergence of "new Leonardos," hybrid creative individuals or teams that will not only develop a meaningful art for our times but also drive new agendas in science and stimulate technological innovation that addresses today's human needs.

For more information on the activities of the Leonardo organizations and networks, please visit our websites at http://www.leonardo.info/ and http://www.olats.org.

Roger F. Malina
Executive Editor, Leonardo Publications

Acknowledgments

The book started as the result of long-term research and preparation for a major presentation of Erkki Kurenniemi work at Documenta 13 in Kassel, Germany, where his work was exposed to a far wider public than ever before. This meant the possibility to reflect upon the enormity of his project and its cultural impact. The preparation for this exhibition also soon compounded the ongoing problem of making sense of the volume and diversity of his prolific output—not least visible in the volume of boxes at the Central Art Archive at the Finnish National Gallery in Helsinki, where a substantial part of his work is deposited—and his use of unruly formats and subject matter. Of course, this was nothing new to many of the contributors to this book, who have shared a fascination with this material for many years and a bewilderment of how best to understand its lasting value. To many, Kurenniemi comes across as a scatty misfit, unable to fit into institutional categories of appreciation someone who actively defies the orthodox categorization of separate disciplines and genres in the spirit of interdisciplinary research. He is both inside and outside written histories.

The book represents our collective attempt to come to terms with Kurenniemi's work in a critical and historical manner, to produce a comprehensive publication for an international readership, and to understand his important contribution to the wider contexts of contemporary (media) art, music, science, and technology. Taken together we understand this as an attempt to write and unwrite (media) art history—in a truly media-archaeological style—excavating a largely unrecognized body of work that offers new insights. It brings together some of the best writing on Kurenniemi over the years with new essays by prominent scholars who have close connections to his work. Many of these relationships are personal as well as professional, and the individual contributions make this apparent for the most part in their chapters, reflecting the way that Kurenniemi's life and work intertwine.

We are grateful to those who made this project possible and apologize in advance for anyone we might have overlooked. Special thanks go to the following individuals: Perttu Rastas and Mika Taanila, who were instrumental in bringing this project into being, and who continue to care for Kurenniemi and his archive; Kai Lassfolk, Mikko

Ojanen, and Jari Suominen at the Department of Musicology at the University of Helsinki; Lars Bang Larsen for his short introduction to the Documenta 13 notebook on Kurenniemi that gave rise to the idea of a larger book project; curator Kati Kivinen and director Leevi Haapala at the Museum of Contemporary Art Kiasma in Helsinki; Pirkko Siitari, curator and former director of Kiasma; Marita Mellais at the Central Art Archive, Finnish National Gallery in Helsinki; Carolyn Christov-Bakargiev, Artistic Director of Documenta 13, who initiated the exhibition; translator Teo Välimäki; assistant copyeditor Mathias Ruthner; Esa Alanne at the Finnish Cultural Institute in Copenhagen for his support for the project at a crucial stage of its making; Doug Sery and Sean Cubitt at the MIT Press for their confidence in the project; and Susan Buckley, Judith M. Feldmann, and others at the MIT Press for their editorial support. We are also grateful to Leonard Rosenbaum for preparing the index. We would also like to thank the following institutions who supported the project: the Finnish Cultural Institute in Copenhagen, AVEK (the Promotion Centre for Audiovisual Culture), Kunsthal Aarhus in Denmark, and Winchester School of Art, UK. Finally, we thank the contributors to the book, and most of all Kati and Erkki himself. The book is dedicated to his visionary ideas, and to algorithms yet to be written that will preserve our memories in future archives.

Joasia Krysa and Jussi Parikka
Aarhus and London
Summer 2014

Foreword: Fragments as Monument

Erkki Huhtamo

Erkki Kurenniemi's career may seem like a textbook case of postmodern incoherence. Here is a partial list of his roles during the past forty-plus years: could-have-been nuclear physicist, electronic music composer, inventor, experimental filmmaker, roboticist, computer graphics and animation artist, curator of popular scientific exhibitions, obsessive recorder of his own life, visionary thinker. One might add "university dropout" and "perennial dissident," both at the heart of his highly uncommon career. Until illness robbed him of the ability to speak, Kurenniemi was an endless talker, a stream-of-consciousness verbalist (yet another epithet). The list of his interests is endless: the architecture of the computer, electronic music, industrial robots, biotechnology, human–machine interfaces of the future, computer games, the prospect of uploading the human consciousness onto a microchip, and life in outer space, to name just a few. Kurenniemi's visions may have been world-embracing, but he expressed them in carefully formulated sentences, shifting from the mundane to the extraordinary in a matter of seconds. In an era when utopias have turned into clichés of the popular media, he has a rare skill to surprise and, if not always to convince, at least to stir the mind.

Kurenniemi is one of those rare individuals who have an idiosyncratic way of perceiving reality. His vision is neither political nor theological. He is not a preacher trying to convince others about the Truth, or to mold others into his image. Neither is he a businessperson (the attempts he's made have proved catastrophic). Kurenniemi is a hybrid of a scientist, humanist, and bohemian artist, who fits uneasily among the banalities of the bourgeois society. He is a citizen of the pataphysical Republic of Immodest Ideas. Yet, wild as his predictions may be, they are not without foundation. Like those of Hans Moravec, they spring from thorough learning and a hyperactive mind. Kurenniemi never gets totally carried away by his visions—humor and skepticism serve as a counterbalance. He readily admits that although science is mighty, it is not almighty—there are things we do not know. Still, being a technological optimist, Kurenniemi would probably add: *not yet.*

It is not inappropriate to call Kurenniemi a techno-utopian; however, he should not be deemed uncritical or naive. Science and technology will inevitably change life on

Earth, but they are not quack medicines that will solve global problems with a sleight-of-hand. Kurenniemi does not offer short-term solutions; he looks further into the future. While doing so, he again and again returns to the imperfectness of the human being. Prosthetic technologies like head-mounted retinal displays (envisioned by Kurenniemi before prototypes existed) function as extensions of the body, yet they fail to solve the key question: mortality. In the end, the human is just a machine molded by evolution. It is the mind, consciousness, that matters, not the rotting meat frame we are forced to live and die in. Even the carnal pleasures of sex will eventually be replaced by something else.

Kurenniemi's quest for survival provides an explanation for the storage and documentation mania addressed in many of the chapters in this book. This growing "databody," it seems, is a necessary stage for reaching immortality. Turning a life into a multimedia database is the first step; uploading it onto a computer is the second. With the aid of an appropriate artificial intelligence algorithm (yet to be invented), Kurenniemi hopes he will be resurrected as a virtual clone after his meat body has turned into dust.

Kurenniemi's plan for transforming himself into a sentient multimedia database is not just a figment of an eccentric imagination. It is logically derived from his understanding of the evolution of digital computing. Ted Nelson, another brilliant college dropout and self-made visionary, is a kindred spirit. Like Nelson, Kurenniemi has tried his hand both in art and in practical applications, restlessly moving from one idea (and institution) to another. Both have created careers that consist of brilliant fragments, beginnings, and dead ends rather than of finished "masterworks" or lasting institutional structures. But while Nelson has always made sure that his main claims to fame—hypertext and *Xanadu*—envisioned in the 1960s, are firmly attached to his name, Kurenniemi's achievements fell into relative oblivion for years until their recent resurrection, as this book seeks to establish. As his mind kept wandering from one area to another, it kept drawing a jigsaw puzzle–like map, with pieces missing or lost, but its general outlines perceptible.

What can be said about Kurenniemi's forays into technological art? As surviving evidence shows, here as well he was unquestionably a pioneer, but lacked either the flair or the circumstances for keeping his achievements in the public eye. He remains a somewhat shadowy figure, known in "high definition" only by a select few. He has experimental films, computer graphics, and animations to his credit—firsts not only in his native Finland, but often on the international scale as well. Similarly, his electronic music of the 1960s was well informed by the latest developments in the field and contributed to them, but was often released in semiobscure ways (as soundtracks for promotional films, for example), when released at all. It has therefore joined the history of technological art and music not organically, but rather retroactively.

Among Kurenniemi's remarkable achievements are the interactive music instruments he designed and built in the early 1970s. They deserve attention, for it is becoming evident that Kurenniemi's creations, like the "video-organ" DIMI-O, anticipated interactive installation and performance art by years. In fact, they belong to the earliest achievements of computer-based interactive art. DIMI-O uses a video camera attached to a computerized organ as an input device. A dancer can thus create an interactive soundscape by the movements of her body. This system anticipated David Rokeby's famous, equally pioneering *Very Nervous System* (1986) by more than a decade, and was developed along parallel tracks with Myron Krueger's early works in interactive media art. Another one of Kurenniemi's devices, DIMI-S (Sexophone), allowed participants, chained together by handcuffs, to create music by touching each other's skin. In the wake of Nam June Paik's and Charlotte Moorman's *Opera Sextronique* (1967), Sexophone contributed to cybersex experiences in the guise of art.

Kurenniemi understood early on that the encounter between human and the ever-smarter media machine provided the key to the future. Learning to cope with machines, teaching them to be smarter, engaging in increasingly intimate interactions with them, perhaps becoming subsumed into them—these are some of the questions that occupied Kurenniemi throughout his entire career. Whether he found any lasting solutions to these issues is in the end less important than having raised them, and being part of and stimulating the process. Brilliant fragments of insight can be as important as accomplished institutionally implemented systems. New technology should not be left in the hands of hard-core technocrats and businesspersons. Artists and idiosyncratic thinkers and tinkerers like Kurenniemi have an important contribution to make. They point out that technological developments are never as prosaic, predictable, and one-dimensional as they may seem.

The fragments eccentric pioneers leave behind tend to solidify into monuments over time. That has been happening to Kurenniemi for some time now. His works have been publicly praised and shown at major art events as Documenta 13 in Kassel (2012). Still, such public monumentalization is always worth questioning and counteracting, and I believe Kurenniemi would engage in such activities himself if he could. Kurenniemi's life's work, with its formidable unpredictable celebration of the fragment, may provide the best argument against efforts to mold the complexity of his profile into that of an idolized technocultural hero.

An earlier version of this text was published as "Kurenniemi, or The Life and Times of a Techno-Visionary" as liner notes for the DVD by Mika Taanila et al., *The Dawn of Dimi* (Helsinki: Kiasma/Kinotar, 2003).

Introduction: Writing and Unwriting (Media) Art History

Jussi Parikka and Joasia Krysa

Erkki Kurenniemi's work is difficult to place. His work defies categorization, which makes any introduction to his work an unenviable task. Kurenniemi is emblematic of the eclecticism of much of post–Second World War media art history in that it is clearly "not just media art" but rather extends over a multidisciplinary terrain that ranges from the aesthetic to the scientific and technical. If one were to choose to apply general terms such as "media artist" or "experimental artist" to Kurenniemi it would preclude other parallel descriptions such as a student of nuclear science, composer, engineer of electronic sound machines, an odd sort of everyday archivist, a curator of science and technology, and a public figure writing about technology, evolution, and the future of the quantum computing universe. The mix is immodest, and much of Kurenniemi's enthusiasm is too.[1] Anyone claiming that by the year 2048 the humankind could be downloaded into a computer memory and recreated in artificial form sounds more like a 1980s science fiction author than a serious scientist. But this does not change the fact that Kurenniemi's media-artistic futurological claims can be taken seriously.

This book is an excavation, a critical mapping and an elaboration of Erkki Kurenniemi's multiple sides as resonating with the central themes of media art history and present conditions. This introduction gives some insight into his oeuvre, and this is the first international book-length assessment of this pioneering, wild, interdisciplinary spirit, which borders on the fantastical and yet stems from serious scholarly research. The two get often infused in each other and make up a unique speculative and insightful combination. That said, our task is not to paint a picture of Kurenniemi as any sort of a singular genius. Instead, we want to approach his writings and his projects as illuminating important points in terms of broader audiovisual innovation, artistic experimentation, and conceptual thinking both in the context of his time and crucially for its impact today. He is a symptom of the development of technological arts and the experimental spirit of the information age. As a symptomatology, this book is about media arts, the culture of sound engineering and emerging media arts, interfaces and archival fever in creative practice, all read through the figure of Kurenniemi. Furthermore, at a time when an increasing amount of practical and theoretical attention is placed on the issue of digital art conservation,[2] Kurenniemi's work strikes a chord: the archive is together a theme, practice, and a material context for much of his activity. This body of art practices and collections provides another way to address the archival in technical media culture.

The book is divided into five parts, which are introduced individually by different experts in the field. The parts address (1) Kurenniemi's archives and the wider context of creative practice in digital archival culture; (2) his standing in the technological arts and also the contemporary art field; (3) his scientific and technological visions; (4) his contributions to music and the sonic arts as perhaps his most important field, where he was recognized internationally and written about in the *Wire*[3] and other magazines; and (5) interviews and conversations with Kurenniemi over the years. We have also added a separate section of images that were selected primarily from his archive deposited at the Finnish National Gallery's Archives and Library in 2006. This serves to emphasize the important role the archive plays for Kurenniemi and for this project: it is not merely a place of storage for objects, images, ideas, and fragmented leads that are gathered in a physical place. The archive is more an active reference point through which Kurenniemi's artistic projects are situated as cultural historical practices. His work elaborates the archive as a place of potential new variations, such as in Kurenniemi Active Archives (http://kurenniemi.activearchives.org/), featured in this book in chapter 9. We feel a visual chapter in this uncompromising form operates in the spirit of Kurenniemi's practice and as a "mnemosyne" to suggest the potentiality of alternative epistemologies.[4]

The introduction offers the contextual glue to key ideas that will be investigated in more detail by scholars from multiple disciplines reflecting the multiple identifications of Kurenniemi's practice. These writers include media theorists, musicologists, software and sound artists, curators, historians, and media archaeologists. Through the chapters it becomes evident why Kurenniemi is now regarded as such an important figure: initially part of the Finnish scene of experimental music (even if less as a composer), as well as the fields of engineering and cutting-edge technology since the 1960s, he gradually became an internationally known figure who offers an perspective on the development of sound and visual arts engineered into future visions of human growth. That he has been neglected in many of the histories (that tend to overlook the non-English-speaking world and especially some geographically less central regions) leads us to redress the imbalance. In many ways the book also builds on the recent attention given to Kurenniemi in curatorial projects, notably his inclusion as part of Documenta 13 in Kassel in 2012. Entitled simply *In 2048* it was the first large international presentation of his work including his pioneering music instruments, such as his DIMI series of electronic synthesizers and electronic music compositions; robots; generative computer animations; short films; and his writings on music theory, computer programs, and his personal diaries recorded in all kinds of formats—audio tapes, video tapes, handwritten texts and those processed by computers.[5]

The year 2048 is an important symbolic reference point for Kurenniemi. By the year 2048, speculates Kurenniemi, quantum computers will have advanced to the stage that, with sufficient memory resources, it will be possible to upload the entirety of a human being onto a digital format, leaving behind our current organic-based slime-reality and thereby live as data-humans in the postsingularity universe.[6] It becomes a futurist-melancholic reference point in the same imaginary universe that other theorists and fiction writers from Ray Kurzweil to Vernor Vinge have suggested. Yet such transhumanist visions get an alternative twist in Kurenniemi's work, and turn from speculative cyberfiction to material reality in his instruments and interface design.[7] What from a discursive perspective is a highly problematic dismissal of the fleshy, gendered organism in favor of a utopia of the databody is resolved in the more interesting side of Kurenniemi's design projects. One apt characterization by Simon Reynolds has been to refer to Kurenniemi as a mix of Karlheinz Stockhausen, Buckminster Fuller, and Steve Jobs.[8] It is easy to see how an interest in Kurenniemi opens up oblique routes into many of the current discussions around the nonhuman, the new materialism, and philosophies of realism. Even in attempting to transcend the human, the contradiction remains that the human is affirmed in the figure of Kurenniemi himself.

But before we advance too much into futurology, let's take a step back and introduce Kurenniemi's work in a more biographical fashion.[9] He was born in 1941 in Hämeenlinna, Finland, to a cultural family: his mother Marjatta Kurenniemi was a famous writer of children's books. Kurenniemi's early interests in technological music, the recording studio, and new sorts of instruments as part of creative expression resonated with international trends that came rather quickly to Finland in the late 1950s. For example, Stockhausen visited the country and his texts were being read there as well.[10]

Electronic sound and its technological development attracted a lot of people in the music field—for instance, Henrik Otto Donner—but also more generally people interested in the new engineered visions of arts and society. Kurenniemi was active as a volunteering assistant at the Department of Musicology in the University of Helsinki in the 1960s, but he was interested in the studio as a technological environment for creative practice. Indeed, Kurenniemi was invited to plan the first electronic music studio for the department. During those years he also worked as assistant at the Department of Theoretical Physics between 1962 and 1973, a period that included him receiving a bachelor degree in physics (1968). However, despite these academic achievements, he characterized himself as a dropout (see the interview with Taanila in this collection). Kurenniemi became not a typical academic but a tinkerer who moved between jobs and disciplines, across the arts and sciences.

In addition to audio, he worked with experimental filmmakers, including a composition with Donner for Eino Ruutsalo's short film *Hyppy* ("The Jump," 1965). Kurenniemi was active in contacting people and studios in other Nordic countries such as Sweden and Norway. His own film *Electronics in the World of Tomorrow* (1964) has an avant-garde science-fictional flavor, but the later experimental film collaboration with Jan Bark stands out for other reasons. With composer Bark, Kurenniemi can be credited for what is likely the first Nordic computer animation, *Spindrift* (1966–1967), screened publicly only a couple of times before the more recent reconstruction by the documentarist and filmmaker Mika Taanila in 2013. *Spindrift* is a classic case of an experimental post–Second World War sensibility: Kurenniemi programmed the animations on the PACE TR-48 analog computer that he got access to through the University's Department of Nuclear Physics. Misuse of scientific equipment produced early art pieces, which in this case meant shooting the animations directly from the computer monitor onto 16 mm film.[11] This fits into the wider context of computer arts: for instance, John Whitney Sr.'s computer animation work that was viewed not only by specialist avant-garde eyes but was also present in the opening credits of *Vertigo* in 1958; and the emergence of computer arts, gaining wider visibility through such instances as the exhibition *Cybernetic Serendipity* at the ICA in London in 1968.

Kurenniemi was deeply interested in digital technologies, too. This was evident in his work in developing the electronic music studio as well as an interest in the early development of synthesizers. He was aware of the international predecessors—for instance, the RCA synthesizer from the 1950s—and became known as the go-to man on the Finnish music scene when one wanted a custom-made instrument. Ralph Lundsten, Osmo Lindeman, M. A. Numminen, and others turned to Kurenniemi for his technical expertise. Even if Kurenniemi's instrument legacy is mostly familiar through his DIMI synthesizers from the 1970s, already the Sähkökvartetti (Electric Quartet) was a testament to some of the later designs that incorporated technologically mediated collective electronic sound making.[12] It included the main unit and individual controllers: a "Melody machine, Electrical saxophone, Violin machine, Drum machine, filter bank controller and a light sword"[13] to distort vocal sounds.

Kurenniemi was also a public persona early on. Already in the late 1960s he was featured on television on several occasions, speaking about the new computer aesthetics and technological tools of sound and art. Such perspectives mixed with views of the coming digital/information society have characterized his interests throughout his career. In Finland and across a range of Western countries, the "future shock" (Alvin Toffler) was a much-discussed topic by futurologists, sociologists, and policy makers. Everything was about the future. Toffler's more popular-audience-oriented *Future Shock* appeared in 1970, preceded by the more policy-styled *The Year 2000—A Framework for Speculation on the Next Thirty-Three Years* as well as some Finnish publications of similar nature. But Kurenniemi's specialty was to tinker with ideas of the future through speculative but innovative gadgets and machines.[14]

He was seen as a specialist of computer music as well as a commentator on computer code and the future of the arts. For Kurenniemi, as for so many others since the 1960s to our current era of digital home music studios, computers made things cheaper and more accessible. The studio, which had only gradually become such a centerpiece of music production in the 1950s and especially the 1960s, was already in the midst of change, becoming a technological milieu in itself and gradually shifting from a physical place to the computer. The futuristic visions of what might change in the coming one hundred years had to do not only with socioeconomic considerations but also with how it would change the status of the musician and the composer. For Kurenniemi, the computerized composer of the future might be less of a music-specialist and more of an industrial designer.[15] This sort of a vision of the industrial engineer of music was connected to his interest in multisensorial experience. As the music notation system allowed music to be read visually, Kurenniemi's vision includes the future engineer of computer sounds to be infused with other sensorial experiences of visuals, touch, and even smell. Computer music was not only music to the ears, but for an expanded sensoria that represented a future multimedia-system, before such terms became part of the mainstream discourse in the 1990s.[16]

Kurenniemi's interest in multisensoriality was evident in his work on synthesizers. The different DIMI projects characterize Kurenniemi's activities in the 1970s. The earlier projects, DICO (1969) and DIMI-A (1970), included a digital memory that was programmable (see "Interaction of Music and Technology: Erkki Kurenniemi's Music and Musical Instruments," by Lassfolk, Ojanen, and Suominen, in this collection). Besides its internal features, such as larger associative memory in the DIMI-A, it also stands out because of its peculiar interface design. Chromatic pitch, volume, and, for instance, sequence jumps and tempo were controlled via a touchpad and two styluses even if, despite appearances, the instrument was meant primarily for studio use only.[17]

Many of the projects since 1970 were collected under the work of the company Digelius Electronics Finland, which Kurenniemi founded together with Jouko Kottila and Peter Frisk. However, the projects were also somewhat supported by public funds including the work toward DIMI-O (1971) by the Finnish Innovation Fund (Sitra). However, none of the machines became a commercial success, explaining why they are now left only as individual machines located across different institutions in Finland and Sweden. Current scholars at the Helsinki University musicology department's studio in collaboration with the Museum of Contemporary Art Kiasma have undertaken reconstruction of his instruments over the past years, resulting also in some of these being exhibited and performed again, most recently at Documenta 13 (Kassel 2012), Kunsthal Aarhus in Denmark (January–March 2013), and most recently at the Museum of Contemporary Art Kiasma in Helsinki (November 2013–March 2014). DIMI-O continued Kurenniemi's 1960s ideas of the future of computer arts as mixing different sensoria: in this case through the optical input mechanisms of the video screen that allowed visuals—including movement—to be integrated into the composition. There is something telling in that the machine was tried in diversely different contexts: from a dancer demonstration video to participation in experimental settings at the Department of Psychology at the University of Oslo, among other places.[18]

The collective sound production idea of the Electric Quartet was carried over to the DIMI-S—or the so-called sexophone. It was a device that involved a maximum of four people forming a human–machine cybernetic loop with the machine (through an electrode/controller held in one hand) and other people, embracing them in any part of the skin they dared—or were allowed—to touch. Currently the machine is part of Kiasma collections; it fits with other performative arts that explore possibilities in collective social experiments that link technology, art performance, and public social behavior. With a later work in 1973, the DIMI-T, a.k.a. the *Electroencephalophone*, Kurenniemi delved deeper than the skin, using the brain and its measured electric patterns as the input for the synthesizer machine. This was indirectly influenced by the biofeedback theories of Manford L. Eaton,[19] but just as much one could have thought Kurenniemi's ideas resonated with Alvin Lucier's earlier projects of brain music: the brain as input for arts and music (e.g., in *Music for Solo Performer*, 1965).

Besides his own company Digelius, Kurenniemi worked as an expert for various big technology companies. During the period 1976–1978 he worked at Rosenlew on industrial robots and automation, and later at Nokia's cable machinery division between the years of 1980 and 1986.[20] One has to emphasize that this was before Nokia became the famous global telecommunications and mobile business it is today. But it is clear that Kurenniemi always lived a double role. He had his day job at such companies, but he also had his other work: a publicly quoted intellectual, talking about automation, robots, and the changing landscape of labor and work (see, e.g., "Robots Go to Work" in this collection). For Kurenniemi, the laboring man is a general automaton, a Turing machine that is also programmable.[21] The machine was present both in the increasingly automating work life of factories and in the creative practices of electronic society.

It is also during his time at Nokia that he promoted the idea of a Personal Communicator. Of course, Alan Kay and the Xerox Palo Alto Lab (PARC) had envisioned many personal solutions that were similar to Kurenniemi's ideas. But Kurenniemi always puts his own twist on things. In 1971, he envisioned a personal device linking information and entertainment in one gadget: radio, audio, post office, bank, typewriter, museum, art exhibition, notebook, calendar, videophone, etc. (see "Message Is Massage" in this collection). The Personal Communicator was returned to again later and discussed in 1987.[22] Some years passed and such an idea was commercialized by Nokia and other companies. The early 1990s saw the commercial mobilization of the term, and Nokia's own Communicator appeared in 1996. But Kurenniemi was imagining something even more comprehensive: not just a mobile device to replace different home media, but "3D4π" lenses worn as eye glasses that project a 360° augmented reality view. As one important instance in the media archaeology of Google Glass, such visions of vision were part of his own history of invention as well as the cybervisionary world of 1980s. Virtual reality hype was growing since the 1980s cyberpunk genre became the lingua franca of the international technological culture. Such vocabulary and themes were boosted in the 1990s especially through the hype created by *Wired* and other related magazines in the United States, including *Mondo 2000*. Kurenniemi performed his role as a Finnish version of tech visionaries by arguing that future humans would be symbiotic beings linked via electronic devices in wearable computing, satellite-connected glasses, and new intimate forms of virtual reality that one then (in 1993) could get a glimpse of through computer games.[23] Perhaps a media archaeology of Oculus Rift could be identified here too.[24]

Working as a consultant and even as Head of Planning for the Finnish Science Centre Heureka between 1987 and 1998 allowed Kurenniemi to continue the visionary side of his interests. Despite the fact that his own company did not succeed commercially and that some of his ideas were considered unrealistic, Kurenniemi was awarded various significant national honorary prizes between 2003 and 2011, including the Order of the Lion of Finland medal from the Finnish President. During the 2000s he returned to working with instruments: together with Thomas Carlsson (UK) he started collaborative work on a new DIMI device. However, in 2005, Kurenniemi suffered a stroke, which had a severe effect on his health.

Kurenniemi's interest in technology and science was always contextualized in the arts. This meant a specific musical interest entangled with his mathematical mind-set: formal theories of sound and musical harmony that he also expressed in theoretical form (see "Chords, Scales, and Divisor Lattices" and "Tonal Theory" in this collection). His archives contain several papers on musical and mathematical theory as well as scientific texts, which since the 1960s all the way to 2000s express a developing interest in a range of topics including cellular automata, theories of evolution, quantum and string theory, artificial intelligence, computational structures and algorithmic thinking, and also, importantly, graph theory (see "Graph Field Theory" in this volume).[25]

Besides writing texts—which were never really published in any significant scientific journals internationally or in Finland—he was a collector. Kurenniemi's archives are full of news clips and copies of texts that revolve around the aforementioned fields. And he was a collector in other ways too: his archival fever[26] become a sort of a combination of his scientific interests with an experimental performative side: to collect and record as much as possible about the world around him. This drive included extensive diary-writing, photo diaries, video diaries as well as collections of paraphernalia: shop receipts, paper clippings, and material objects, including, oddly, even samples of pubic hair that can be found in his archive, demonstrating the carnal side of his interests (see Paasonen, "Fleshy Intensities," in this collection). Like many of his other activities there is also something prescient about the obsessive recording all aspects of his life that reminds us of the use of social media (see Røssaak, "Capturing Life: Biopolitics, Social Media, and Romantic Irony," in this volume).

The archival activity of Kurenniemi is particularly compelling and reflects his machinic sensibilities. The archival side of his art practice is featured as a recurring part of several chapters in this collection and in the collection of images reprinted from the archives. There are two main ways how he explained this drive for collecting that feed his personal archive: on the one hand, he saw such material as necessary "playback" material for the future databody humans who don't have firsthand knowledge of the world in flesh; on the other hand, such material was to be the source for a reconstruction of himself in 2048, when such reconstructions were made possible by quantum computers. This is in line with some of the singularity writers predicting the point at which artificial intelligence machines will become self-aware. Clearly there are further links to be made to the fantasies of artificial life and the dreams of reanimated life through cryogenics of Walt Disney, but perhaps this book can also be thought of as an act of reconstruction of some of these fragments. Besides afterlife, it is the archival that expresses a wider transversal theme in twentieth-century science-arts. Buckminster Fuller's Dymaxion Chronofile contained 200,000 documents and hundreds of card indexes and was a similar project to what Kurenniemi later envisioned: to collect in regular intervals snippets of life into a multimedia scrapbook of sorts, an activity which for Fuller continued for decades from 1920s to 1983 as a record of one particular human period recorded by means of media:

> If somebody kept a very accurate record of a human being, going through the era from the Gay '90s, from a very different kind of world through the turn of the century—as far into the twentieth century as you might live. I decided to make myself a good case history of such a human being and it meant that I could not be judge of what was valid to put in or not. I must put everything in, so I started a very rigorous record.[27]

Peter Weibel characterizes this project as "the product of a life of excessive self-observation, the product of an immense effort of immortalization in order not to be forgotten, but remembered,"[28] which echoes Kurenniemi's desire. Fuller's rationale for the project is clearly something that might have influenced Kurenniemi. Of course, similar themes emerge in more recent life-logging projects; Gordon Bell's Microsoft Research Labs-project MyLifeBits[29] especially offers something like a systematic version of Kurenniemi's—and Fuller's—aspirations backed up by a major corporation. The meticulous recording of traces becomes an index of the digital archive of life that in such Microsoft projects leads to the total annotation of a holistic archive; computational techniques of databases and search become attached to an analytics of life management, a version of the Quantified Self and self-monitoring. The archive for reconstructing the self is then actually a way of constructing the self in a continuous feedback loop.[30] Besides Vannevar Bush's Memex (the memory extender) plans from the mid-twentieth century or figures such as Ted Nelson and Douglas Engelbart, we need to add Kurenniemi as part of the media archaeology of such analytics of the self, of the Quantified Self.

Many of the themes addressed briefly in this introduction are elaborated in the chapters of the book. We want to underline, however, that the book is not merely about Kurenniemi but is an attempt to understand some important cross-sections of the art, sound, and computational fields arising at a particular point in time in Northern Europe outside of the Anglo-American perspective on media art history. We want to highlight the transnational connections of Kurenniemi's interests, the multidisciplinary articulation of arts through engineering, and the public role and rhetoric of future technological society that grows from current scientific innovations.

In terms of media studies, there is a temptation to think of Kurenniemi as the Finnish McLuhan, as both were engaged in provocative analysis of media change brought about by electronics. However, one should not stretch such comparisons too far: McLuhan stemmed from a literary background, Kurenniemi from an interest in music technologies; McLuhan's analysis was grounded in periodization of the media historical epochs of the past, whereas Kurenniemi was facing forward to the future to speculate about what would be waiting around the next corner. Kurenniemi's method could also be referred to as "situated speculations," which alludes to how artistic expressions are at the same time products of their own generations and contexts, even as they constantly strive to surpass them in imaginary ways. Kurenniemi's perspective on the past was less of a historian: he was interested in time-scales of evolutionary biology (see "Origins of Life, Intelligence, and Technology" in this collection). His reference points were those of not literary or media historians as much as those of scientists of evolution who showed more interest in the long-term narratives concerning human perceptual and cognitive capacities. Indeed, he did not consider himself an artist—a group he referred to as "cognitive amateurs."[31] One could hence think that his interest was in the cognitive and affective modifications of humans, their sensoria, and their society.

More to the point, his ideas and experiments were constantly evolving and future-facing—but also often left unfinished.

One way to focus on this artistic practice is to elaborate on the notion of the unfinished. It is a recurring feature in a lot of his activities from some of the engineering projects to the unending nature of the archival collecting. Kurenniemi was also engaged in a project to write a science fiction novel—entitled *2048*. An interesting personal letter of his from 1993 reflects on this side, and introduces the concept of *unwriting*. Perhaps this short passage reveals a lot about his relation to poetry and science, as well as a certain computational drive to exact formulation that is, as one knows in the Turing age, automatable:

> Writing a letter, even an unmailed one, could be part of the therapy. I have always had the urge to become a SF writer, have some files titled "2048," in Finnish only, sorry. I wouldn't be the first one to turn personal madness into poetry. But that's not my inner ambition, cf. above, and much too much has been written already. Have to find a way to unwrite things. Mathematics would be the ideal language for compact ideas. Think of Newton's formula F = ma. Four letters which govern almost everything we see. Well, not quite. For that you also need Maxwell's equations. I think Maxwell needed something like a full page to write his equations because the compact vector notation was not known in his time. In vector notation you need four short lines only. Then Einstein rewrote the Maxwell equations in four or five letters. Today, they need not be written at all because they follow automatically from the gauge condition. This I mean by unwriting poetry.[32]

Unwriting and undoing stand out as interesting ways to understand Kurenniemi's own back-and-forth movement between the arts and sciences. A constant back and forth of writing and unwriting, of sketches and fragments, of projects and ideas characterizes his attitude to the media arts. It produces a long tail of projects and leads, which are followed through, commented on, and elaborated in this book.

Notes

1. Here we allude to Huhtamo's short piece on the immodesty of Kurenniemi, "Fragments as Monument," the foreword to this collection.

2. See, e.g., Bernhard Serexhe, ed., *Digital Art Conservation: Preservation of Digital Art: Theory and Practice* (Cologne: ZKM, 2013).

3. Harri Uusitorppa, "Finnish Digital Music Pioneer Erkki Kurenniemi Almost Passed into Obscurity, Until His Rediscovery by This Year's Avanto Festival," *Wire* 225 (November 2002).

4. "'Mnemosyne" refers to Aby Warburg's "Mnemosyne Atlas of Images," produced in 1927, that consisted of forty wooden panels with a thousand pictures from books, magazines, newspapers, and other daily life information sources. These pictures were arranged according to different themes, somewhat like chapters.

5. Erkki Kurenniemi's *In 2048* was exhibited as part of Documenta 13, June 9–September 16, 2012, http://d13.documenta.de/#participants/participants/erkki-kurenniemi/. Joasia Krysa was a contributing curator, and further commissioned a related work by Constant referred to in this volume in the chapter "Archiving the Databody."

6. For discussion of this idea that pops up in fragments across Kurenniemi's writings and thoughts, see Mika Taanila's film about Kurenniemi, *The Future Is Not What It Used to Be* (2002).

7. Indeed, many of his other visions have been incorporated into the innovation labs of contemporary engineering corporations: Kurzweil works for Google as director of engineering; Kurenniemi was at Rosenlew and then, significantly, at Nokia.

8. Simon Reynolds, *Retromania: Pop Culture's Addiction to Its Own Past* (London: Faber & Faber, 2011), 389. See also Perttu Rastas, "Erkki Kurenniemi—Finnish Hybrid of Stockhausen, Buckminster Fuller, and Steve Jobs," panel talk at *RENEW: The Fifth International Conference on the Histories of Media Art, Science, and Technology*, Riga, October 8–11, 2013.

9. The Wikipedia entry on Kurenniemi is well curated and kept up to date by Finnish specialists of media art, including Perttu Rastas: http://en.wikipedia.org/wiki/Erkki_Kurenniemi (last accessed April 2, 2014). Much of the bibliographic details we provide here are in debt to their online entry.

10. Petri Kuljuntausta, "Elektronimusiikki ja zeitgeist," *Musiikin Suunta* 35 (3) (2013): 54–55.

11. Mika Taanila, "Spindrift—utopia puhtaasta televisiosta," *AVEK-lehti* (2014): 16–20.

12. "Ungdom för helvete/Sähkökvartetti esiintyy," Yle-television, September 2, 1969, http://yle.fi/elavaarkisto/ (last accessed April 3, 2014).

13. Jari Suominen, "Erkki Kurenniemi's Electronic Music Instruments of the 1960s and 1970s," In *Kurenniemi—Man from the Future* (Helsinki: Finnish National Gallery, 2013), 9, http://www.lahteilla.fi/kurenniemi/a-man-from-the-future.

14. Herman Kahn and Anthony J. Wiener, *The Year 2000—A Framework for Speculation on the Next Thirty-Three Years* (New York: Macmillan, 1967). Jyrki Siukonen, "Dead Computers Tell No Tales—Remarks on the Futures behind Kurenniemi's 2048 Resurrection," in *Kurenniemi—Man from the Future*, p.10.

15. "Kahdeksan tahtia tietokoneelle," Yle-television, October 27, 1967, http://yle.fi/elavaarkisto/.

16. "Ihmisen uudet mahdollisuudet/Tietokonekoodi on uusi nuottikirjoitus," Yle-television, March 3, 1969, http://yle.fi/elavaarkisto/.

17. Suominen, "Erkki Kurenniemi's Electronic Music Instruments of the 1960s and 1970s." See also Lassfolk, Ojanen, and Suominen, "Interaction of Music and Technology," in this collection. Mika Taanila's film *The Dawn of DIMI* (2003) focuses on Kurenniemi's Digital Music Instrument series' projects.

18. Suominen, "Erkki Kurenniemi's Electronic Music Instruments of the 1960s and 1970s," 22.

19. Ibid. For general information on Kurenniemi's instruments, see http://www.lahteilla.fi/kurenniemi/en (accessed April 3, 2014).

20. http://en.wikipedia.org/wiki/Erkki_Kurenniemi (accessed April 4, 2014).

21. Erkki Kurenniemi, "Ihminen ja automaatio," *Matemaattisten aineiden aikakauskirja* 43 (6) (1979): 347–352.

22. Erkki Kurenniemi, "Henkilökohtainen kommunikaattori eli Peeseellä kyberavaruudessa," *Kauppalehti*, Data Maailma, no. 108B (June 8, 1987), 14–16.

23. "Keinotodellisuus on tulevaisuuden tori," Yle-television, October 13, 1993, http://yle.fi/elavaarkisto/ (last accessed April 4, 2014).

24. Oculus Rift is a virtual reality headset company and the 2014 purchase by Facebook triggered a renewed interest and media debate about VR-devices in gaming and other applications. See, e.g., http://time.com/37842/facebook-oculus-rift/.

25. His archives, at the Finnish National Gallery's Archives and Library, contain several versions of the text "Graph Field Theory," which can be considered his main mathematical/scientific writing. It stems approximately from early 1990.

26. See Jacques Derrida, *Archive Fever: A Freudian Impression*, ed. Eric Prenowitz (Chicago: University of Chicago Press, 1996).

27. Buckminster Fuller, "Oregon Lecture no. 9," July 12, 1962. Stanford University Library, R. Buckminster Fuller Archive. Quoted in "Dymaxion Chronofile," Wikipedia, http://en.wikipedia.org/wiki/Dymaxion_Chronofile.

28. Peter Weibel, "The Digital Oblivion: Towards a Material History of Media," in *Digital Art Conservation*, 185.

29. http://research.microsoft.com/en-us/projects/mylifebits/ (accessed January 27, 2015).

30. See Gordon Bell and Jim Gemmell, "A Digital Life," *Scientific American*, March 2007, http://www.scientificamerican.com/article/a-digital-life/.

31. As quoted by the Kurenniemi expert and archivist Perttu Rastas in "Kurenniemen vempeleet ja visiot Kiasmassa," Yle-television, October 25, 2013, http://yle.fi/elavaarkisto/artikkelit/erkki_kurenniemi_digitaalisten_nakyjen_nakija_100032.html#media=99957 (last accessed November 12, 2014).

32. Erkki Kurenniemi, letter to Bronwyn, July 20, 1993 (computer file, Kiasma).

I Archival Life

Foreword

Perttu Rastas

Ha ha ha, fantastic … tis-m-. Mm organizational thinking is increasing and constantly becoming more significant. We're not talking of some kind of clean-up, no … but systems. The briefcase and its meticulous organizing so that everything inside has its own place. On the left a recorder, between it and the left side a slide rule, in the front a microphone and empty cigarette boxes scribbled full of ideas, ready to be set on the mantelpiece once one gets home. On the right some A4-sized material. The last ones on the bottom, the recent ones on the top. A book, some tissues and a card organizer. A film container, which turned out an excellent conveyor of small cards—some hundred cards, which my pockets are filled with anyway. The cards themselves organized: white for long-term memory, yellow for short-term memory, blue for projects, red for components … or something like that. At least the last one isn't yet quite clear.

—EK, diary, August 1975

The cultural significance of Erkki Kurenniemi is connected with his early experimentation in both music and building digital instruments as well as his extensive intellectual and cultural endeavors into digital culture. One of these early processes was his enthusiastic recording of everyday life, which can be said to have started in the beginning of the 1960s when he made his first 8 mm and 16 mm home movies; or at the latest in the beginning of the next decade when he continued to observe and record his surroundings using a portable cassette recorder, used to create about one hundred cassette diaries (between the years 1970 and 1975). At the time, he also started to furiously record his life in written diaries, and to collect different paper documents from his everyday life. From the 1980s onward he continued his passion through photography and video recordings, which he produced at an increasing rate until his illness in the early 2000s.

Even though this is not particularly unique in itself, Kurenniemi had a broader objective in mind from a very early stage: to provide an archive of his own history for future generations interested in observing human life in the aftermath of the World Wars and at the time of the formation of the postindustrial society where digital technology in its many forms gradually became the crucial factor in culture, industrial management, and engineering.

The archive project was interrupted in the end of 2005 when Kurenniemi suffered a serious stroke. Because of a string of unfortunate coincidences, he was brought into care too late, and his use of cigarettes, alcohol and several different prescription and recreational drugs over an extended period probably didn't aid the healing process. The stroke especially affected the speech and language centers of his brain, which has resulted in severe communication problems.

Because he was adamant about keeping all his collected archives and documents intact, the archive of the Finnish National Gallery negotiated a deposit agreement with Kurenniemi. It dictates that the extensive and multifaceted collection be archived and can be indexed, digitized, studied, and used as exhibit material. Among others, the archive has partnered with the Department of Musicology at the University of Helsinki, where Kurenniemi built a production studio for electronic music and his first electronic synthesizer in the beginning of the 1960s. The department is involved in several related research projects, and is able to provide a small but important collection of early electronic instruments for use by current and future academics and musicians. Some of the instruments are still partially functional and in use.

Why Was Kurenniemi a Lifelogger?

"Lifelogging" is currently an umbrella term for all everyday documentation connected with different media technologies—for which the Internet has provided a fertile breeding ground. Still, Kurenniemi has never written seriously or extensively about his archiving philosophy and its roots even though he diligently indexes his own recordings. Yet he must have known that many of the thinkers and artists he idolized—from Buckminster Fuller to Ray Kurzweil—were also ardent archivers of their own material.

In an interview with Mika Taanila in the early 2000s, Kurenniemi stated that people are interested in history because of a psychological need to reanimate the past to answer questions about themselves and their origins. According to Kurenniemi, historical reconstructions are significant on a broader scale than just in terms of museums, because the same techniques can be used to design the future and create predictive models for the coming era when Earth will become just one of many potential living environments for humanity.

Yet there is something rather obsessive about the manic pages of Kurenniemi's written diaries, his audio diaries from the beginning of the 1970s, and his later photo and video diaries. They are completely uninhibited, full of mathematical reasoning that operates almost as a separate language for him. The shifts between public and private content are rapid and merciless.

Strange how one becomes blind, almost sleepwalking through these turning points of one's life. This is connected to that and so on. At some point today I stumbled onto Koestler's name in passing. Maybe he's here today. During today's bus rides I noticed the stunning yellows of the leaves, this year's autumn foliage has been impossibly wonderful. So much is left un-erupted, hopefully in the fundaments of mathematics. Take this as your new image, believe that finding the essence will require exactly this, giving your all. It might have some logic to it, not an unbelievable notion in light of what's to come. Something rarely simple seems to be present. It might be connected with the understandability of mathematics. On the other hand, it might be just bollocks. Anne just went to bed, coughing. There's something irritating in the air, we've not vacuumed for months. (EK, diary, October 10, 1986)

Home Movies

From the beginning of the 1960s to the early years of the 1970s, the earliest "home movies" of Kurenniemi are another interesting example of lifelogging. Even though he was mainly documenting events from his environment and the pace of life around him, he was quick to comment on subjects such as technology and the advance of computing and the imagery surrounding it—which gathered little interest from other cultural activists at the beginning of the 1960s.

The similarities between Andy Warhol's and Kurenniemi's documentary practices have remained relatively unnoticed. Both made films in the 1960s, although Warhol was more serious about it, even attempting actual movie production. Kurenniemi's movies are perhaps more authentic home movies that represent the popular roots of filmmaking in the Godardian sense.[1]

Recently we've also learned that Warhol recorded the everyday life of his "culture factory" and collected related items in cardboard boxes. He referred to these boxes as "time capsules" and amassed a collection of 612 capsules during his career. The principle behind the contents of a capsule was simple: "What you should do is get a box for a month, and drop everything in it and at the end of the month lock it up. Then date it and send it over to Jersey."[2] Since Warhol's death, the contents of these boxes have reached a wider audience and are gradually being opened up and studied. Warhol was also passionate about keeping diaries and recordings from the middle of the 1950s onward, and called his tape recorder his "wife," explicitly stating: "when I refer to 'us,' I mean myself and my recorder."[3]

Anyone thoroughly familiar with Kurenniemi's later recordings from the 1980s to more recent years has come in contact with not only reports of everyday life, but also a vast array of his intimate encounters with several women, where the camera records everything that happens between the two people in real time. His partners were aware of the recordings, often even contributing actively to the plot. Kurenniemi himself refers to this boldly as "porn," a genre he considers to be a pivotal force in our lives.

Often lasting several hours of long, unedited recordings—including what happens before and after the sexual act—they are fairly boring to watch, which Kurenniemi himself acknowledged in one of his video diaries, asking: "Who would want to watch this unedited stuff"?

This part includes some of Kurenniemi's texts that deal with these concerns of life-logging—"Oh, Human Fart" (2004), "Relative Life" (2003), and unedited excerpts of his recorded diaries from 1971, "Audio Diary C4008-1"—as well as a chapter by Susanna Paasonen, "Fleshy Intensities" (2014), that deals with Kurenniemi's archiving practices and a connection to body.

Notes

1. On home movies, Godard says: "As for me, I've become aware, after 15 years of cinema that the real 'political' film that I'd like to end up with would be a film about me which would show to my wife and daughter what I am, in other words a home movie-home movies represent the popular base of the cinema." Quoted in Dave Kehr, *When Movies Mattered* (Chicago: University of Chicago Press, 2011), 151.

2. Andy Warhol, *Time Capsule 21* (Pittsburgh: The Andy Warhol Museum, 2003).

3. Jean Wainwright, *Art Monthly* 254 (March 2002).

1 "Oh, Human Fart"

Erkki Kurenniemi
September 5–8, 2004

Man is a machine. A machine produced by evolution. I find it impossible to think that for mere nostalgic reasons, such a slime-based system would be preserved ...
—Erkki Kurenniemi

Me and Computers

This article is a short premature self-obituary, relating some of my activities with things happening in the world of technology, decade by decade.

Fifties: Electronic Music and the Electronic Computer

I was five when the ENIAC electronic computer was started. During the fifties, as a schoolboy, I read about computers and electronic music. Max Mathews used the computer to generate music. With my father, I visited the Bull computer factory in France, and I was sold.

Sixties: Analog versus Digital

I started studying mathematics and theoretical physics and got my first job at the Institute of Nuclear Physics, programming an analog computer. Soon came the first digital computer, the Swedish-made Wegematic 1000, with vacuum tubes, a drum memory, and a thirst for kilowatts of power. At the same time, I had an unpaid job at the Institute of Musicology, developing their electronic music studio.

I read about the Buchla and Moog voltage-controlled synthesizers and started developing my own circuits, determined to learn the spiritual life of transistors and operational amplifiers. Voltage-controlled circuits are just specialized analog computing modules, and patch-cord patching is the natural way to program them, a technology borrowed from the old manual telephone exchanges and Hollerith punched card machines.

I began developing an integrated analog/digital music studio with combined voltage and digital control. Digital signals were used as triggers or gate signals, and also as square-wave sound. The final musical pieces were still edited the traditional way, by cutting and splicing analog full-track audio tape.

As integrated circuits became available in the mid-sixties, it was only natural to use them for memory and processing in sequencers and other musical instruments, but the main thing, a computer, was still unreachable to ordinary individuals.

I tried several "architectures" for pre-computer music instruments. At the 1968 conference on computer music in Florence, Italy, I read a paper entitled "The Music Terminal," presenting the idea of a graphic terminal, with local A/D–D/A converters for a sampled sound interface, but still connected to a central mainframe.

Seventies: Computer on a Chip

The first instrument in the DIMI series, the DIMI-A (Digital Music Instrument with Associative Memory) was constructed in 1970. It was a 2-channel non-processor digital sound synthesizer, with a serial MOS shift register memory of 100 16-bit words, built from about 200 TTL MSI circuits. Analog circuitry was only used for two 8-octave band-pass binary-selectable output filter banks, giving 127 timbral qualities for both channels.

The idea of an associative computer memory was attractive (alternatively known as a CAM, Content Addressable Memory). Most probably I got the idea from Teuvo Kohonen, the leading Finnish theorist in self-organization and neural computing. In the DIMI-A, all of its 16 bits circulated at high speed in their CCD or "bucket brigade" shift registers. Half of the 8 bits determined the time code for the instruction. Of the remaining bits, 4 determined the parameter, and 4 determined the value for that parameter. For example, a 4-tuple of 4-bit memory word nibbles (a, b, c, d) might mean: at bar a, at its beat b, set c (e.g. the pitch (mod 12)) to value d (e.g. #F).

This meant that to use the DIMI-A was to write machine code. That didn't turn out to be such a good idea.

The DIMI-A was followed by a series of Dimi instruments:

The DIMI-O was an electronic organ (with 1536 bits of MOS memory for 32 polyphonic steps) controlled by a video camera plus 1-bit video digitizer input and a 4-octave polyphonic keyboard.

The DIMI-S or "sexophone" was another user interface experiment. Four players, connected to the instrument by wires, touched each other and generated up to six-voice parallel sequences, by repeatedly tapping each other's bare skin.

The DIMI-T or "electroencephalophone" was not really digital. Two clips at earlobes registered one component of the brain's electrical activity. The signal was amplified about a millionfold and filtered to eliminate all but the delta-alpha range of EEG. With some more processing the signal was used to modify the pitch of a voltage-controlled oscillator.

The last and the most unsuccessful DIMI instrument was the DIMI-6000. Intel had introduced the first fantastic "computer on a chip" or "microcomputer," and we obtained the first unit in 1972. The microcomputer-controlled synthesizer consisted of two subsystems: (1) the processor cards CPU, ROM, RAM, RS232 (for the display terminal), FSK modem (for C-cassette data storage), 8-bit control voltage DAC; and (2) the audio cards: VCO cards, VCF cards, analog multipliers for modulators, FET crosspoint switching matrices, etc.

Digelius Electronics, the company founded to manufacture and market digital instruments, crashed, and I moved to industrial robotics. Jukka Ruohomäki, a Finnish pioneer of electronic music, wrote a sophisticated piece of software called DISMAL for the DIMI-6000. It was in effect a music assembly language. But then the world was not interested in fiddling with code. It wanted to twiddle knobs instead and pound keyboards.

Eighties: Personal Computers

I bought my first Apple II computer in 1980, with a stupendous 32 kilobytes of memory. The built-in BASIC language was good for writing "epic algorithmic art": simple sounds, graphics animation, and textual game-like interaction. Later came the Forth language, which allowed "lyrical algorithmics," compact programs whose structure was more beautiful than their output.

The Midi interface arrived and changed the way music was made. The Midi interface (a local area network) and the Midi file standard introduced a new way to encode music, not as sound or as written notation, but as a real-time record of performer actions and gestures. The capacity of home computers was not yet sufficient for sampled sound.

I also considered Midi a tragedy for music. It bound us to an equal-tempered pitch scale for more than two decades. The computer would have been the first truly scale-independent musical instrument, but that chance was lost by bad formulation of the Midi standard.

For the 10th annual exhibition of the Dimensio group of experimental artists, at Kunsthalle Helsinki (1982), I constructed Master Chaynjis, the swearing robot.

It was a human-looking head rolling around the exhibition hall on two wheels, limited by an umbilical cable connected to the ceiling carrying power and serial data. It had four collision sensors at each corner of the base frame, and a limitless vocabulary speech synthesizer onboard. It performed a fractal-tree-shaped dance on the floor and every time it collided with a wall or a visitor's foot, it swore. The worst thing it said was something like: "Oh, human fart."

The aggression worked better than I had anticipated. Two days before the end of the exhibition, a religious fanatic was told by his god to destroy the swearing robot. The body of M. Chaynjis is gone, but his immortal soul is safe on a 5¼ inch floppy disk, waiting for resurrection.

The virgin era of home computers ended when the Macintosh appeared in 1984. Still too slow for sampled sound without extra audio cards, it nevertheless quickly established itself as *the* art machine No. 1. Windows later copied the Mac GUI, but Apple originally plagiarized it from Xerox Palo Alto Research Center, the place of magical alchemists who invented things like the personal computer, the mouse, the graphical user interface (GUI), the Ethernet, and the Internet, and later ubiquitous computing.

Computer games emerged as something more forceful than any previously known form of art. Nineties: Internet, mobile phone, laptop. It is too early to try to assess what really happened during the nineties. It has produced a kind of "Garden of forking paths."

The era of John Cage took over and ended with the ultimate composition: a stretch of stationary white noise at 0 dB level, starting at minus infinity in time, and going to plus infinity in time. Interestingly, some results in mathematics hint at the possibility that this grand random stream is unique! (Try to Google "Bernoulli shifts.")

The laptop emerged as the most amazing musical instrument to date, with sufficient computing power to generate anything a human ear could differentiate.

Noughties: Present and Future

The mobile phone will become the ultimate musical instrument. Networked from Bluetooth to the Net, it will again allow spontaneous jam sessions, global and/or local, with or without acoustic local channels. Music will return to one of its main roots as a form of many-to-many social group communication, freed from the strict chain of command God → Composer → Performer → Audience.

A message to Nokia. The keys of the mobile phone must be larger, with longer travel and pressure-sensitive, for music playing. And they should be all around the device. The phone also needs the 6D positioning sensors, with 3 spatial and 3 angular coordinates. A name for this generic instrument? Phonephone, phonyphone, or simply phone?

But still more ultimately, the hardware of musical instruments will disappear.

Combining silicon acceleration transducers mere millimeters in size with Bluetooth-like wireless, possibly inserted into fingertip bones, will give all the manual dexterity ever needed and make mechanical things like mice obsolete. For visual perception, tiny video projectors, implanted inside eyeballs, will project images directly onto the retina.

Both technologies need a good power source to be developed, such as an implantable biobattery extracting electrical energy (or ATP) from the surrounding living cells.

Very soon, LED-sized and shaped (and priced) video chips with integrated optics will literally cover all the walls and ceilings around us. Farewell, Orwell! "Privacy" will be an obscene word (it already is, in some circles). Cells isolated from their surrounding social control may develop into terrorist or cancer cells.

I keep on working with tonal music theory, in terms of whole number divisibility and simple numerical ratios, in the footsteps of Leonhard Euler (1707–1783). My present problem is to identify the triad chords in the divisor set of the "Donald Duck" number 345600.

Having been mostly apolitical, except for slow drift from mild *Zen-Marxism* to still milder *liberal anarchism*, I now have a cause. The ideology of *sustainable development* is too slippery, because it does not specify absolute limits to change. I want to sharpen sustainable development into the *Museum Planet Earth* idea.

Briefly, this says that in step with the transhumanist fall into singularity, in less than 100 years, we should turn the planet Earth into a museum. This means an asymptotically stopped change in the resident human population, biodiversity, biosphere, environmental chemistry, climate, and so on.

But, everything will be allowed in space: economic expansion, population explosion because people will no longer age, genetic science and nanotechnologies of unimaginable power, warfare, and worse. A deal?

To make this Utopian future more acceptable, I shall briefly describe its econopolitics. In 2100, for example, print 10 billion "Earth licenses" and distribute them to all the then-living humans. No more licenses will ever be printed. Licenses can be sold. This way, the people who want long life and long-lived children can have them, but only by migrating into space. This will be cheap, because there will be people wanting to stay down here, purchasing Earth licenses at a price that will amply cover the price of the lift to orbit for the seller.[1]

Notes

First published in *Framework* 2, 2004, FrameFund, Helsinki.

1. Created by *Mathematica* (September 21, 2004).

2 Relative Life

Erkki Kurenniemi

The hard sciences are successful because they deal with the soft problems; the soft sciences are struggling because they deal with the hard problems.
—Heinz von Foerster, 1971

Life as a Relation between System and Environment

While considering the problem of defining life in the SETI context[1] I got stuck in the following attempt:

Definition D1 A subsystem L of a closed dynamical system M is *living* with respect to M if, and only if, from almost all initial conditions L grows to cover almost all of M.

I don't care about being too specific about the nature of a dynamical system or a subsystem at this stage. Instead of a traditional dynamical system, M could be a fiber bundle, a cellular automaton, or a graph with additional structure.

If the initial L is empty, we have the problem of ontogenesis. Then M should possess some fertility, excitability, or Langton lambda parameter. Instead of "closed system," "sterile substrate" might be a better term. At least for terrestrial life and computer simulations, thermodynamic disequilibrium is essential. I don't consider evolution under selection or genetic algorithms as basic. GA [genetic algorithms] may just be a "quick and dirty" optimization technique found by biology. Eukaryotic cells may indeed already use far more advanced methods.

Although one can immediately find absurd counterexamples, the relativity of the definition makes them more palatable. For example, a drop of ink in a glass of water should be considered living because it diffuses to cover all of the water volume. Why not, if the glass really were a closed system? In a thermodynamic system, D1 makes disorder (entropy) "alive" and temperature differences "dead." If the temperature is below the melting point, a crystalline structure would be living. In Conway's "Game of Life," on a finite field, almost all random configurations eventually die out: death is alive.

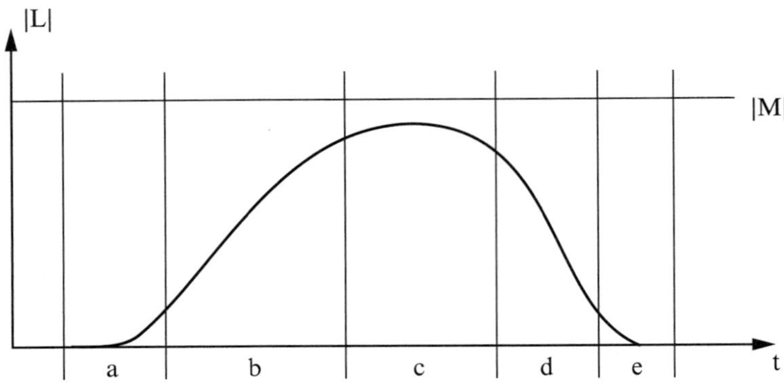

Figure 2.1
A nonoscillating life cycle.

Life, as a relation between two systems, seems to be transitive: if L_1 grows to cover all of L_2, and L_2 grows to cover all of L_3, then L_1 also will cover L_3. The definition allows us to compare competing forms of life: if a piece of L_1 and a piece of L_2 are grafted on the same substrate, several outcomes are possible, among them the cases that one definitively wins over the other.

Relativizing life brings forth a temporal perspective. In figure 2.1 the phases of life are demarcated, for a nonoscillating life cycle as: a—genesis, b—growth, c—saturation, d—decline, e—extinction.

The definition allows life to die out. Frank Tipler in his brilliant and shrewd *The Physics of Immortality*[2] elevates the immortality of life as a cosmological postulate and then proves the existence of God and resurrection of humans. According to D1, Tipler's God (the completion of the future c-boundary of spacetime) would be the only really living entity. What he dismisses is the possibility that humanity will be wiped out by a more advanced civilization, which decides not to resurrect us as virtual reality simulations.

The Boundary of Life

The process of growth postulated in D1 makes the advancing, receding, or stagnant boundary of the living system a focus of interest. Terrestrial life seems to have reached the saturation phase (c) long ago. Despite adaptation to quite "hostile" habitats from deep sea vents to Antarctic rock cracks, there are significant volumes devoid of life. Is there a common feature demarcating the boundary of life?

My impression is that there is not. A combined effect of convection in fluids (atmosphere, oceans, desert sand, snow, glaciers), lack of energy sources, and lack of key nutrients may be the key. Temperature limits for biological life are quite strict. Space technology and future self-replicating nanotechnology will push the boundaries of terrestrial life. It remains to be seen whether firewall methods will be effective against computer viruses.

Evolution and Complexity: Major Transitions

John Maynard Smith leaves open the question between two major alternatives:[3] 1. In the absence of physical changes … evolution would gradually slow down and stop. 2. Even in the absence of physical change, coevolution would continue indefinitely.

Thus, it is not known whether evolution is driven or autonomous. One possibility is given by Bak's hypothesis of self-organized criticality: the tendency of spontaneous motion toward the edge of chaos.

"There is no theoretical reason to expect evolutionary lineages to increase in complexity with time, and no empirical evidence that they do so." Nevertheless, during evolution complexity has increased. Maynard Smith and Szathmáry attribute this increase to what they call "major evolutionary transitions."[4] They list eight such transitions:

1. Replicating molecules to cells
2. Unlinked replicators to chromosomes
3. RNA to DNA + protein
4. Prokaryotes to eukaryotes
5. Asexual clones to sexual populations
6. Unicellular to multicellular
7. Individuals to colonies
8. Primate societies to human societies

The authors note three features common to several of these transitions:

1. Entities that were capable of independent replication before the transition can only replicate as parts of a larger unit after it
2. Division of labor (specialization)
3. Changes in language, information storage, and transmission

It is interesting to note that these properties are rather similar to the general qualities of object-oriented programming languages: encapsulation, association, abstraction, polymorphism.

Abstraction: The Semantics of DNA

In an A-Life computer simulation, it is generally preferable to make use of as high level of abstraction as possible, just to save computer time. In a functional or object-oriented programming language, one goes on building layers of abstraction on top of previous layers.

In biological life shaped by evolution, one would expect efficiency to pay off. If abstraction plays a part, where is it in a living cell? In the hypothetical RNA world there is little hierarchy (fig. 2.2a). In prokaryotes, the molecular transcription/translation chain can be seen as a two-level abstraction hierarchy with genes as "abstractions" of proteins. Considerable economy is attained by separating DNA-to-RNA transcription from RNA-to-protein translation.

In eukaryotes, the nuclear plasma membrane allows for one additional level of abstraction. RNA posttranscriptional splicing means that genes can be interpreted in several context-dependent ways. Because we now have a well-developed mechanism for DNA processing and expression control, the fundamental information storage and replication mechanism gets conserved and new levels of abstraction arise as more flexible layers of DNA code regulation and interpretation. The upward arrow in figure 2.2c represents the incredibly complex regulatory network.

Three other mechanisms that make the DNA language more "high-level" are: methylation, tertiary (spatial) structure, and transcription factor binding. It may be that these mechanisms have already saturated the geometrical degrees of freedom available in a three-dimensional space. Further speculation may be idle before more is known about the noncoding DNA, which recently has been shown to possess a footprint of "language." Were it true that the geometry has been saturated and consequently conserved, more abstraction levels would be needed both in the upward and downward pathways.

In other words, the layers in figure 2.2 do not represent hierarchical abstraction levels. The highest-level abstraction may actually occur somewhere in the middle layers. The problem now arises: how do the higher levels communicate between cells or the distant parts of the organism? Compare the situation with the orderly ISO-OSI model of computer communications (fig. 2.3). Although all communication ultimately takes place at the lowest physical level, the control hierarchy forms virtual links between all levels of abstraction. Such virtual communication may be also essential for efficient A-Life.

Sweeping aside the still intractable problems of morphogenesis associated with embryonic development, I am astonished by just the maintenance of bodily symmetry in an adult. I have a harmless wart in my right thumb. Then I found a similar one in the left thumb, though not identically located. Recently, I found a third one on my right big toe. I keep eagerly fingering my left big toe …

I cannot believe that such a long-distance communication network could be maintained with simple molecular concentrations. From data communications theory one would suspect a highly structured, compressed, and even encrypted distributed system, packet switched, circuit switched, or both. As the central nervous system is associated

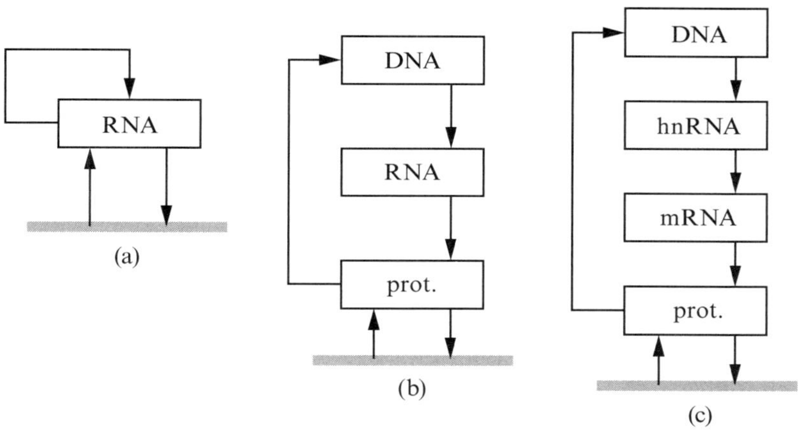

Figure 2.2
Molecular hierarchy in an RNA world: prokaryotes and eukaryotes.

and possibly saturated with movement control and sensory tasks, the morphogenetic system (of organisms without CNS) could naturally use, as fast signaling routes, the fluid circulatory system. To test this hypothesis, one should perform relatively fast correlation analyses for concentrations of conventional blood-signaling molecules and associated regulatory proteins from symmetrically taken sample pairs.

The hypothesized encryption would not be directed there against parasites. Its function would simply be to make the global signaling patterns invisible (seemingly randomized) and thus biologically neutral during the transmission.

Topological Cellular Automata

Si un hombre nunca se contradice, será porque nunca dice nada.
—Miguel de Unamuno (quoted in Schrödinger, *What Is Life?*)

My current utopian dream is that one might be able to define mathematical life in an as general a setting as the following. As the initial state take a random graph. Consider it to be a cellular automaton with the edges as the computational cells and the state of a cell to be the valencies of its two end-vertices. Consider an edge to be alive if its two end vertices both have the degree 3. Vertices of degree (number of incident edges) 2 and those of degree greater than 3 would correspond to the two phases on the boundary of which the living phase exists. The actions available to a cell are local transformations of graph topology.

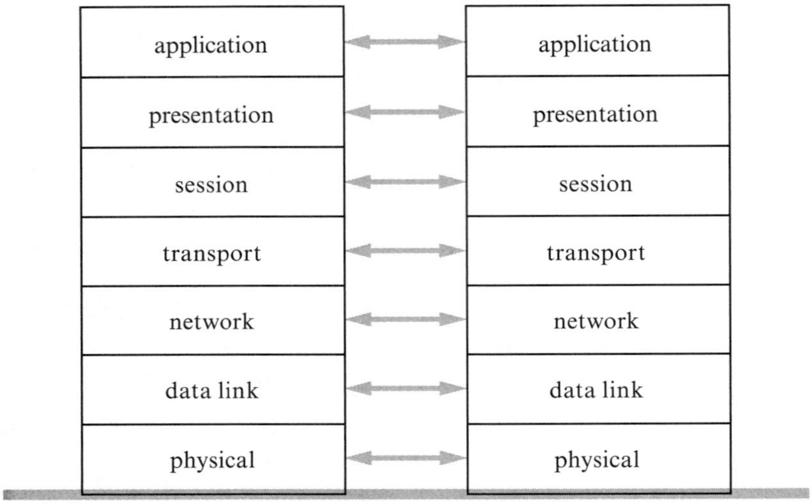

Figure 2.3
The ISO-OSI model of computer communication.

I would like to exclude the two simple actions of splitting a vertex and contracting an edge because they do not conserve the number of edges. This leaves us with two kinds of actions: (1) Split a vertex without creating a new edge between the daughter vertices. (2) Reconnect an incident edge to the other end vertex of a given edge. Actually, any combination of moves 1 and 2 could be allowed. Moves of type 1 are essentially irreversible.

Despite some effort, I haven't been able to write a consistent cellular automaton rule that also tends to a trivalent (vacuum) state. This failure prevents me from speculating that a vacuum is the ultimate living substance.

Notes

Unpublished text, 2003, Heureka, Vantaa, Finland.

1. E. Kurenniemi, "Elämän, älyn ja teknologian alkuperä," in *SETI: Vieraan älyn etsintä*, ed. Oja Heikki (Helsinki: Ursa, 1994), 151–162.

2. Frank J. Tipler, *The Physics of Immortality* (New York: Doubleday, 1994).

3. John Maynard Smith, *Evolutionary Genetics* (Oxford: Oxford University Press, 1989), 302.

4. John Maynard Smith and Eörs Szathmáry, *The Major Transitions in Evolution* (Oxford: Freeman, 1995).

Additional Literature

Kurenniemi, E. "Origins of Life, Intelligence, and Technology." In *SETI—Search for Extraterrestrial Intelligence*, ed. Seppo Linnaluoto and Jouko Seppänen, 157–163. Espoo: Finnish Artificial Intelligence Society, 1993.

Szathmáry Eörs, and John Maynard Smith. "The Major Evolutionary Transitions." *Nature* 374 (1995): 227–232.

Von Foerster, Heinz. *Observing Systems*. Seaside, CA: Intersystems Publications, 1982.

3 Audio Diary C4008-1

Erkki Kurenniemi
December 3–6, 1971

Translation by Reija Meriläinen, 2012

(00:00:00) (*Click click, radio signal, blows in the microphone five times, click, blow*) One, two, three, puppadadud. Fuck, fuck, fuck, this is sensitive. There we go. (*Blow*) Yeah, a dreaming computer … will be the last human invention. Well not the last one, but … the last invention. Because a dreaming computer will already have dreamt up everything. Prior unconscious. Well, no. Dead computers may only be in two spaces: in an idle loop waiting to be interrupted or in a conscious space receiving and handling external information, printing it. A sleeping computer is not in an idle loop. Yeah, well, of course it is, it does ask questions and wakes up when needed, but otherwise it dreams. It is organizing its files, optimizing, associating, organizing, thinking, planning. And only when called upon, it interrupts its sleep for a little while to answer a question. (*The sound of the microphone being touched, cut*)

(00:02:22) (*The voice is closer*) A dreaming computer, a dreaming computer is … A dream computer is the first, a dreaming computer is the last human—a dreamt-up computer is the last one made by man, a dreaming computer is the last invention that makes a human. (*Cut*)

(00:02:46) (*Voice comes from further away, very silent*) And finally the Earth gained consciousness. (*Whispering*) Another month went by, the Earth went to sleep. It makes that noise, until the space troops came and stirred it from its slumber. (*Cut*)

(00:03:16) (*In English*) Compare the extreme cases that completely parallel processor that makes one address. The next one, two address, four address … And so on. (*Cut*)

(00:03:35) (*A hard blow, the voice is very near; whispering*) The rumble of the wind, the hiss of the stove. Water dripping in the sink and people fucking at the other end of the apartment. (*Pause, the sound of adjusting the microphone, two cuts, distant female moaning*) (*Cut*)

(00:04:41) (*In the car*) Yeah, yeah, hell, yep. Time and space, time and space. (*Cough*) Yep, time and space. December 3rd, '71 … And … the time, hold on … 12:10 p.m. Just woke up, driving along the shore. Esso coming up. (*Yawning*) Aaah, it's about three or four degrees Celsius. Still there's some white frost on the ground. It's damp, gray, and cloudy. The sea has yet to freeze over. There are quite, quite, quite a lot of buoys. Buoys, buoys, buoys. (*Singing*) Well, goddammit. (*Speaking*) What, yeah, yeah. I went to bed at two, had quite a productive cloud. Had, had some thick, complicated, and painful dreams. I think the Hong Kong flu was waging war inside of me last night but it hasn't been able to push through yet. I'm still in good health. I hear that the influenza epidemic is at its peak in Helsinki. Goddammit, Kaivo-, Kaivopuisto, Kaivopuisto Park, yeah. It is so gorgeous, there is this, this, this (*singing*) infinitesimal trace of snow everywhere that only just … plom, plom, plom, plommm. What is it, what is it? (*Snarling*) Ugh. (*Continues singing*) This morning is, shall it be—like so many mornings are—dedicated to hate. Every morning I hate someone. (*Speaking*) And these days, nowadays, it is so nice to have a well-deserving target. That fucking bitch. (*Singing*) Fucking bitch. (*In a low nasal voice*) Haa. Ha ha ha haa … Haa, ha ha ha haa. Haa, ha ha ha haa. Haa. Oh, why can't such a morning last for all eternity? (*Yawning*) I'm shivering (*singing*) in the car, is there anything sweeter? (*Speaking*) Yeah, so I have thought up this plan that henceforth every now and then—unless maybe if it's fucking freezing out or something—I will have my tape recorder with me in the car on my way to work, all the while recording as I talk bullshit. One makes all kinds of plans. This is already an old plan. And (*yawning*) today, this right here is the first realization. After a year, a year from now, this nervous, fumbling, vapid chitchat will have changed quite a bit. There will be systems—I don't know what kind yet. This thing (*clears throat*) will have branched out, and it will of course have this, these moments and, and then it will also have these new—I especially expect that it will also have these consequential moments. This is a splendid setup, I am noticing how fun it feels to have people looking at me there, pedestrians … we have to, like, wait here and stand there, so of course you can see it from the outside, a weirdo ranting into a microphone. But of course, everything can be explained. (*Chuckle;* (*singing*) Oh, my heavens. (*Speaking*) Well then, Senate Square, Senatstorget on the right, the University on the left, post office, police station, er, government, government palace. And this Snellmaninkatu here, this sure is a fine street. And the House of Nobility, they are taking down the coverings from when it was all wrapped up not so long ago. I don't think I'll take this thingy with me in the café. Now wouldn't that be a farce, if I started just babbling to this recorder wherever I went. (*Singing*) Oh yeah, oh yeah, oh yeah, yeah, yeah. In any case, I am noticing that this is a good output channel, this is a damn good output channel for anyone, although it might turn out, turn out to be unrealistic to think that this kind of deliberate output … Oh, no but yeah, yea, yea, yea, yeah, that's it, that is, that is, that is, oh dear Marshall. (*Singing*) Everything is but a mosaic, everything is but a mosaic, this is all getting to be such a fucking mosaic, gone to mosaics, gone all mosa, ow, and that mosa, moza, mo-o-o-o-o-o-oza. Well, time to sign off for now, I'm trying to find a parking place to get … oh and there, there's one, just freed up. Bye. (*Cut, blow, "pah," cut*)

(00:13:35) (*In the car*) And Saturday. Fourth of December, I think. The time is 11:30—no, 11:40 a.m. And I am starting at my, um, house. The streets are wet, the sky is gray, future gray, ambitions wet. Primary ambition: coffee, juice, and lingonberry pie. Then again, this kind of a Saturday makes you feel the whole ambivalence, wanting to go downtown to idle about and browse in bookshops and check out girls in fancier bars. But then again, duties call to, er, test the bladder and everything. And then, like, just to go to that same old coffee shop, Liisa's Café, musty old place. (*Yawning*) I'm by the shore. Ohh. How beautiful the Finnish nature is, all around me. Last night, (*singing*) last night, what did I get up to? I took a nice bath and then-nn-nnn-nnn-n thought about all kinds of stuff. Yeah, well, I started forming this kind of a theory, intelligent algebra. See we have this datatandandin-information, but it is kind of like this type of raw material. Intelligence is meaningful, and it is always carried by some data, maybe. But it can only be meaningful in relation to some other intelligence or data. So this brings us, on the whole, to this issue of interaction between one's own and others' information. And I would maybe assume that intelligence is just a form of this type of interaaaaaaaaaaaaction. (*Speaking*) And oh, see, they've finished that building that is apparently in front of the Mannerheim museum, or it's almost done, there are still some rental coverings hanging off the balcony. (*Yawning*) Oyoyoy, a sweet, sweet pile of shit there. Ilmatar, Silja Line, a ship with white sides. A bit of road E3 there. Police car floating there on the shore. Why, there's ice there, more formations that have trickled down from the rocks. Hakkapeliitta beer commercial, every other building is a bank, every other building is not a bank. Hakkapeliitta, Hakkapeliitta, Hakkape, Hakkapeliiliiliitta, Hakkapeliitta. Am I overdriving this shitty tape? No, I'm actually rather underdriving it. And piece, piece … Yeah (*clears throat*), the thing is, though, whether this kind of an audio tape can kind of, like, convey that moment's feel. And maybe I'll then concentrate and, um, fall into some kind of a trance. (*Yawn*) Into slightly embarrassing consciousness. Damn, I'd so like to just go downtown but it's all jammed down there. Should I go see Kärlekens språk, andra del today? Here I am, courteously letting pedestrians cross the street to the market hall … There, zap zap zap, zap zap zap. (*Singing*) Light blue blazing in my eyes; is it the City Hall shining in front of me? Perhaps. And here, waiting in the Market Square traffic lights. These lights have become a sort of temple for me over time. Many a morning, I have sat here in this car and at this obligatory moment of pause, I can't help wondering about eternity. The ultimate questions that don't give me peace until sometime in the afternoon when I get, like, busy with work. And eventually, that week is then ruined. We'll see whether there's a problem with me driving straight to Katajanokka from here, it's kind of my base camp. For fuck's sake, where am I going right now, should

I go to (*singing*) Liisa's Café, Liisa's Café? Anyway, we're off to the Great Square, Great Square, umm, what's it called, Senate Square, Senatstorget. What is the Great Square? What is the senate? Hmm. Behind me the university library, Snellmaninkatu. We're on the wrong lane. (*Cough*) Don't you, bitch, gawk at me. Hallituskatu crossing, traffic coming in from the right. Get on with it, boys! Italian, French wagons. There's also an English one. (*Singing*) Da te dadaeda dadadadadada daeda dada dae. (*Very low*) Taa dae dae. Oh, quelle blondengue là sur la street. O-oo, A-aoo. (*Sniffs*) Sorry, boysenberry. Hohoho haa, haa, haa. I ammm, umm, guess where, on Vironkatu, turning onto Mariankatu. Here I am driving past my old apartment building. Oh my, what cute da-, grandpa and little girl. Both so small. Restaurant, restaurant Savanna Savanna. Zaza zaza vanna. Oh fuck, fuck, I need to take a trip this weekend and see into these people's thoughts. Yeah, well now we're turning onto Liisankatu, hope I can fit there. After that, dear diary, farewell, farewell, till next time. It's fucking crowded. Well, I'll just have to squeeze in. Bye-bye. (*Cut*)

(00:24:30) (*In the car*) Afternoon of the same Saturday at about 5:15, driving home from Digelius at Katajanokka. It's dark, raining. Bought food by the railway station. Before that, I spent an hour in the Suomalai-, er, Akateeminen bookstore, but I didn't buy anything. Had a LISP manual in my hand for a while and there (*yawning*), in the bookshelf, by the bookshelf, the Hong Kong flu got me—zong!—in the throat. Now it's a battle of life and death. I'm gonna go home, get stoned. I'll go see if Paavo is home. I'll read books, do whatever, and then there's bread and butter and tea and all kinds of sweets and stuff. So that, so that everything is kind of prepared. Someone just said, was it maybe Uffe, that he has a friend who works at an, umm, bacterial or virus laboratory of some pharmaceutical company or something, and according to Uffe, he had claimed that all the people working there are fully convinced that this Hong Kong flu is a deliberately sent message. (*Singing*) Oh, thank you Mao for sending me this gift. Tadaedadae da dae dadadad. (*Speaking*) Now I am driving up a hill, by the shore, up like that and down like this. Shoot. I was at INP today. I was punching Heke's schemes up some cards. And then (*clears throat*), Siegfried came and wanted us to pick up some grind lists from the state's data center. Hannu and me, we went to help her. And shit, was that a place or what! White dreary staircase and no doorbell, just these, kind of fire doors everywhere, not to mention the data sitras. Yeah, and all of the sudden—goddamn I'm glad I went—I got to go to the engine room of UNIVAC. Wowee wow (*singing*), wowee wow, what records! The record computer, it had quite a fascinating drum memory. Boing! So, we're approaching Merikatu and steering is getting to be quite arduous without switching, but maybe I can manage to uphold this recording situation closer to—all the way home. Shit, it's damp and dark, as it does tend to be in the city at this hour. Fuck, it's difficult to manage the car with one hand, steers inevitably onto the sidewalk. They have brought a new barrel of oil there but I guess I'll wait until tomorrow morning to go pump gas, unless Paavo is all finito. So, here we are, home, I'll turn off the recorder. Bang, bang. (*Cut*)

(00:29:46) (*In the car*) Ohh. Sunday, the 5th, 1:45 p.m., just woke up. Driving along the shore. People walking around, so many of them, taking leisurely Sunday strolls. It's gray, all over. It's above zero, so it's warm, quite fantastic. Last night I dreamt a lot, but I can't remember much of it. But I remember where it ended; it was some old corner room of the Institute of Nuclear Physics, next to the engine room. Yes, the engine room. And there was some lump a bit like Enetra, an IBM 3060, and what was interesting about its build was that it had this solid case that could be lifted up when necessary. What surprised me was that the ceiling of the room was so low that it couldn't be taken off completely. On the bottom right side there was a red start switch, and Petri Laurikainen was operating it. Click—the computer started and made this awful noise for a while before settling down. The blare was apparently some burst for the fan, which slowly turned on. I don't remember what happened then. Well, the computer started. Anyway, at this point, they started throwing people out and that was pissing me off because there was this beautiful blonde girl who wasn't quite what you'd expect there but considerably more, um, perfect. (*Yawning*) I tried to give her a kiss there when she was ushering people out. It was sad. But other people came to this room too, and i-i-ivy, by the ivy, or whatever plants they were, there was this punched paper, paper tape punch. Mäkelä asked me to mark something in the weekly calendar as soon as, yeah well, as soon as something else had happened or come, or driven. And (*clears throat*), but I don't know, I don't know, I think I'm mixing up the essentials. Anyway, seeing inside that computer, it was alternating between sparse and dense. Hmph. Frankly, this, this idea about these ponderings each morning feels a bit strange now. Oh well, I'll stop now, brak-. (*Cut*)

(00:34:34) (*In the car*) Hehe, I was just feeling a bit timid with having an army transporter truck in front of me. Now we're heading off downtown. Next, um (*yawning*), Pohjois-Esplanadi Street. I'm at the Unioninkatu traffic lights. (*Cough*) And onwards. (*Cough; singing*) There is something so delicate about this city here. (*Speaking*) Was I supposed to tell you about last night when han, han, um, Mara came to talk to me? Why, there's a strange sneak. We were considering taking a trip today, but I'm not sure whether I'm in the mood or not. There's a lot of other things to be getting on with and it's already, I think, past two soon. Anyway, I'm heading north here from Mannerheiminkatu, we're passing the statue of Finland's Field Marshal Carl Gustaf Mannerheim. We see that there is a new traffic sign pointing toward the Finlandia Hall. We see that we are approaching red lights so we shall stop at them. We find that the National Museum has a sharp-edged roof, which is silhouetted dark gray against a light gray background. It is broken only by a few cables stretching between streetlights. Yes, and what can we say about this? Well, where is the Finlandia Hall, anyway? Now that it's, oh I see, you go around from there. There are a lot, a lot of lights, traffic lights. Naturally. Aha, there's the grand Finlandia. And it looks, it looks quite a bit like a locomotive. Very, very pleasant. Good, good. Shit, such waste of tape. Smart filter, senior. Haha. I see, and here they have erected some new hotels and whatnot, the first of which is gray and at the moment looking quite nasty, and another one that is white, it looks quite a bit nicer. What could I say about it, is it, where are those windows from? From here? I don't know, I do not know. (*Stronger wind, echo*) I must say that (*yawning*) I would like there to be some massive restructuring of the city, taking down all buildings except for these new white ones and these, maybe these old yellow ones. Because buildings like this Exhibition Hall here ... Although maybe this is better than—better to keep this Exhibition Hall and take down a dozen residential blocks and build something, oh, wild instead. Again, I find myself wondering where I am heading. I think maybe I'll go get some coffee at Kolme Kreiviä. And now I'm heading off. I'm just driving around. Maybe I'll head out to see some countryside outside Helsinki. Although, what's there to see? Now I just worry whether I can produce any interesting material here to keep up the interest of (*yawn*) the listeners. Ha, they only get the final solution. Um, no, no, no, can't do it, no point in trying. And this reminds me of the fundamental function of this talking and recording. This is an exercise for breaking away from the matter. One minimal moment of this exercise is probably realized in not being frugal with the tape. Because suddenly, we find ourselves at the point where—well, maybe not there yet, well no, um, well yeah, we are at the point, wherever we are, that no matter how much I record and never overwrite, the consumption of tape doesn't really, really affect me economically. Well, we can see where this monetary limit lies if (*yawning*) this, at this moment it can cost, um, 5 markkas per hour—not so much, 5 markkas for one and a half hours, with some discounts. Well, let's say 10 markkas for two hours, yeah. So twenty-four hours a day would cost 120 markkas per day, which undoubtedly is a bit much. So if the price of a cassette is cut down to one-tenth, then, considering that no fucker would record more than six hours a day ... (*Yawning*) Well, in any case, we have now arrived at Ruskeasuo (*clears throat*) somehow. I might try to, ah, find some gas station and pump some gas. Some cars driving up this way. The whole, um ... Now I'm tired. There's a Teboil there, service and bar. Teboil with its service bar. Making a mental note, I guess I could go there ... on the way back. Now I gotta step on it to make it before the light turns red. This is, what is this, Turku, Björneborg, Hämeenlinna. So, we're on the Hämeenlinna-bound road. The Haaga boroughs are there, close by. Should turn on the headlights, there. See, there's some strange boxy house there with some decorative trimming around the edges of the balcony. There's a bit of light in all those houses, which is quite cold and gloomy at this time of day. It seems they like to have their lights on in the

South Haaga apartments. Then there on the right, there's some of that wild region. Laaturakenne has just put up a house there and then a bit onwards, there's that kind of an old-fashioned red one, and next to it a long gently sloping mansard roof. And now we are, we're passing below the train tracks, through an underpass. And on the right there is a wide expanse of, what is that called, a bog field. (*Cough*) I guess it's that Ruskeasuo. Damn, why don't people walk there? There in Ruskeasuo, what was it, that dream that took place on those wetlands? And now, there's a self-service Esso station there but … I suddenly notice that the Haaga vocational school is an exceedingly gorgeous building, very nice. I see, there's more of the swamp on the right, how have I not noticed it before? And there, there are no buildings there yet. Oh, what a grand wide expanse of swamp and certainly places that are wet or snowy at the moment, at least wet. Pisses me o-, now we drove up to a traffic light where there's this little street crossing. And another Shell there again, of course. Anyhow, we notice that all service stations are still open at this hour. Yeah, but it's getting a bit monotonous, this chatter, I'll turn off the recorder and drive on in solitude. Goodbye! (*Cut*)

(00:48:37) (*In the car*) Yeah, on the same trip, driving back, at the Mäkelänkatu traffic lights, got some gas. Seven, seven, five, five, eight, Shell. Saw Korhonen's new furniture store and almost felt like popping in. Have been having many thoughts. The time is, let's see, half past two. (*Cough*) Have been thinking about Anna, about how nice it would be to drive around with Anni-Saara, maybe on a Sunday afternoon like this, and show her places. That's the way it goes. And this here again, this, this (*chuckle*) Mäkelänkatu formation. Three cars in single file on the center lane and one on each side lane, as sort of wings. There's another squad coming up after (*yawn*), after them. We'll have to see where I get my morning coffee today. The most important questions in my life begin with what, where, and when, having anything to do with eating. It matters if it has to do with eating and answers the question of what or where or when. Juice at Kolme Kreiviä Café at two thirty. Oh my. Maybe I'll meet my friend there (*in a nasal voice*), engineer Pekka Suomalainen. Perhaps, perhaps. Perhaps I'll read some Burroughs there, *The Wild Boys*. I have it in my bag with me. I feel like I'm getting all stuck in my ways, getting all pedantic and reclusive. Hahaha, fantastic, tis-. Mm, getting more organized by the day. I'm not cleaning or anything, no … It's about systems. Organizing the suitcase, having an assigned place for everything in there. Recorder on the left, a slide rule behind it, a microphone in front, and empty, filled-up cigarette cases to be placed on the mantelpiece when I get home. Some A4 material upright on the right side. Lost ones on the bottom, current ones on top, topmost. A book, some tissues, and a box for cards. A case for film, photographic film, that with small modifications turned out to work great as storage for the small cards, about a hundred of them that otherwise fill my pockets. There's a system for the cards, different colors: white for long-term memory, yellow for short-, short-term memory, blue for projects, red for components, and so forth … or something. That last one, at least, isn't quite clear yet. Oh, this is not the lane yet. Apparently, I drove up to the Hämeenkatu–Mäkelänkatu junction and I don't mean to get onto Hämeenkatu but to jump over it, over and down. And I'm just about to—ha, yellow and green. (*Car accelerating*) Is this Lautatarhankatu where … Lautatarhankatu or Lautatarhantie? Can't see a street … And here (*yawning*), we'll step on it, ha, aah, oh, I see, something has changed here, there's a one-way street there, we'll have to (*clears throat*) make a little Umleitung here. But that's not, this is of course some kind of a dairy area here, or what is this—no, slaughterhouse area, I mean, production halls. But there's not a lot of factory romance in the air. This is Työpajankatu, no, what the heck is it? And now we're on our way downtown, Tynnyrintekijänkatu and a gasworks on the left. What is that shitty building there? Dilapidated, waiting to be demolished, I guess. A small triangle of a gasworks, some buses. Now there is this feverish feel in the air again, driving through this passage of getting in and out of the city, even though I'm not even (*yawning*) in a hurry. There's still this sense of racing somewhere … argh, f-f-fuck. I'll just quit this jibber-jabber (*chuckle*) for now. A pile of coal on the left, the future ahead. Bye! (*Cut*)

(00:56:22) (*In the car*) And what, may I ask, is the declaration of this Independence Day? It's around, around noon (*yawning*), ha-aah, holy shit. Driving in Lauttasaari. It hasn't been so wet lately but there's still this same gloom. And the same temperature. So, what will I talk about today? Yesterday, yesterday … whatever happened yesterday? Most of it was spent in the engine room of INP. A lot of program runs. Smoothed some out, ran Heke's lists, ran Nelli. Instead, life is still in bud, no code for it yet. Yep. But so, here we are, is it Jorvaksentie that we're passing? Lauttasaari and its landmark church. A heath and then this curved ramp onto the motorway. Not going very smoothly. (*Pause, car accelerates.*) But I think we made it without getting run over by that Volkswagen behind us. The sea is calm, just a couple of birds swimming there and a fair amount of people strolling along the shore. There's some smoke swirling up from somewhere in the horizon. Peculiar water towers, peculiar place. Ugh. Well, today I woke up this early partly because Philip, Philip Donner, called. The other Philip is in the midst of war, perhaps no longer though. And … Now that I don't know where to go, should I keep on driving along Tarvantie or head toward Leppävaara-Alberga and Otaniemi? I think I'll turn back right about now. It's this pain of a holiday morning, whether to drive far, whether to drive far, far away? Or to just turn and head back to the city? That is the question. The birch trees are all white in this part of town. It's as if … I cannot verbalize everything … that I'm feeling. Six markkas to Leppävaara, seven to Helsingfors, one to Otaniemi. (*Chuckle*) Hehe. So, at work last night, I found an issue of *Electronics* I hadn't read before. Or I'm not sure if I had read it, but I can't remember it at all. But so, there was a long article in it called "Computers for 70s" which I didn't have time to finish; I hardly got halfway reading it. Umm … But yeah, it made me think of all kinds of stupid little things. And now I'm carrying an *IBM Systems Journal* with me, which is very well suited for rifling through while drinking my Independence Day's morning coffee. Now if I only knew where to go. There's the Lehtisaari shopping center, but it doesn't exactly look like there's an open and welcoming café there. Is Kuusisaari the next place, then? Oh, what a cool looking building. (*Cut*)

(End 01:03:31)

4 Fleshy Intensities

Susanna Paasonen

As the video file (dated February 5, 1988) begins to play, I am looking directly at the crotch of Erkki Kurenniemi, who is stroking his limp penis and lifting up his testicles for the camera. The rest of the room is visible behind Kurenniemi, and his female partner is lying on the bed, witnessing him play. Kurenniemi moves away from the camera, then gets closer again, and positions his genitalia firmly at the center of the frame. I am now looking at the tip of Kurenniemi's penis, now at his scrotum in extreme close-up: genitalia keep on moving in and out of focus and almost touch the lens of the camera. The image is fuzzy, both from the magnetic noise of the original VHS tape recording and from the naked human body standing just a bit too close to the camera. The visual landscape is pink-beige, hairy, noisy, and full of skin, while the soundscape consists of casual remarks between Kurenniemi and his partner, his body moving and knocking into objects, the camera zoom being operated, some laughter, and Bob Dylan's "When You Gonna Wake Up" playing in the background. Soon, Kurenniemi moves to the side and zooms in on his partner, who squints her eyes in the bright light, then moves over and starts to pull down the sheets. She looks directly at the camera, then at him, pulls back the covers and complains that she suffers from "acute video allergy." Kurenniemi answers, "Well, let's deal with it by turning off this camera, then. Or?" As she murmurs "yes," the scene ends at 01:52.

This video diary clip is one of many documenting Kurenniemi's mundane sexual play—or attempts thereof—with the camera both recording the play and being very much an active agent in the network of actors and the dynamics of desire that comprise the scene. The camera is an organic, perpetually present, if not necessarily inconspicuous part of Kurenniemi's everyday life, as mediated by his diary logs. His partner's bout of "acute video allergy" is telling of everyday intimacy that is both saturated and disrupted by the presence of the camera: in the logs, the camera both amplifies the fascination of sexual play with its possibilities of editing and replay, and takes on the role of an antiaphrodisiac antigen.

Sexual fantasies, memories, and acts are perpetually present in Kureniemi's multimedia diary logs recorded on analog and digital video, handwritten diaries, digital files, and voice diaries on cassettes from the 1970s to the 2000s. In these mundane records of events, fantasies, and observations, Kurenniemi's attention constantly revolves around, focuses and clusters on the embodied and the sexual. Pornography abounds as homemade videos, images harvested online, clips and still images grabbed from films, television, and magazines. The fleshy and the pornographic characterize—and even animate—these personal logs. This chapter maps the fleshy dynamics of Kurenniemi's diary logs by asking how they have been put together and what intensities they may facilitate, encapsulate, or mediate.

Horizontal Threads

In Kurenniemi's written logs, the details of theories ruminated upon, exercises at science fiction writing, people encountered, erections and ejaculations achieved, sandwiches and bottles of wine consumed, commutes and travels made, defecations planned and performed, hash smoked, casual yet systematic observations made of friends and passersby, things planned, imagined, and remembered—all follow one another and overlap, with just occasional asterisks separating one strand of thought from another. Analyses of scholarly articles merge with remarks on intimate bodily odors while video material of streetscapes intermeshes with candid shots of neighbors going about their everyday lives.

March 1987

Now it's getting a little hard with these systems, I'm writing on a Kawasaki that's only installed to my system Fancy Fonts, I wonder what future generations will make sense of this? …

I'd like to debauch in my head with all the options but it's up to the imagination. I notice how strongly our computer culture is still attached to paper. Time for the last drops [of wine].

Let this document be strongly in the spirit of Kawasaki. I dare to write. I bought the new US edition of *Penthouse* and wonder if it could inspire any new, even more incredible debaucheries. First X, now also I have given my pubic hair a punk cut, both of us are left with just vertical stripes. Stubble starts to grow on the sides. If you consider the level of visualization in *Penthouse* you have to wonder: is that all?[1]

There is no Goffmanian backstage, no simple division intended between the public and the private, the micro and the macro in these logs. Details of reading *Science* and *Nature* are presented alongside those describing the most recent issues of *Penthouse* and *Razzle*, and observations on fractal theory are provided in the same detail as recurrent plans and experiences of shaving pubic hair.

March 11, 1987, 23:59

(((2 3)(2 3))((1 3)(1 3))((1 2)(1 2))).

(I shaved lots of my cock hair again with a new scalpel knife. A little blood. The core of that problem seems to be only a small push away. Now I'm drunk again. Beautiful fragile Mohawk from the cock to the belly button. Oh so thin, the same thing for X soon. I'm a hairless man, plus impotent. In addition, happy.)

January 1, 1989, 22:59

1.5 hours of VHS tape from the morning, too bad that two of X's orgasms were left out, in the afterglow I dug her asshole with my middle left finger and it led to something. X complained that I've neglected her ass. Nice to hear …

Sex life in good shape. Such orgasms last night. The hash won't last for even another week.

23:41 went to jerk off to *Erotiikan maailma* [*a porn magazine*]. Now an opened Côtes [*Côtes du Rhône, one of Kurenniemi's regular brands of affordable red wine*]. Light in the kitchen but empty.

On the basis of his diary records, Kurrenniemi's daily activities have been driven by three intermeshing key interests, or passions: *a passion for theory* (physics, biology, and the mathematical theory of harmony), *a passion for technology* (synthesizers, robots, computers, and consumer electronics such as cameras and mobile phones), and *a passion for fleshy intensity* (sex, pornography, alcohol, drugs, and combinations thereof). Any boundaries between art, science, science fiction, pornography, and the documentation of everyday life remain porous at best. As I argue, for Kurenniemi, both theory and joys of the flesh involve potential pleasure, affectation, and intensity of experience: they are all about playful experimentation and some kind of a potential high.

March 11, 1987

Zeus = ((1 1 2)(1 2 2)) = (((1 2)(1 2))((1 1)(3 3))((2 3)(2 3))). Will not be solved today. Will probably dream tomorrow night. Please note that I don't want to rush. It'll surely come and I want to play with this pleasure. These are my orgasms. Now I'll go to bed, and since I've already jerked off today, there ought to be enough time.

May 10, 2000

Now a light high and a Halcion [*sleeping pill*]. I may try to think a little or stew the tonal chord generator or better, the polyphonic generator. With that I'll soon get the best easy stuff from emergent simple tonal theory. Difficult to stop and turn the thing off. The pleasure of this keyboard is so big. I could probably write until morning.

The overall horizontality of Kurenniemi's personal archive can be explained through its principles and aims of documentation premised on the inseparability of perception and consciousness. In accordance with his transhumanist theory, Kurenniemi wanted to record his consciousness for future access. Memories, impressions, and theoretical speculations were understood as data to be stored and processed. For this database to be accessible, the algorithmic structure of one's persona would need to be mapped and relocated, or installed on some computational substrate such as a computer or brain tissue grown from stem cells.[2] To a degree, this hypothesis—whether it be considered a futuristic theory or a vision of science fiction—resonates with Henry Bergson's conceptualization of consciousness as inseparable from sensory experience, and of the immediate data of consciousness as lacking any juxtaposition of events. For Bergson, consciousness is a multiplicity of horizontal threads and vibrations that resonate according to a particular personal rhythm. This rhythm applies both to perception and to recollection as embodied processes of different kinds, in that perception is about affectation with or by external images whereas memory is a more introspective activity:

> But, if you abolish my consciousness, the material universe subsists exactly as it was; only, since you have removed that particular rhythm of duration which was the condition of my action upon things, these things draw back into themselves; and sensible qualities, without vanishing, are spread and diluted in an incomparably more divided duration. Matter thus resolves itself into numberless vibrations, all linked together in uninterrupted continuity, all bound up with each other, and travelling in every direction like shivers in an immense body.[3]

The traces and records of Kurenniemi's life stored in the archive may resonate with those of the people browsing through them, but they can never vibrate as they would in his consciousness, defined by its specific rhythms, layered histories, desires, fixations, and drives. What can be resurrected from the archive is by necessity another creature altogether: an assemblage or patchwork strung together from more or less random documents that are largely lacking in metadata. Traces and resonances may linger on, but they will do so at the speed and frequency of another person's embodied consciousness—they are different kinds of shivers in a different body.

Following Kurenniemi's stroke and his ensuing loss of ability to form sentences, the connections between the different items in the archive have been drastically cut. Rather than a systematic archive, what remains is disjointed assemblage of highly edited works and random flotsam. This can be seen as unavoidable in the sense that horizontality was Kurenniemi's underlying principle of accumulation: he regarded *everything* as potential data, from the brands of toilet paper to the ticket stubs, receipts, advertisements, Christmas cards, and business cards that accumulated over the years. The ambition was to laterally save and record the material traces and banal incidents of his everyday life, as well as the trajectories of his thought, fantasy, and speculation.

June 27, 1987, 0:50

Tomorrow I'll tell X about my fantasies ... These are connected to the porn magazines I mentioned. Large steel ellipsoid in Kaivopuisto. Flying saucer of the same shape. Copper coils and nobs in Kurenniemi. Copper sheets in the fortress storage room. Killer robots and a Chinese rat cage. Me cut up into a skeleton, only shreds of flesh left. Cuntless porcelain women. Girl guards and me behind a desk. Concentration camp mentality.

Now I guess I'll smoke a little dope, otherwise I'll get too drunk. When X was last here and I then smoked some dope I got all merry and raucous and blabbered on about Y. X got furious the next morning and told me about her shared venereal disease with Y.

Ykä [*pseudonym for a male neighbor*] is probably in some bar picking up someone he's currently prepping for a joint show-off for me. My cock begins to stir already. On the phone with X, I scratched some odd flakes off my cock. Is it some virus, or bacteria? Z probably won't call anymore … Animals or women are screaming on the street. The street was empty when I last looked. Animals.

A Noisy Archive

Kurenniemi himself would probably call his archive not an archive, but a database: a collection of data not to be used for studying the life of Erkki Kurenniemi but for accessing his perceptions and consciousness, and indeed for living his life. This "life" can be understood as the sum of perceptions translating as consciousness, or as a more elusive intensity—"a current," as Kurenniemi himself put it. Kurenniemi granted the media used for remediating his current ideal transparency as the means to an end. At the same time, the media formats involved in this were far from instrumental, and when browsing through the archive, their particular affordances and tactile materialities grow pronounced. A typed diary in PDF format allows for easy access and searchability. Reading the same diary in handwritten format—the same in the sense of containing the same textual "data"—affords additional explorations into the materiality, texture, style, and temporality of the documentation. The backs of the diaries are stained by tobacco smoke, their covers have been worn from use, some have been hand-painted, and others have been doodled on. Inside the diaries, drawings, lists, graphs, photographs, receipts, clippings, bits of paper, and occasional wads of genital hair are glued in to accompany the notes. Some of the diaries involve a multimodal, collagelike aesthetic, while others provide more straightforward textual notes. Kurenniemi's handwriting alters according to the situation and the amount of intoxicants used, and shifts from neat composition to large, restless scribble. All this disappears when the notes are transcribed to electronic format. The PDF files are stylistically much more calm.

Despite the volume of Kurenniemi's logs, years of his notes remain missing. There are long gaps in the chronology of the digital diary files, and the decay of magnetic tape during digitization has generated additional gaps. On a more fundamental level, the affordances of different media condition and dictate that which can be accessed, seen, and heard. In the case of the voice diaries recorded on cassettes, it is often impossible to distinguish noise from information (Kurenniemi's own voice, radio shows, music, and ambient background sounds are all heard). The soundscapes are indistinct and lacking in focus. His 8mm and 16mm films are silent (with soundtracks added later), the VHS tapes come with considerable visual and sonic noise, and many of the digital videos shot with 1990s mobile phones have notably poor image quality and sound resolution to the degree that the audiovisual "data" are barely accessible. The balance and ratio between the signal and noise is constantly unsteady and the processes of mediation remain far from smooth, transparent, or neat. To the degree that the media formats and the visual, textual, and sonic information they provide are impossible to decouple, the archival traces are specimens of media specificity.

For media theorist Friedrich Kittler, what "remains of people is what media can store and communicate," and what "counts are not the messages or the content with which they equip so-called souls for the duration of a technological era, but rather … their circuits, the very schematism of perceptibility."[4] Kittler argues for the primacy of media and their specificities beyond any personal recordings or impressions: what matters for Kittler is that which the media render perceptible. In the context of the Kurenniemi archive, such specific circuits and perceptibility of media matter, for they condition what remains. The scenes, scenarios, moments, dialogues, monologues, fragments, objects, and people conveyed in the recordings are nevertheless of equal importance. The style, feel, and modality of the archive owe equally to both form and content: both are crucial to how the recordings reverberate in our acts of sensing and making sense, and neither can be posited as mattering more than the other.

Kurenniemi defined humans as organic slime machines, the memory functions of which are interchangeable with, and bound to be fused with, the technological. While he positioned himself as a masterful designer, builder, and operator of technology, the relationship was manifestly a prosthetic one. As prosthetic, externalized memory reserves, the material particularities and affordances of storage media limited and oriented how he was able to observe the world and later recall past events.[5] Kurenniemi's autobiography drafts make evident how gaps in his media recordings translated as gaps in his memory. In addition to its main function of data preservation, the archive functioned as both a technology for recollection and a source of enjoyment. Kurenniemi regularly detailed revisiting his earlier diary inserts, reworking and editing them, transferring them to different formats, masturbating to and watching video recordings alone and together with partners, going through his archives of still images, and digitizing and manipulating them on his computer.

December 20, 1987, 19:35

X had a Christmas party on Friday … Finally, I couldn't escape Y. Last night I passed out early on hash until Z called me completely drunk, came over and first pulled my pajama bottoms down and shouted, "Show the balls." Soon I got the camera going so will not need to explain here in writing. Too bad the tape ran out without me noticing. Lost the entire episode with the screwdriver; the screwdriver in her cunt, she's a technically talented girl, me pushing with a vibrator. Today we slept late, watched the videos …

May 10, 2000, 22:54

Third tea and a few slices [*of hashish*]. My mind immediately speeds up for the final sprint of the night. No need to jerk off, guess I already took care of that in the morning, no that was the afternoon, good. Oh when I get to scan old porn pics from my books, especially the stereo images. And then even the rest of my girlfriends will get on the Zip drive as 9 giga backup copies. I then need to add the text from Autobio so that this work becomes the second definitive version.

Recording mundane events and revisiting them added to the overall intensity of experience: paraphrasing Bergson, it was not only the images and bodies of the external world that affected Kurenniemi, but also the acts of recording and reconfiguring images, watching recorded images of his own body and the bodies of others. Acts of recording and rewatching the video logs involved the simultaneous externalization and internalization of perception and memory. Affective feedback loops formed, accumulated, and centered on sex and pornography as the topics that Kurenniemi most carefully tagged, edited, saved, and revisited. If perception, memory, and consciousness are all considered matters of particular rhythm and intensity, then the diary logs make evident the pivotal role of the sexual in and for Kurenniemi's particular personal rhythm, for his individual way of being.

Before digital video cameras became affordable, Kurenniemi shot his video logs on VHS in much more selective manner. The large volume of porn video clips, photos, and textual accounts in the archive speaks of his priorities and interests, and of his focus congealing on the particularities of human bodies and their fleshy intensities. Kurenniemi describes the act of connecting a computer to a network as "small heat," a moment of intensity similar to watching porn or working on a theoretical dilemma. As different as such actions and interests may seem, they all involve an intensity of experience resulting in an affective rush, or at least providing the potentiality of one. It is the more or less playful quest for intensities, affective rushes, and sensory highs that characterizes the flow of Kurenniemi's diary logs.

Differences in Kind

The perceptions and observations that Kurenniemi recorded in order to reproduce his consciousness are media-saturated and inseparable from the technologies used to record them. The possibility and pleasure of capturing mundane events and gestures seemed to intensify both his sensation and his perception. It is therefore something of an understatement to note that the practices of documentation affected how Kurenniemi experienced his everyday life—just as "what is no longer archived in the same way is no longer lived in the same way."[6]

In the video diaries, passersby on the street are seen reacting to Kurenniemi's camera; the urban landscape he is shooting grows into a mediated, reflexive space of seeing and being seen; and dialogues are enacted especially for the camera. Sexual scenarios recorded become performative actions in themselves as partners display their bodies for the camera to zoom in on. The presence of the camera is acknowledged with direct visual contact—in fact, Kurenniemi recurrently asks his partners to face its lens rather than to have eye contact with him during sexual play. This produces a kind of doubling of the sensory where attention and presence are constantly split between the physical act of sex and the static "objective" lens of the camera. Temporality is similarly split between the present acts and the future tense of a replay, as the present is always folding into the future, the revisited and the reedited.

Autobiography 3.fm file, chapter 2: A letter addressed to X

January 31, 1990, 1:29 ... Now I'm making my life into a tape, "Video Verité Totale" ... when I bought a camcorder after you left, I got an idea to shoot everything 24/7. That would be boring and for the moment too expensive but the idea is feasible. For now I've even used the principle in a deconstructive manner, I won't see people who're allergic to camcorders. Like this Y from Pori. Z is game, but she's found a way to tease me: when the tape ends, she joyously spreads her cunt, and only then. This is what relationships degenerate to, if you will, because of our genes. ... The camera is more important than you or me, since it constantly makes imperishable history of both of us. We feel "the wing of history" touching us and we go crazy.

As Jacques Derrida argues, an archive is stored for the future, but people engage with it in the ever-shifting present, with specific aims and purposes in mind.[7] When Kurenniemi revisited, edited, and made comments on his diary notes, reflexivity extended back and forth in time, both toward his past self and more ambiguous future potentiality of his consciousness; both to reflections of finitude (of his own life) and immortality (of his virtual, algorithmic life). The camera, more than any other medium, lies at the heart of this split temporality as the instant generator of historical records that will linger on to be accessed and relived. The year 2048, as the locus of both Kurenniemi's transhumanist project when his consciousness is to become available in machine form, and of the science fiction narratives he imagined and composed throughout the years, remains the end point of the archive's explicit futurity. As a site of fantasy and theoretical speculation, the year 2048 marks the moment when machines will forever revisit earlier memories and records made by human slime machines: as Kurenniemi dryly notes in Mika Taanila's documentary film, *The Future Is Not What It Used to Be* (2002), he expects them to be mainly watching porn.

Archives are defined and driven by the dynamics of forgetting and extinction. The futurity of the archive is conditioned by nothingness and death, just as accumulation and preservation of the archive assumes the threat of effacement and eradication. Or, as Derrida puts it, "There would indeed be no archive desire without the radical finitude, without the possibility of a forgetfulness."[8] Kurenniemi's "archive fever" is fueled by an awareness of imminent loss. His processes of accumulating and storing records of everyday events have been a means of warding off erasure and the limits of human existence—their temporality is geared simultaneously toward both annihilation and eternal life (in 2048, and after).

Rooting for a machine future, Kurenniemi remained firmly fascinated by the dense materiality of the human body. The bodies in his video logs are regularly stylized, but rarely in ways reminiscent of the neatly groomed flesh of commercial pornography. Kurenniemi and his partners are shown painted, cross-dressed, covered in ashes, having sex, loitering about, chatting, sleeping, drinking, smoking dope, defecating, and shaving each other's pubic hair. Body hair, menstrual blood, piss, and shit feature regularly in the diary logs as both visual and textual themes. The tempo is slow, action meanders, tape often runs out, and highlights are missed. These are not bodies as aestheticized surfaces but ones that constantly engage in sexual play and ooze bodily fluids. There is an obvious, tenuous tension between the visceral and detailed fleshiness of Kurenniemi's logs and his idea of uploading them in machine-readable format to be used by future machines, cybernetic or cloned human organisms. For how would a machine sense and make sense of the acts and sensations accounted? What data would the machine be able to read, and how?

September 9, 1974, 11:07

I walk home, accidentally bump into X ... I finish my work up to a point and leave for a walk on the town and eat at the Go Inn pizzeria. Coincidentally Y is there, another encounter, together with Z. From there to Tea House, I play chess with Z. ... A comes later and then X. ... The old game again, women around me... Late bath and heavy sleep. It's raining again, rugged clouds and an early winter feeling, as my sensations increase the time feels longer. I think I understood that Y's boyfriend is in Lapland. What a great partner. Next to, or on a par with B, such incredible breasts, such a fantastic small person's body. I treated X with aloofness. Could be nicer about it. Only waiting to get a squeeze at Y's tits. The hardship of life's battle makes the bones stronger. I take clean clothes. Finishing the Dimi looms ahead.

February 8, 1999, 20:57

I shoot X from each side and explain how important pedantic work is when generating urine. Virtual world digs no dead angles. I've shot the cabin and the toilet and all the nooks and crannies. It'll be easy to interpolate and texture the surfaces.

For the first time, I have an extra feature in Sony that locates the camera and inserts the geo data on the same tape. With it, it'll be quick to make a 3D model. X digs after another cherry-sized piece of hash from her bag ... X lights a cigarette and passes it over to me. I inhale a few times and clean the tip of the cigarette from ash by rolling it against the bottom of the ashtray. I lift the glowing cigarette end against the chunk of hash. The hash gets stuck on the cigarette's hot tip. I lift it close to my mouth ... X quickly learns the trick, and as we've both inhaled five times, we burst out laughing.

Considering the inseparability of the human bodily capacities of sensing and making sense, I argue that such accounts translate poorly as data to be processed by artificial intelligence. While processes of storing and accessing memory are common to human bodies and intelligent machines, the precise processes and forms of memory in question are radically distinct. As Jean-François Lyotard points out, human thought "doesn't work with units of information (bits), but with intuitive, hypothetical configurations": it "isn't just focused, but lateral too"—much like the Kurenniemi archive.[9] For Lyotard, the complexity of human thought and cognition is inseparable from the specificities of human embodiment. Similarly, a camera sees differently than the human eye, and a microphone records sounds differently than the human ear hears them. These affordances and limitations are particular and distinct. A human sees and hears without knowingly looking or listening; sees and hears with acute intensity, and fails to look and listen as her attention constantly shifts, oscillates, and reorganizes. Contrary to the contingency of the human sensoria, cameras and microphones steadily record the audiovisual within the confines of their technological makeup and configuration and without any moments of heightened intensity (unless the operator sets the camera to zoom in or angles the microphone accordingly). Video footage of an event is, then, experientially fundamentally different from a particular person's perceptions of the same event, and a record of a mundane encounter is always dissimilar to a lived experience thereof. Paraphrasing Bergson, these are differences in kind rather than degree.

This fundamental tension, or incongruity, between human and machine forms of memory and perception has a constant, ironic presence in the Kurenniemi archive. The diary logs aim to mediate the rhythms of Kurenniemi's consciousness, perception, and memory in order to reproduce them in algorithmic form. At the same time, these rhythms mesh with, are reconfigured by, and remain inaccessible to the affordances and modalities of the technologies used for storage. Following Lyotard, the rhythms are reconfigured by the inhuman modes of memory that differ in their materiality, organization, and access from human ones. As these human and nonhuman materialities resonated in Kurenniemi's acts of recording and performing his everyday life, the rhythm of his consciousness similarly oscillated and changed. Consequently, the rhythm of his consciousness would be impossible to separate from the technological networks and prostheses that it constantly moved with.

In Kurenniemi's logs, the fantasy of an uploaded consciousness that can continue revisiting the times past for all eternity is much less pronounced than the pleasures of archival practice and the intensities of recording, editing, and revisiting them. Here, the joys of theory and defecation are cut from the same fabric of embodied practice, potential affectation, intensity, experimentation, and play. Living, again, becomes an archival practice of ambiguous temporality that is driven by a quest of affectation, where the past and the future constantly fold into the present. Despite Kurenniemi's transhumanist fantasy of overcoming the slime-based human embodiment, it is the characteristically human fleshy intensities that animate his archive.

Notes

An earlier version of this article was published as "Slimy Traces: Memory, Technology, and the Archive," in *Kurenniemi—A Man from the Future*, ed. Marita Mellais, 32–56 (Helsinki: Finnish National Gallery, Central Art Archives 25, 2013).

1. Translations of the diary excerpts from Finnish are by the author. All names, apart from those given by Kurenniemi to people he did not know, have been removed and replaced with letters X, Y, Z, etc. To preserve anonymity, no particular letter is associated with any individual person.

2. Erkki Kurenniemi, "Aivojen vaihto," *Futura* 2: 51–57.

3. Henri Bergson, *Matter and Memory*, trans. Nancy Margaret Paul and W. Scott Palmer (New York: Cosimo, 2007), 276.

4. Friedrich Kittler, *Gramophone, Film, Typewriter*, trans. Geoffrey Winthorp-Young and Michael Wutz (Stanford: Stanford University Press, 1999), xl–xli.

5. See Celia Lury, *Prosthetic Culture: Photography, Memory, and Identity* (London: Routledge, 1998).

6. Jacques Derrida, *Archive Fever: A Freudian Impression*, trans. Eric Prenowitz (Chicago: University of Chicago Press, 1996), 18.

7. Ibid., 68.

8. Ibid., 19.

9. Jean-François Lyotard, *The Inhuman: Reflections on Time*, trans. Geoffrey Bennington and Rachel Bowlby (Stanford: Stanford University Press, 1991), 15.

Bibliography

Bergson, Henri. *Matter and Memory*. Trans. Nancy Margaret Paul and W. Scott Palmer. New York: Cosimo, 2007.

Derrida, Jacques. *Archive Fever: A Freudian Impression*. Trans. Eric Prenowitz. Chicago: University of Chicago Press, 1996.

Kittler, Friedrich. *Gramophone, Film, Typewriter*. Translated by Geoffrey Winthorp-Young and Michael Wutz. Stanford: Stanford University Press, 1999.

Kurenniemi, Erkki. Aivojen vaihto. *Futura* 2 (2001): 51–57.

Kurenniemi, Erkki. *Diaries*. Erkki Kurenniemi's archive, Central Art Archives, Finnish National Gallery (EKA, CAA, FNG).

Lury, Celia. *Prosthetic Culture: Photography, Memory and Identity*. London: Routledge, 1998.

Lyotard, Jean-François. *The Inhuman: Reflections on Time*. Trans. Geoffrey Bennington and Rachel Bowlby. Stanford: Stanford University Press, 1991.

II Visual Archive

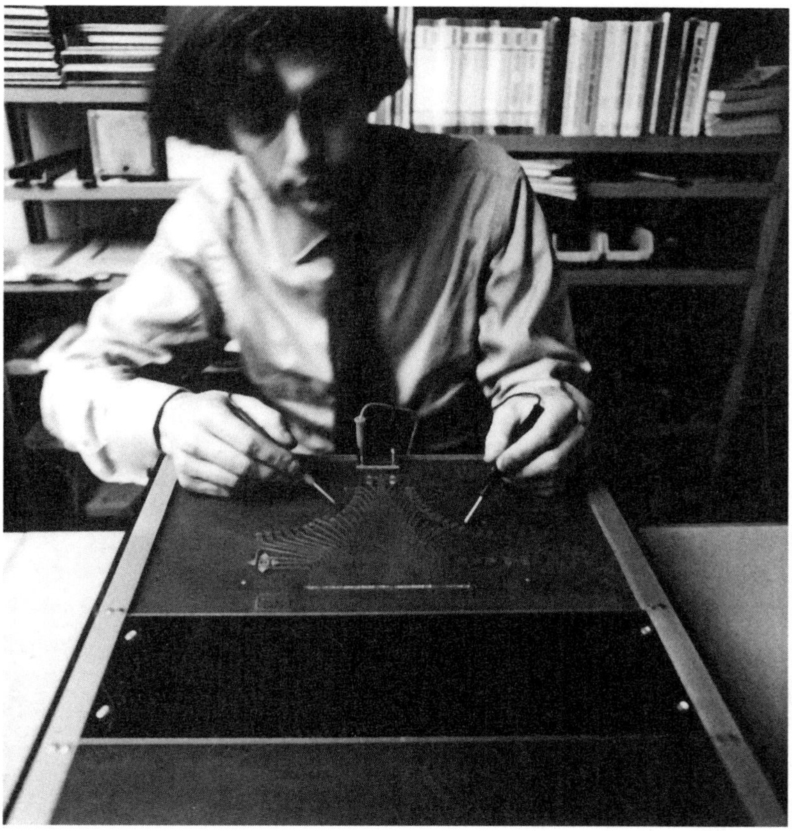

Figure II.1
Erkki Kurenniemi playing a DIMI-A synthesizer (Digital Musical Instrument with Associative Memory), ca. 1973–1974.

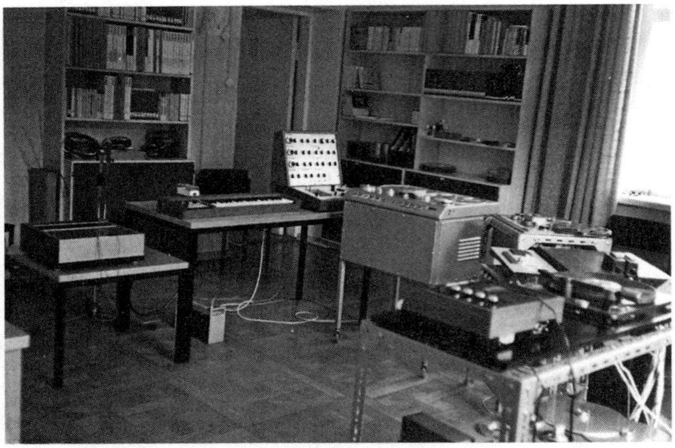

Figure II.2
The Music Studio at the Department of Musicology, University of Helsinki, including Erkki Kurenniemi instruments, 1972–1975.

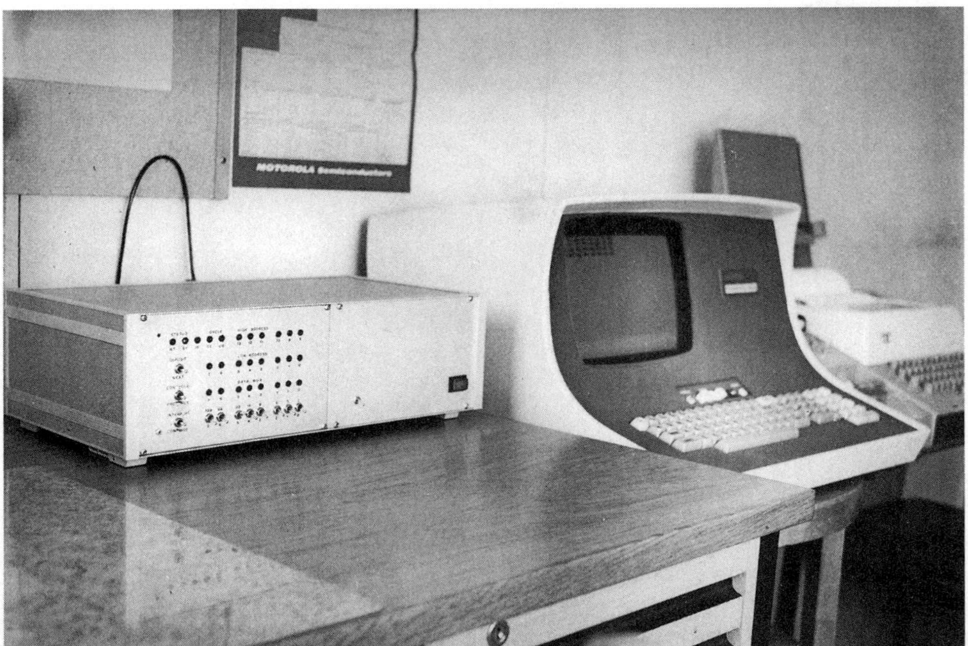

Figure II.3
Digelius Electronics Finland Oy, a company set up by Kurenniemi in 1970 to manufacture DIMIs.

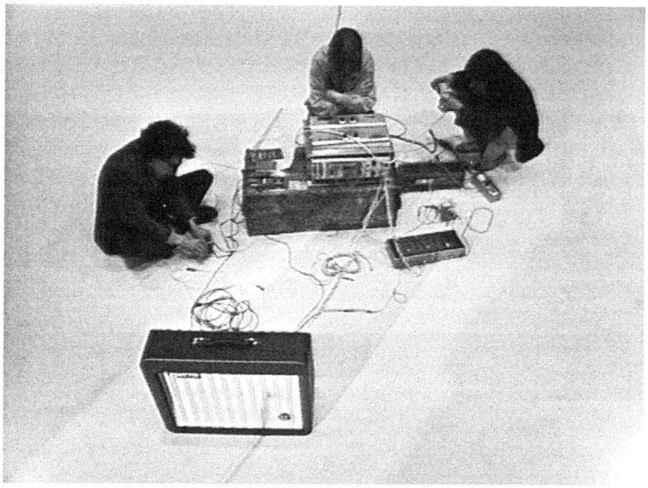

Figure II.4
Sähkökvartetti (Electric Quartet), live performance, operated by Tommi Parko, Peter Widén, and Arto Koskinen.

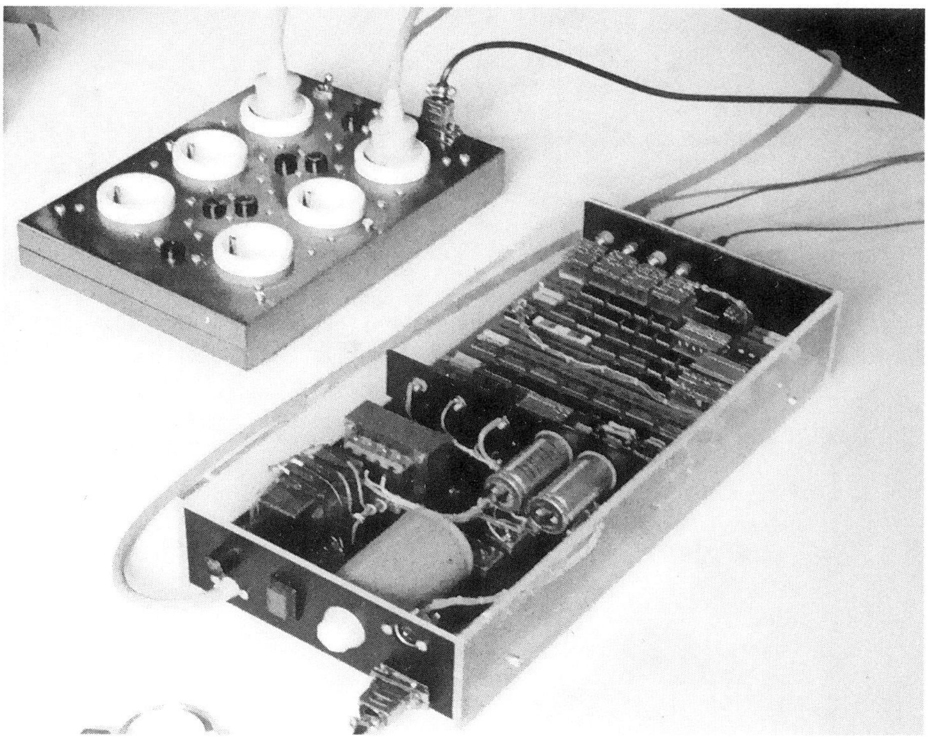

Figure II.5
DIMI-S (a.k.a. Sexophone, or Love Machine), 1972. DIMI-S playing instructions: 1. Ask three people to join you; 2. Ask each of them to pick one iron ball; 3. Touch each other's skin; 4. Play!

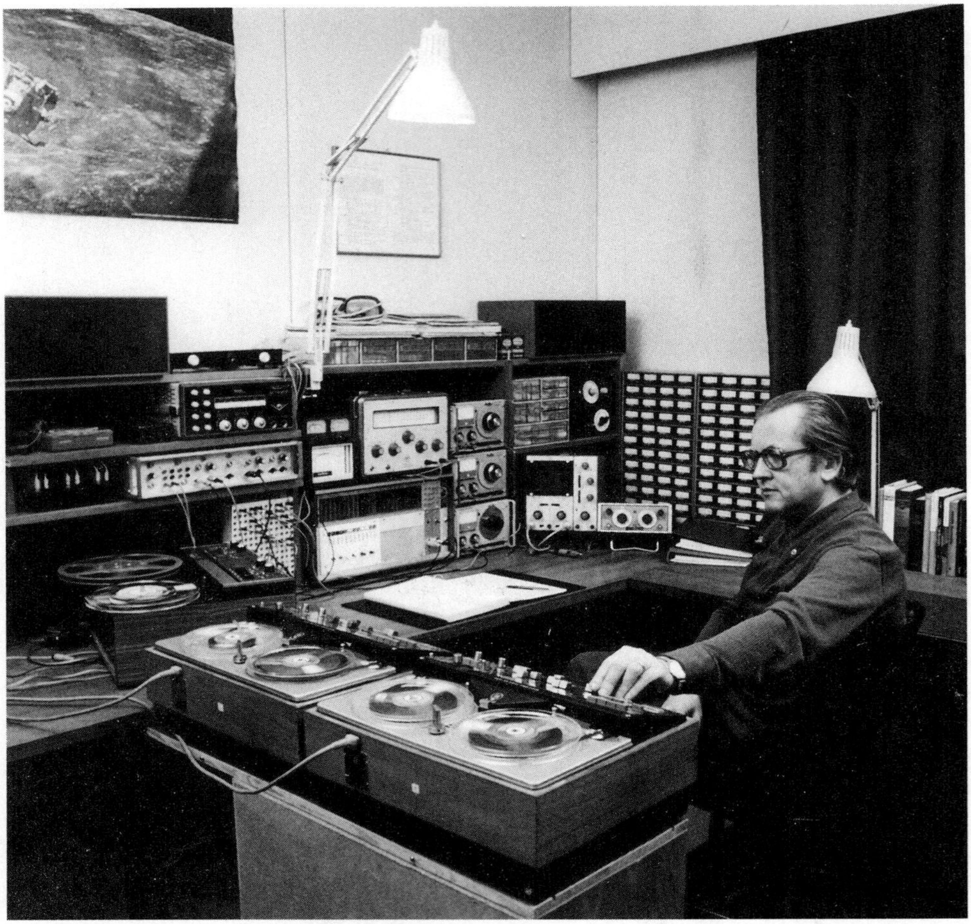

Figure II.6
DICO (Digitally Controlled Oscillator) in Osmo Lindeman studio, Finland, 1972.

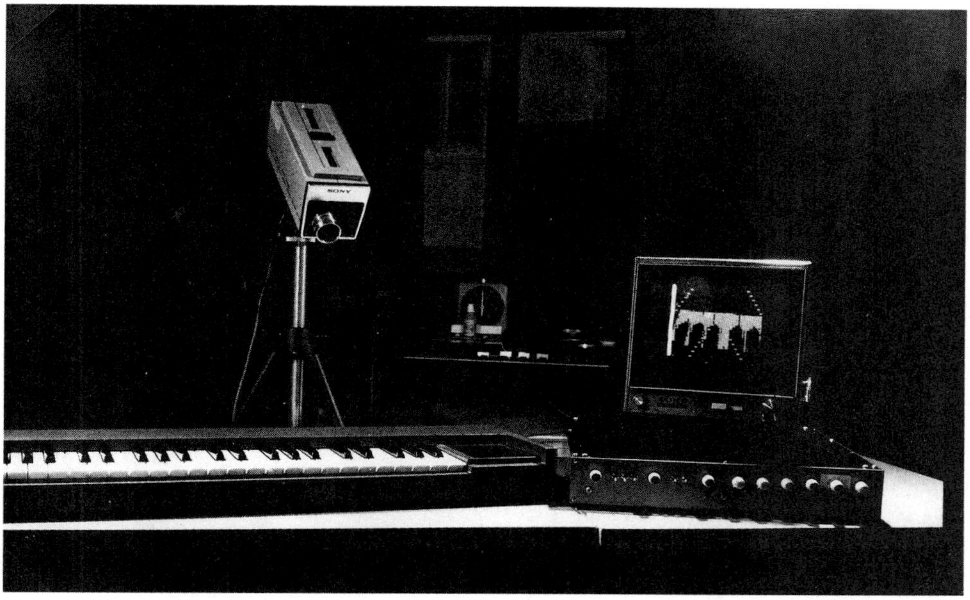

Figure II.7
DIMI-O (Optical Organ), 1971.

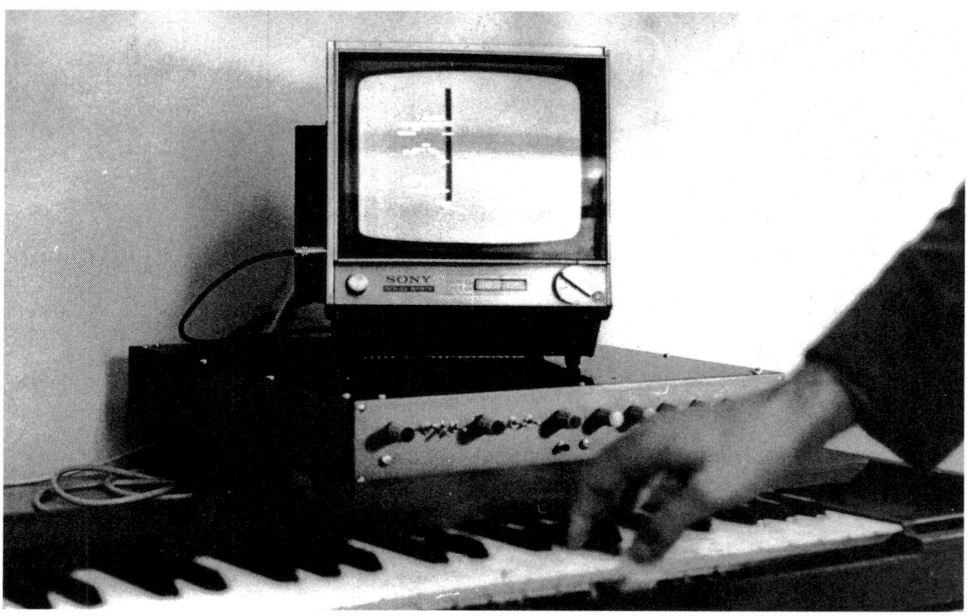

Figure II.8
DIMI-O, 1971.

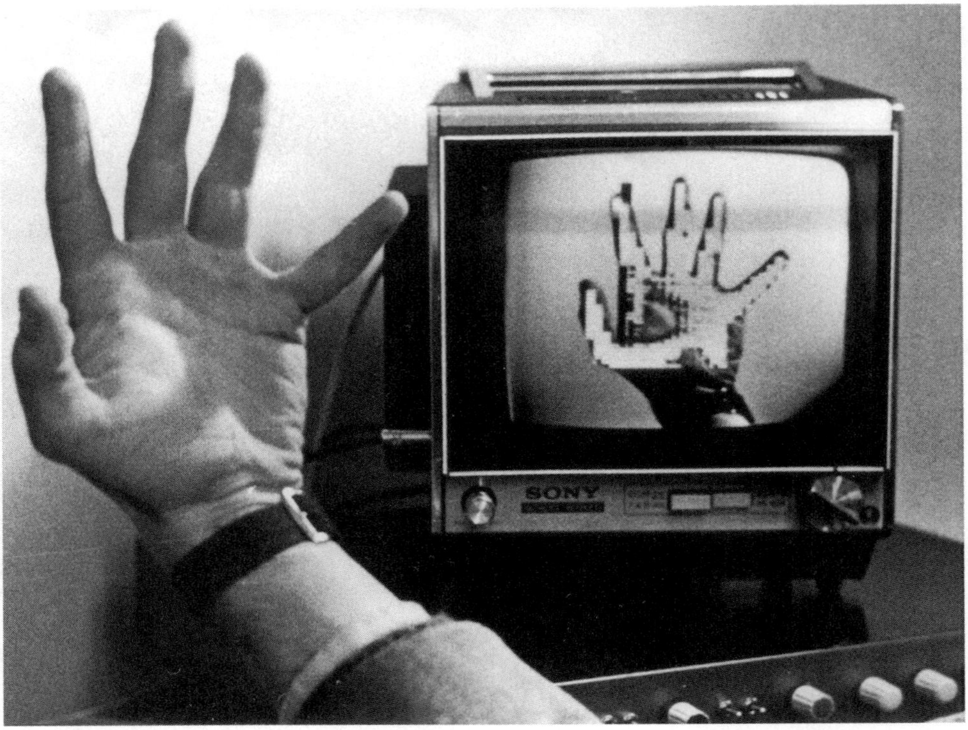

Figure II.9
DIMI-O, detail, 1971.

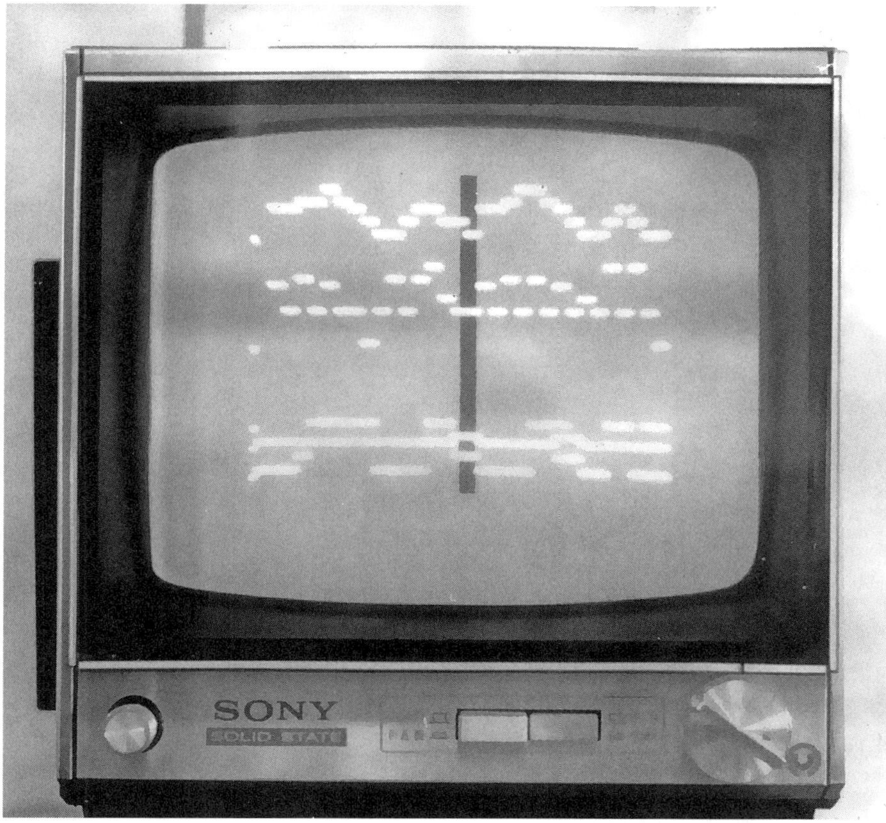

Figure II.10
DIMI-O, detail, 1971.

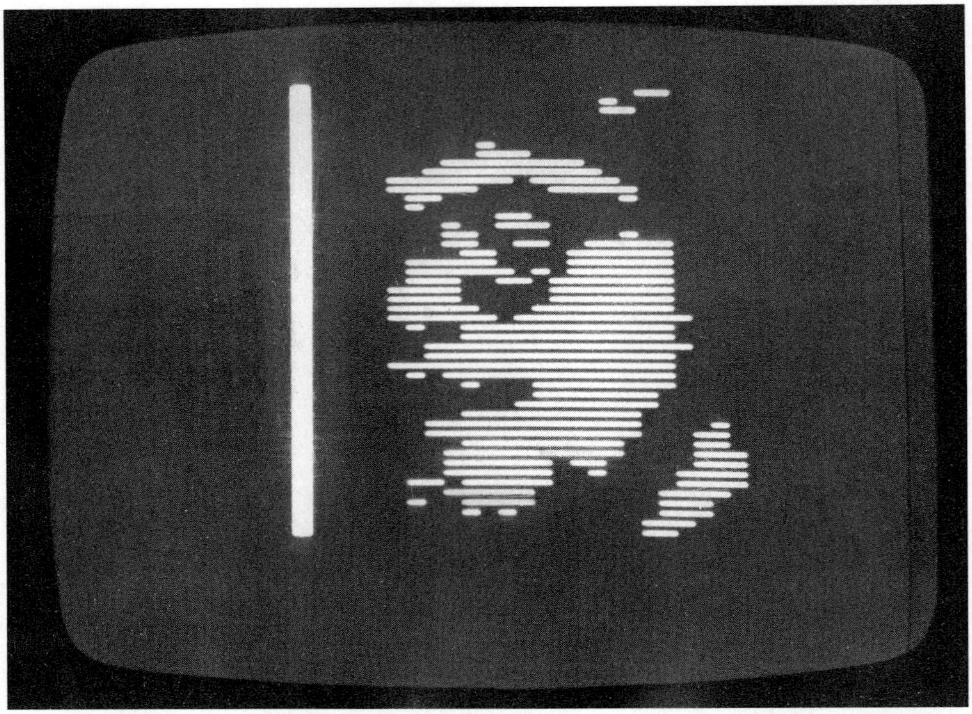

Figure II.11
DIMI-O, detail, 1971.

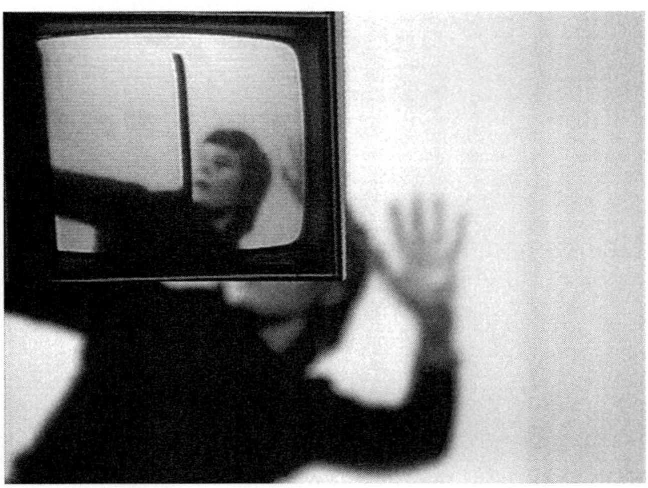

Figure II.12
DIMI-O machine. Still from the documentary *DIMI Ballet* produced by Finnish Broadcasting Company YLE, Helsinki, 1971.

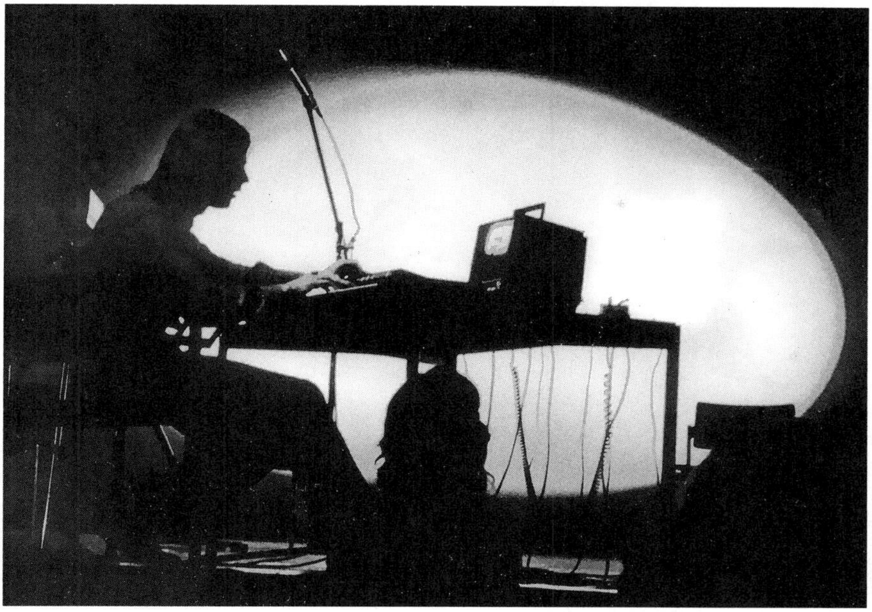

Figure II.13
Erkki Kurenniemi performing with DIMI-O, concert at Vanha Ylioppilastalo, Helsinki, 1971.

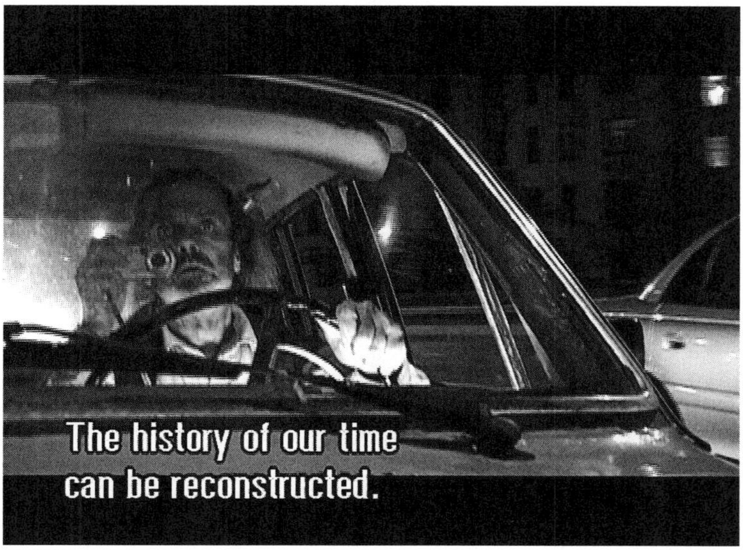

Figure II.14
Still from Mika Taanila's film *The Future Is Not What It Used to Be* (2002). © Kinotar.

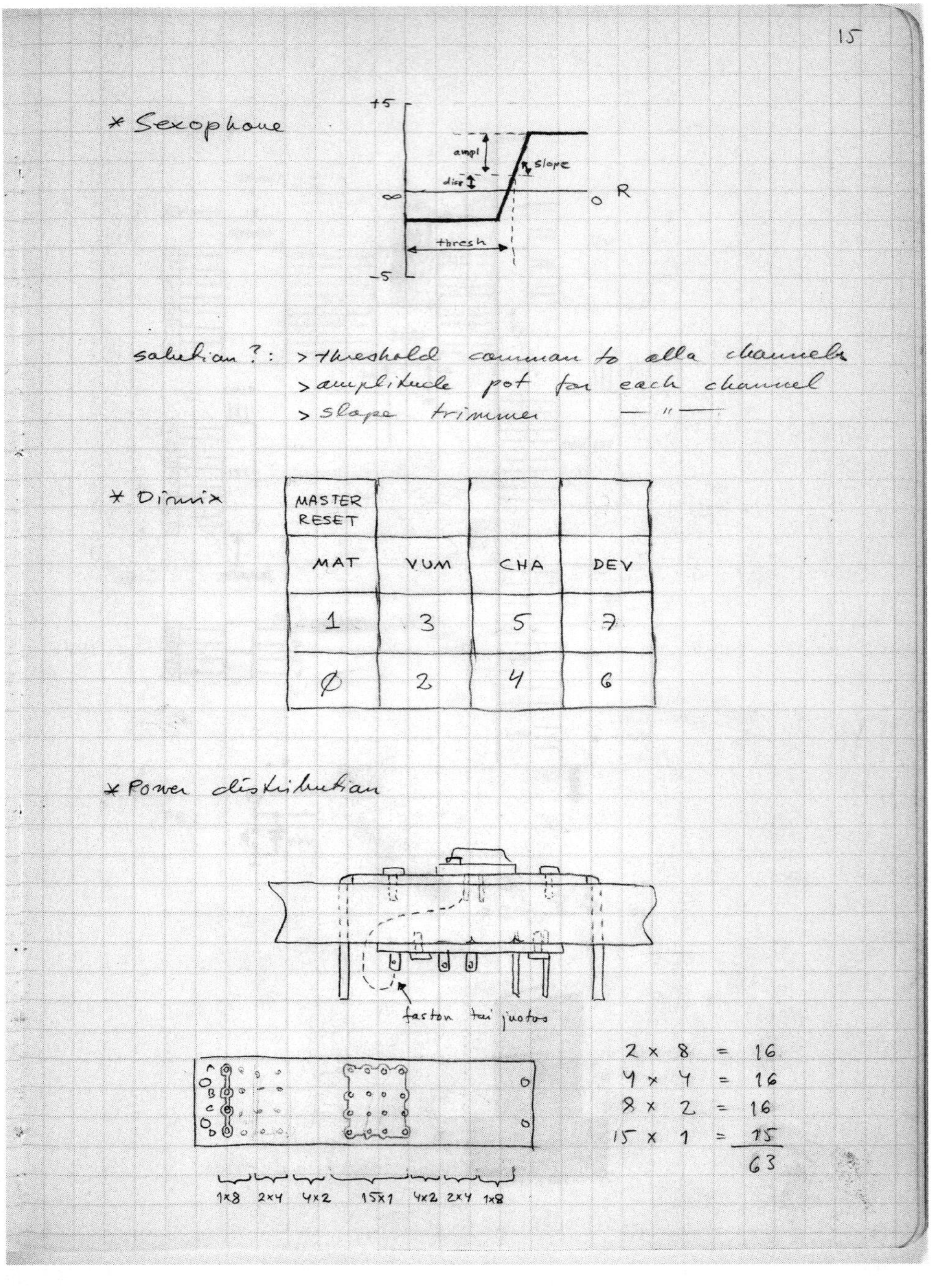

Figure II.15
Erkki Kurenniemi's sketches and notes for instruments. Unpublished.

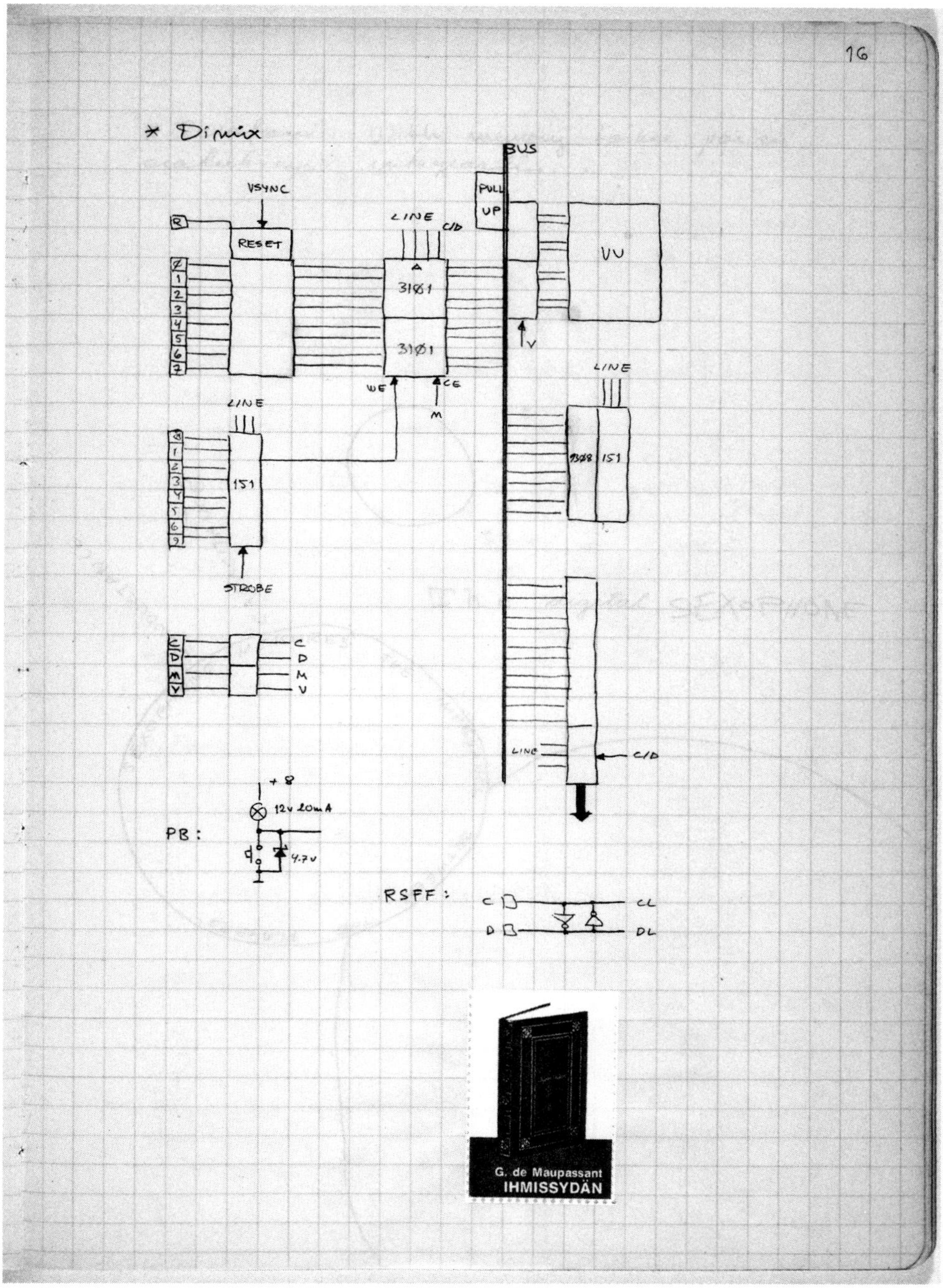

Figure II.15 (continued)

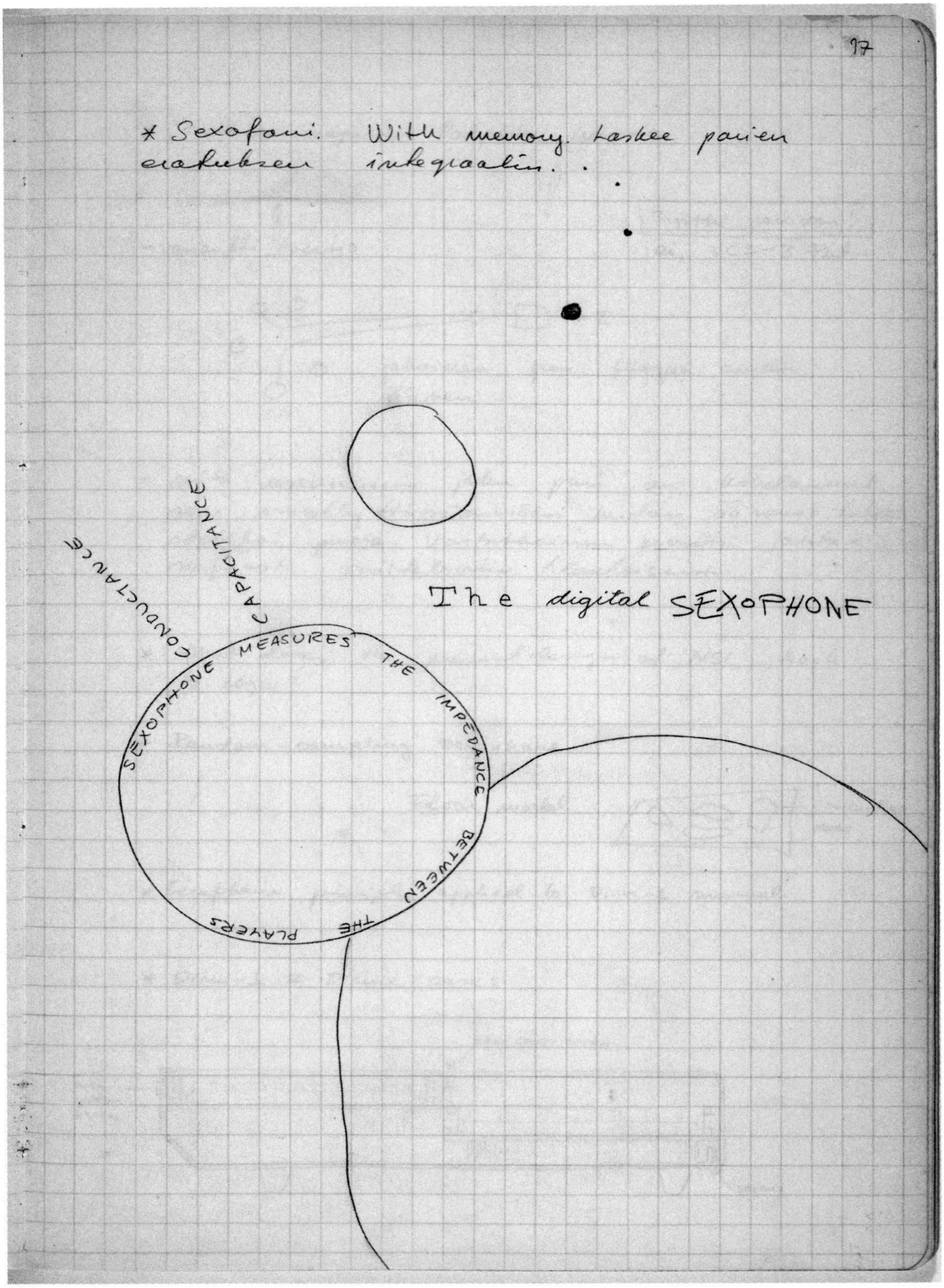

Figure II.15 (continued)

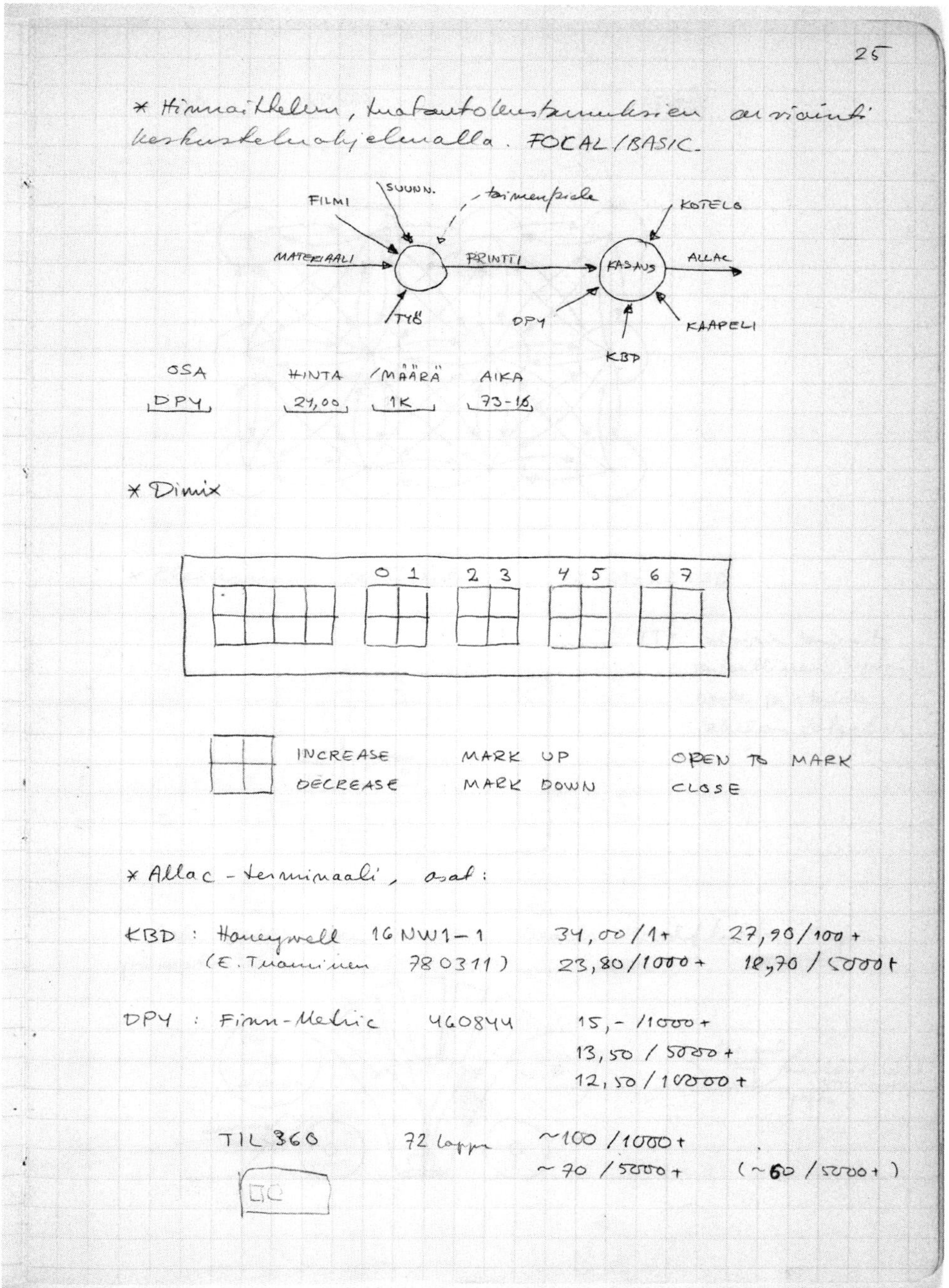

Figure II.15 (continued)

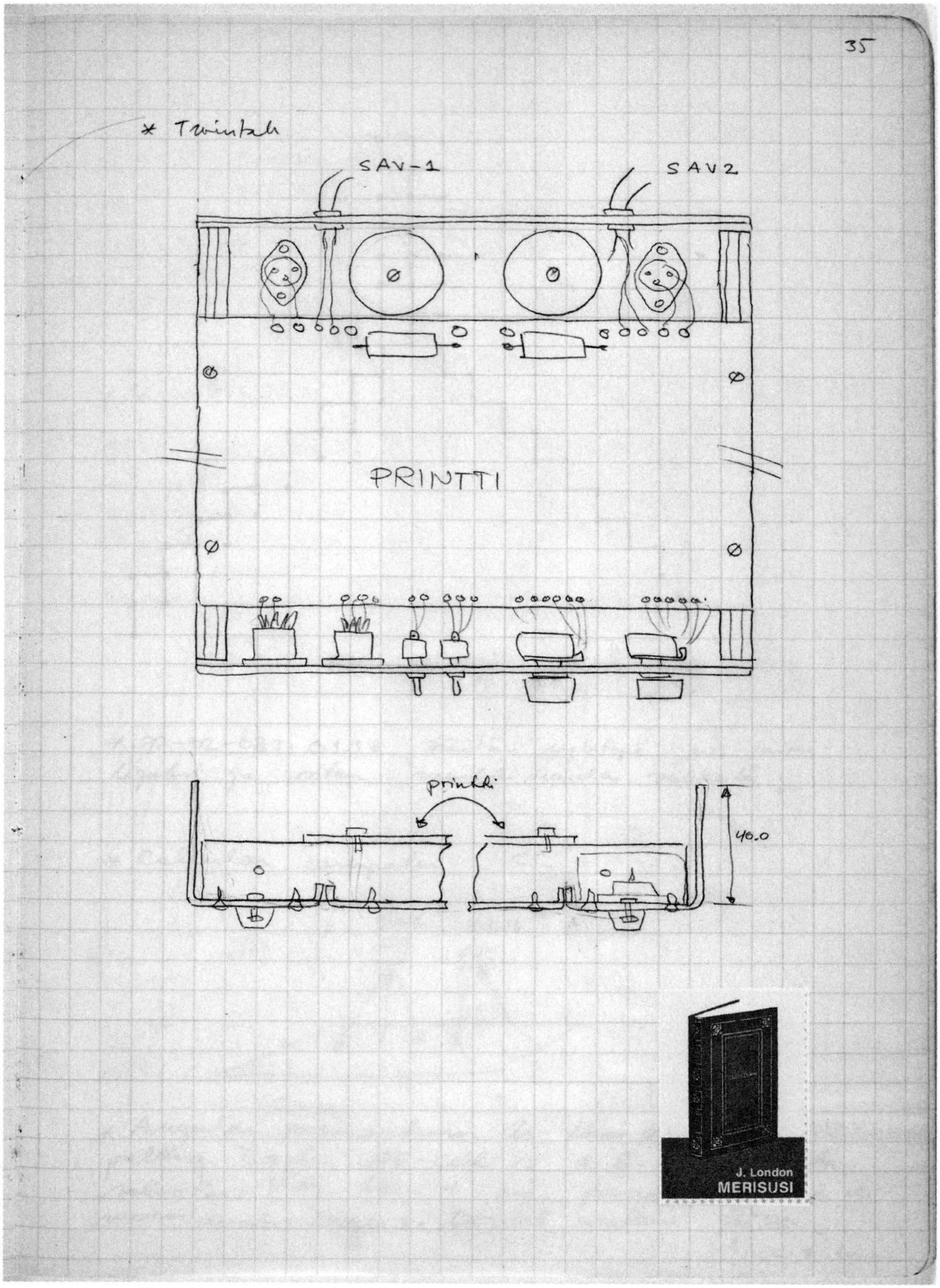

Figure II.15 (continued)

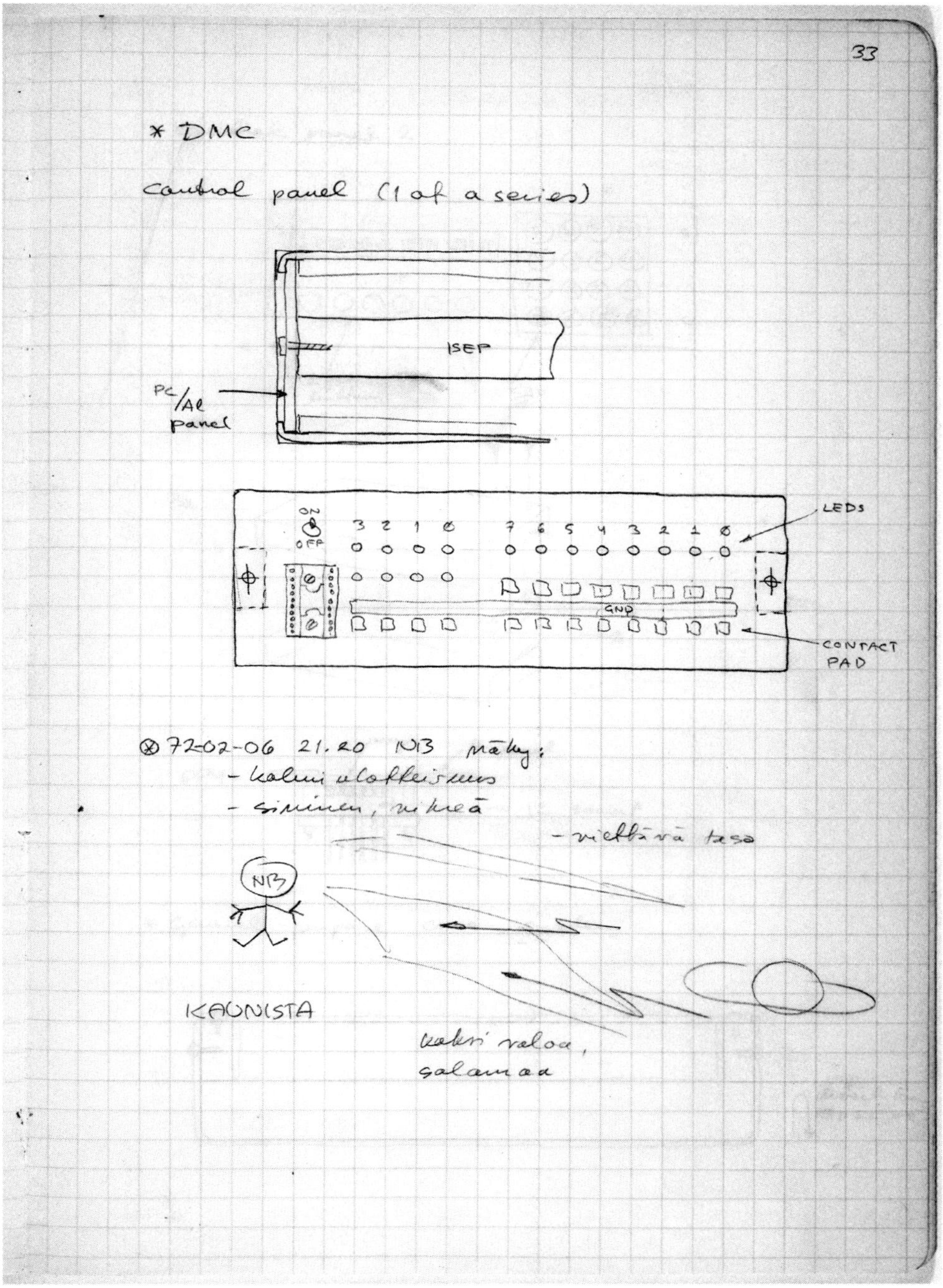

Figure II.15 (continued)

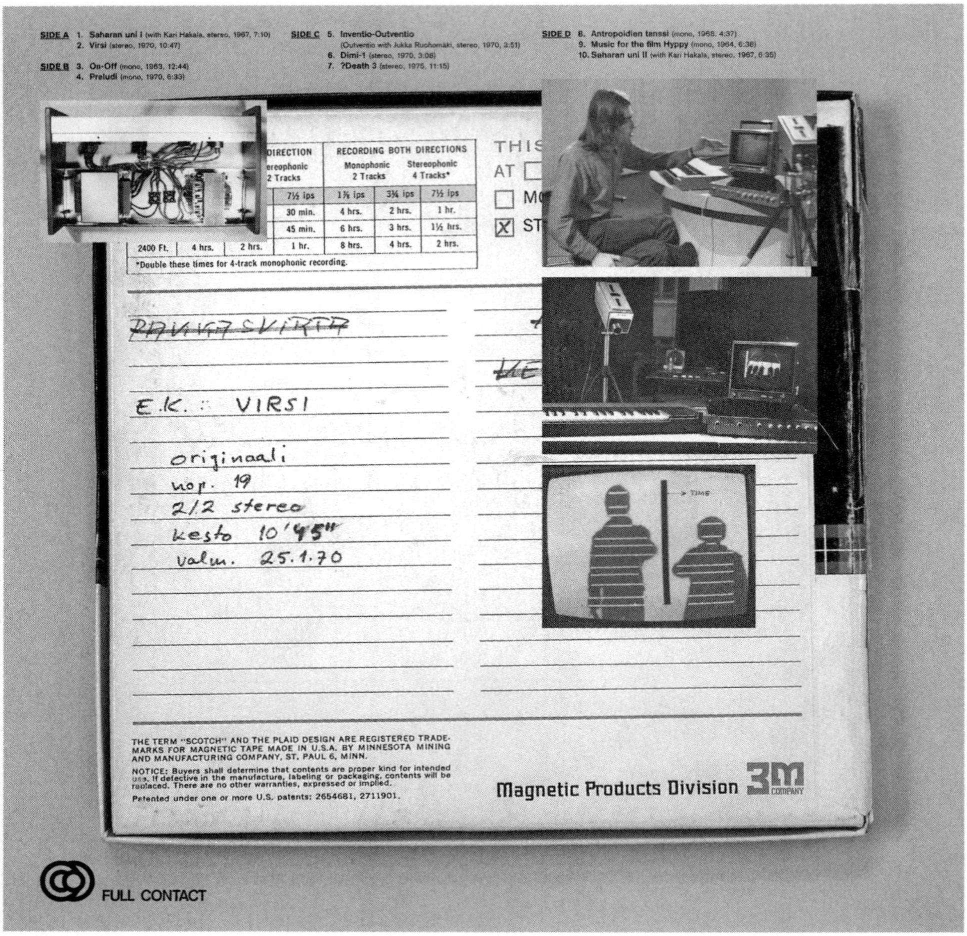

Figure II.16

Erkki Kurenniemi, *Rules*, double LP, Ektro Records, Krypt-022, released June 2012. Back cover: DIMI-O sample sequence; jacket design Musta kirahvi; photo Mikko Ojanen.

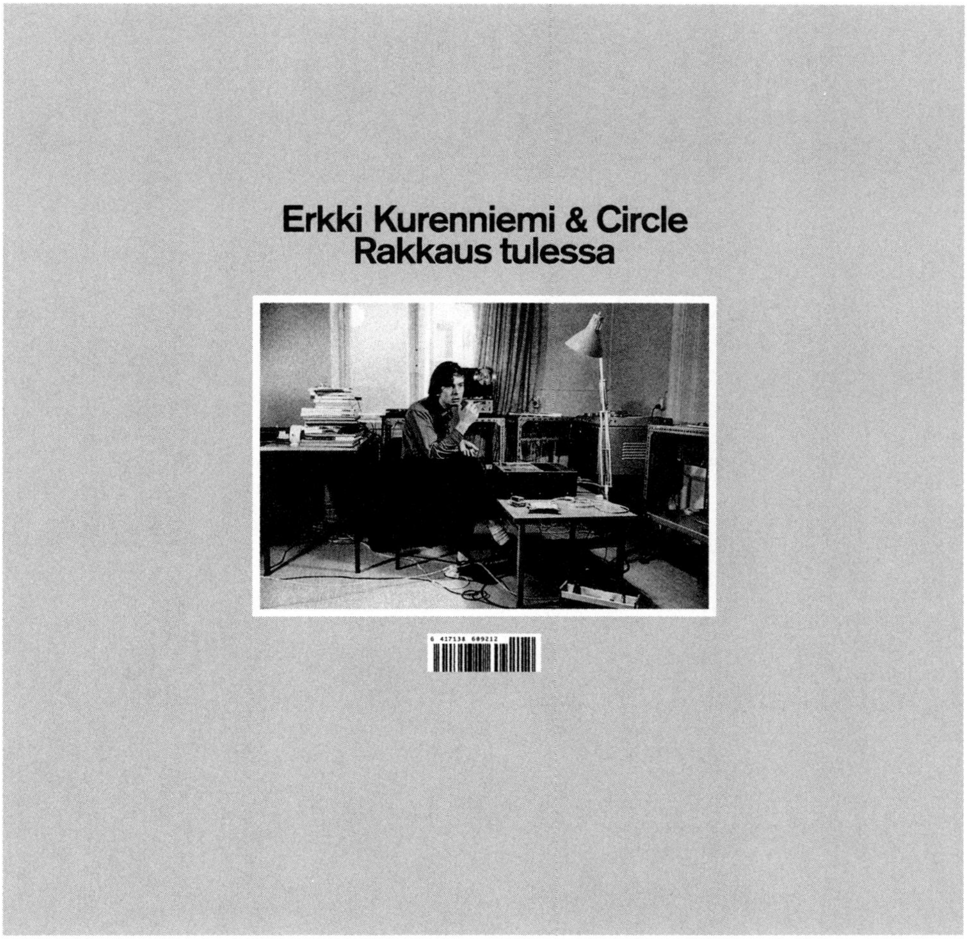

Figure II.17

Erkki Kurenniemi and Circle, *Rakkaus tulessa*, LP, Ektro Records, Krypt-014, released July 2011. Jacket design Musta kirahvi; front cover photo Martti Brandt.

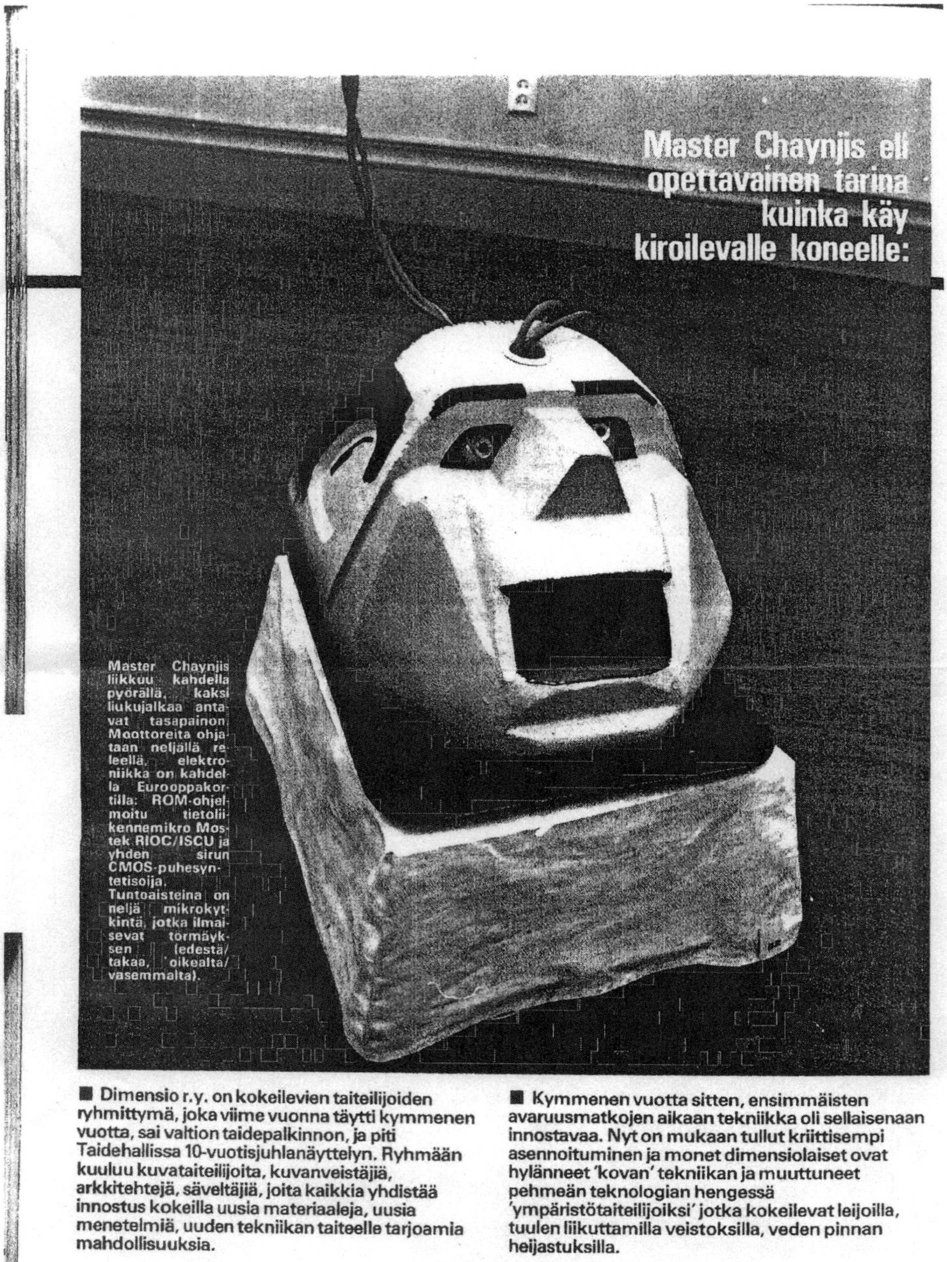

Master Chaynjis eli opettavainen tarina kuinka käy kiroilevalle koneelle:

Master Chaynjis liikkuu kahdella pyörällä, kaksi liukujalkaa antavat tasapainon. Moottoreita ohjataan neljällä releellä, elektroniikka on kahdella Eurooppakortilla: ROM-ohjelmoitu tietoliikennemikro Mostek RIOC/ISCU ja yhden sirun CMOS-puhesyntetisoija. Tuntoaisteina on neljä mikrokytkintä, jotka ilmaisevat törmäyksen (edestä/takaa, oikealta/vasemmalta).

■ Dimensio r.y. on kokeilevien taiteilijoiden ryhmittymä, joka viime vuonna täytti kymmenen vuotta, sai valtion taidepalkinnon, ja piti Taidehallissa 10-vuotisjuhlanäyttelyn. Ryhmään kuuluu kuvataiteilijoita, kuvanveistäjiä, arkkitehtejä, säveltäjiä, joita kaikkia yhdistää innostus kokeilla uusia materiaaleja, uusia menetelmiä, uuden tekniikan taiteelle tarjoamia mahdollisuuksia.

■ Kymmenen vuotta sitten, ensimmäisten avaruusmatkojen aikaan tekniikka oli sellaisenaan innostavaa. Nyt on mukaan tullut kriittisempi asennoituminen ja monet dimensiolaiset ovat hylänneet 'kovan' tekniikan ja muuttuneet pehmeän teknologian hengessä 'ympäristötaiteilijoiksi' jotka kokeilevat leijoilla, tuulen liikuttamilla veistoksilla, veden pinnan heijastuksilla.

90

prosessori 2/1983

Figure II.18

Master Chaynjis robot, 1982, featured in Erkki Kurenniemi with Esa Laurema, "Voi Ihmisen Peru," *Prosessori* 2 (1983).

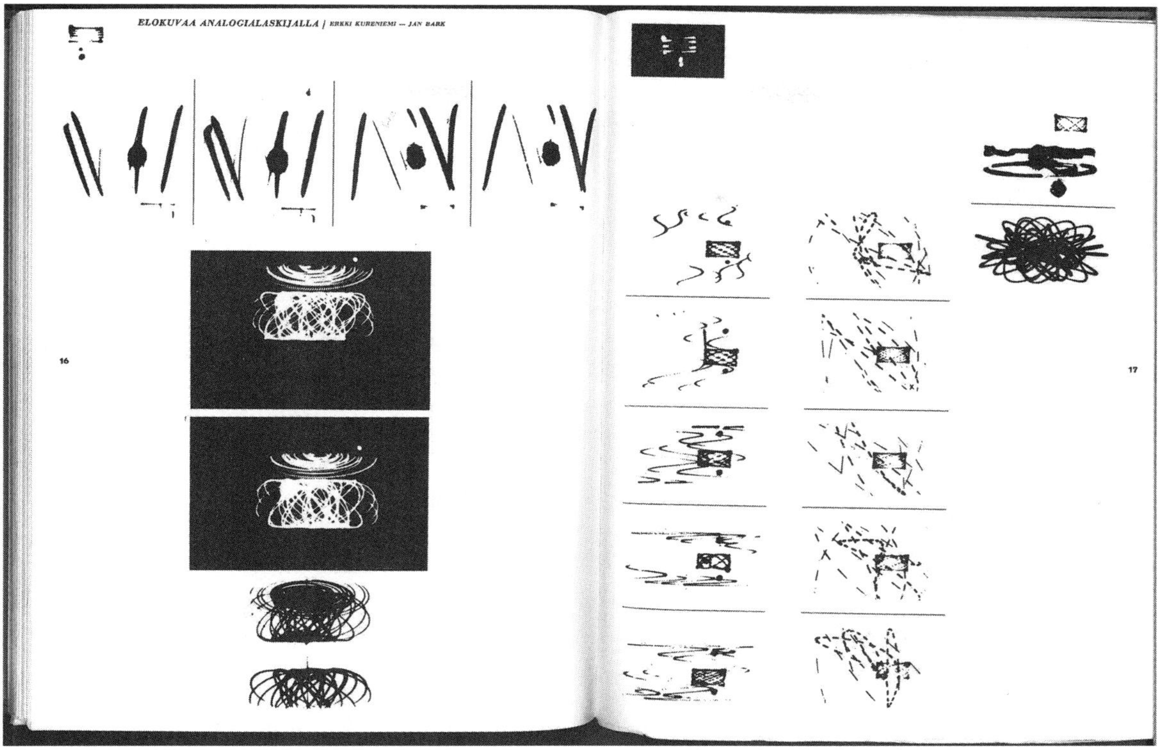

Figure II.19

Erkki Kurenniemi and Jan Bark, tests for *Spindrift*, Finnish magazine *Teekkari* (1963). Scans by Finnish National Library.

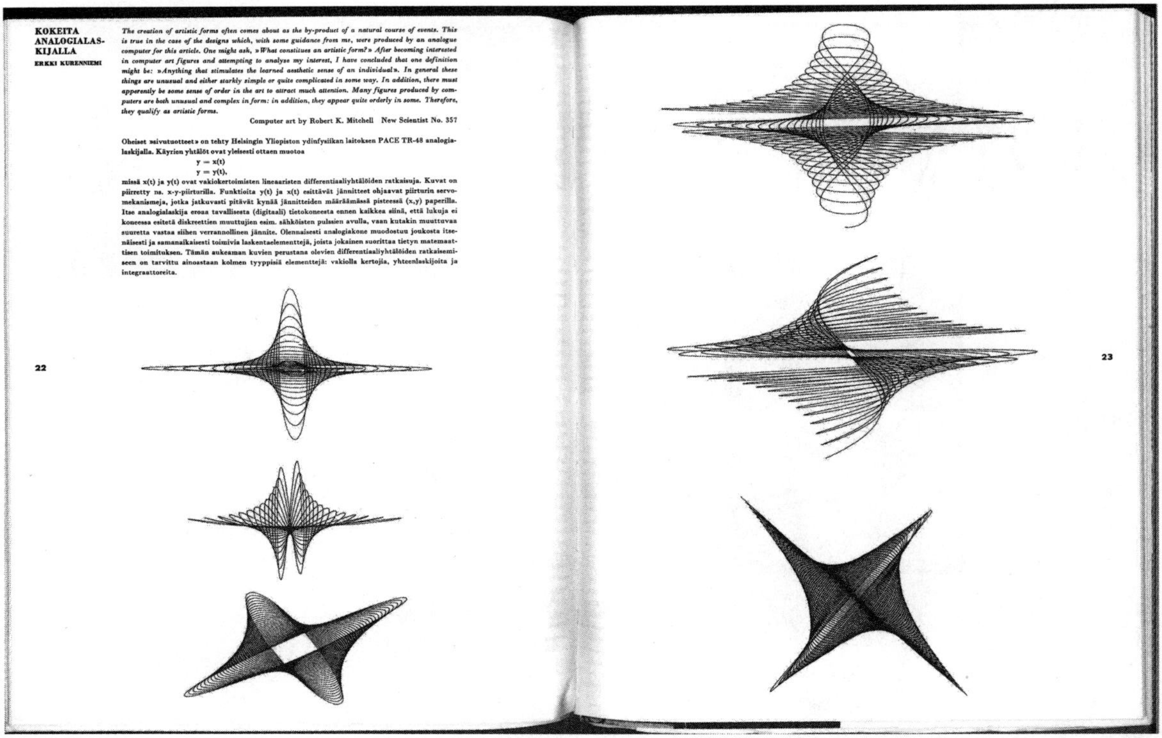

KOKEITA ANALOGIALASKIJALLA

ERKKI KURENNIEMI

The creation of artistic forms often comes about as the by-product of a natural course of events. This is true in the case of the designs which, with some guidance from me, were produced by an analogue computer for this article. One might ask, »What constitues an artistic form?» After becoming interested in computer art figures and attempting to analyse my interest, I have concluded that one definition might be: »Anything that stimulates the learned aesthetic sense of an individual». In general these things are unusual and either starkly simple or quite complicated in some way. In addition, there must apperently be some sense of order in the art to attract much attention. Many figures produced by computers are both unusual and complex in form: in addition, they appear quite orderly in some. Therefore, they qualify as artistic forms.

Computer art by Robert K. Mitchell New Scientist No. 357

Oheiset »sivutuotteet» on tehty Helsingin Yliopiston ydinfysiikan laitoksen PACE TR-48 analogialaskijalla. Käyrien yhtälöt ovat yleisesti ottaen muotoa

$$y = x(t)$$
$$y = y(t),$$

missä x(t) ja y(t) ovat vakiokertoimisten lineaaristen differentiaaliyhtälöiden ratkaisuja. Kuvat on piirretty ns. x-y-piirturilla. Funktioita y(t) ja x(t) esittävät jännitteet ohjaavat piirturin servomekanismeja, jotka jatkuvasti pitävät kynää jännitteiden määräämässä pisteessä (x,y) paperilla. Itse analogialaskija eroaa tavallisesta (digitaali) tietokoneesta ennen kaikkea siinä, että lukuja ei koneessa esitetä diskreettien muuttujien esim. sähköisten pulssien avulla, vaan kutakin muuttuvaa suuretta vastaa siihen verrannollinen jännite. Olennaisesti analogiakone muodostuu joukosta itsenäisesti ja samanaikaisesti toimivia laskentaelementtejä, joista jokainen suorittaa tietyn matemaattisen toimituksen. Tämän aukeaman kuvien perustana olevien differentiaaliyhtälöiden ratkaisemiseen on tarvittu ainoastaan kolmen tyyppisiä elementtejä: vakiolla kertojia, yhteenlaskijoita ja integraattoreita.

22

23

Figure II.20

Errki Kurenniemi and Jan Bark, tests for *Spindrift*, Finnish magazine *Teekkari* (1963). Scans by Finnish National Library.

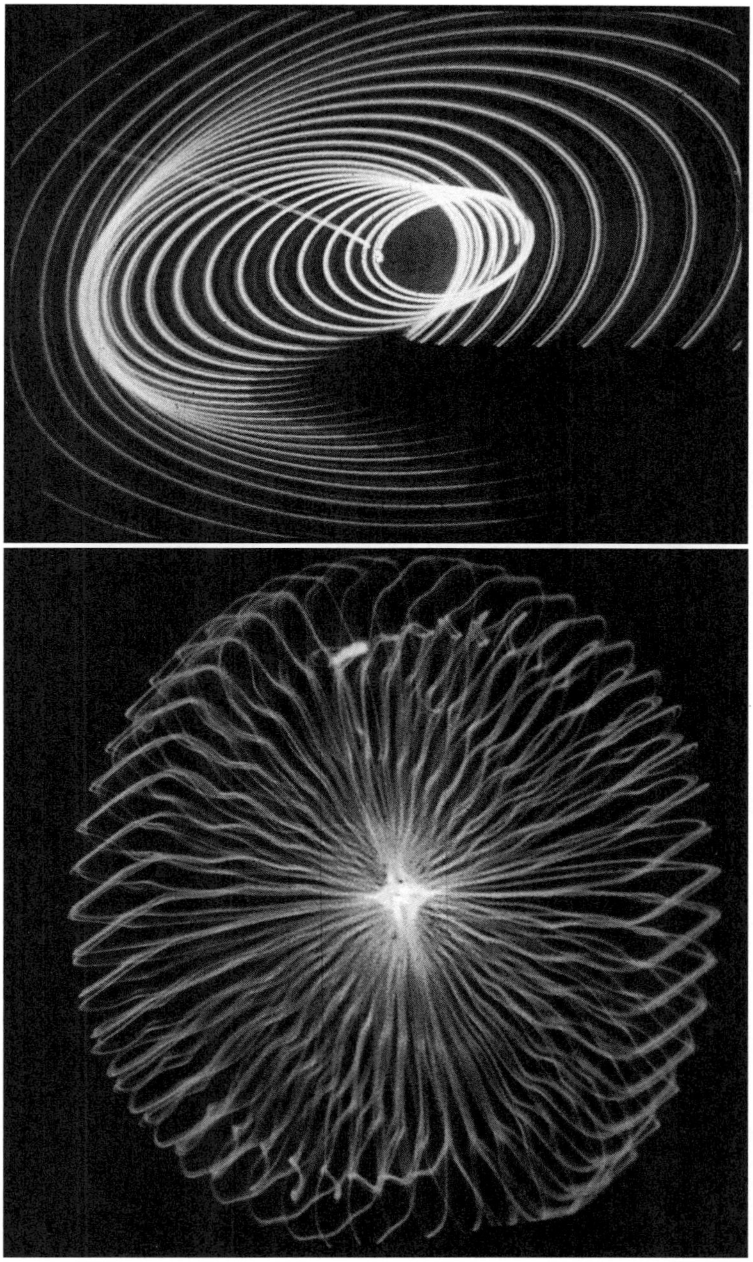

Figures II.21–24

Stills from the short film *Spindrift* (1966–1967) by Jan Bark and Erkki Kurenniemi. Courtesy Museum of Contemporary Art Kiasma in Helsinki.

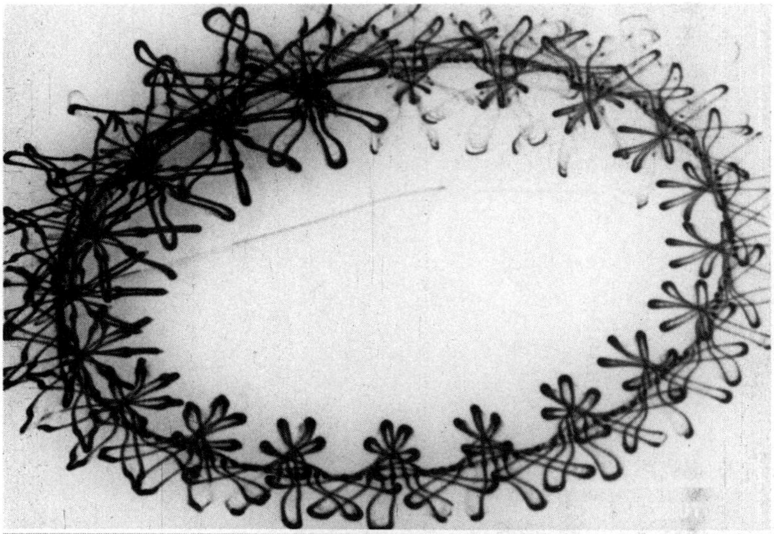

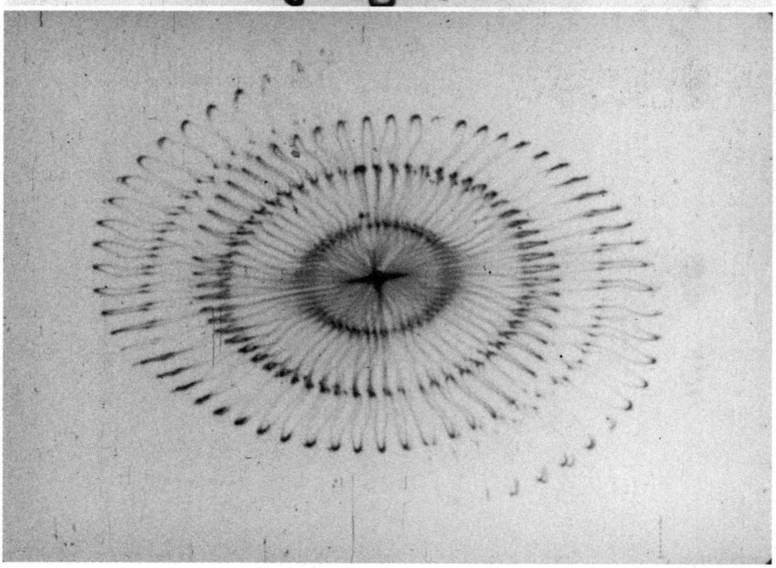

Figures II.21–24 (continued)

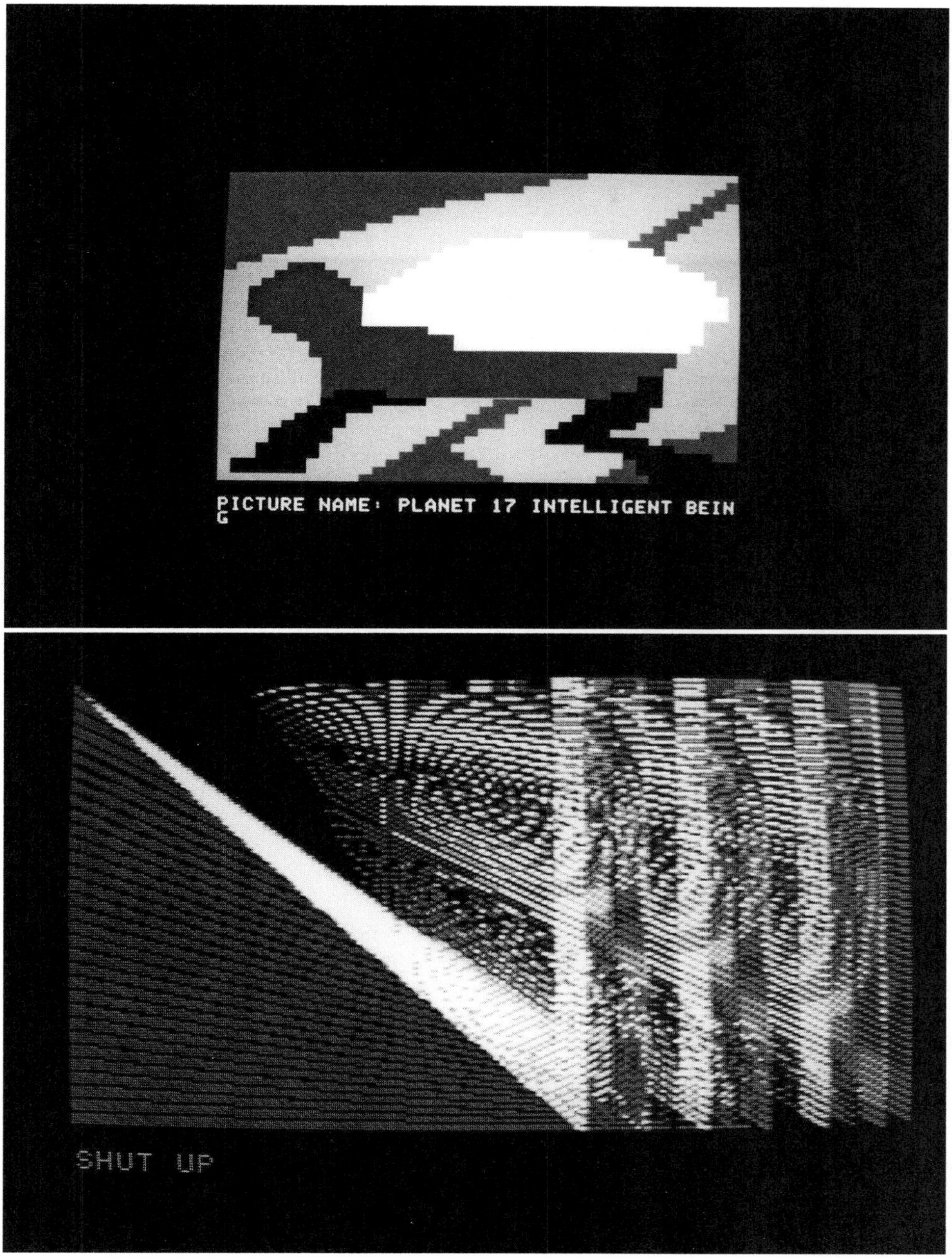

Figures II.25–27

Screenshots of basic animations and computer graphics, produced by Kurenniemi at the beginning of the 1980s using an Apple II computer.

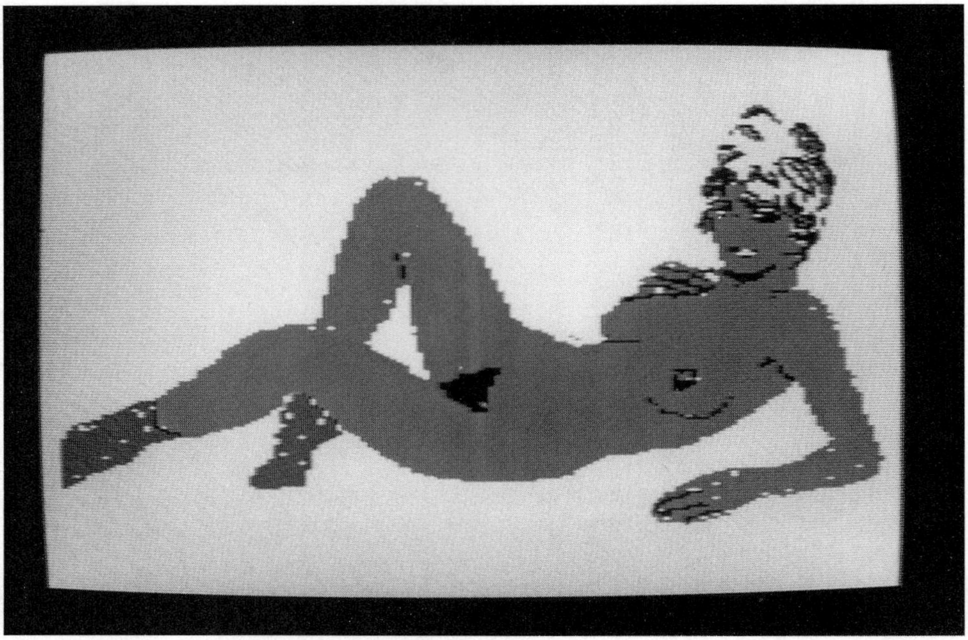

Figures II.25–27 (continued)

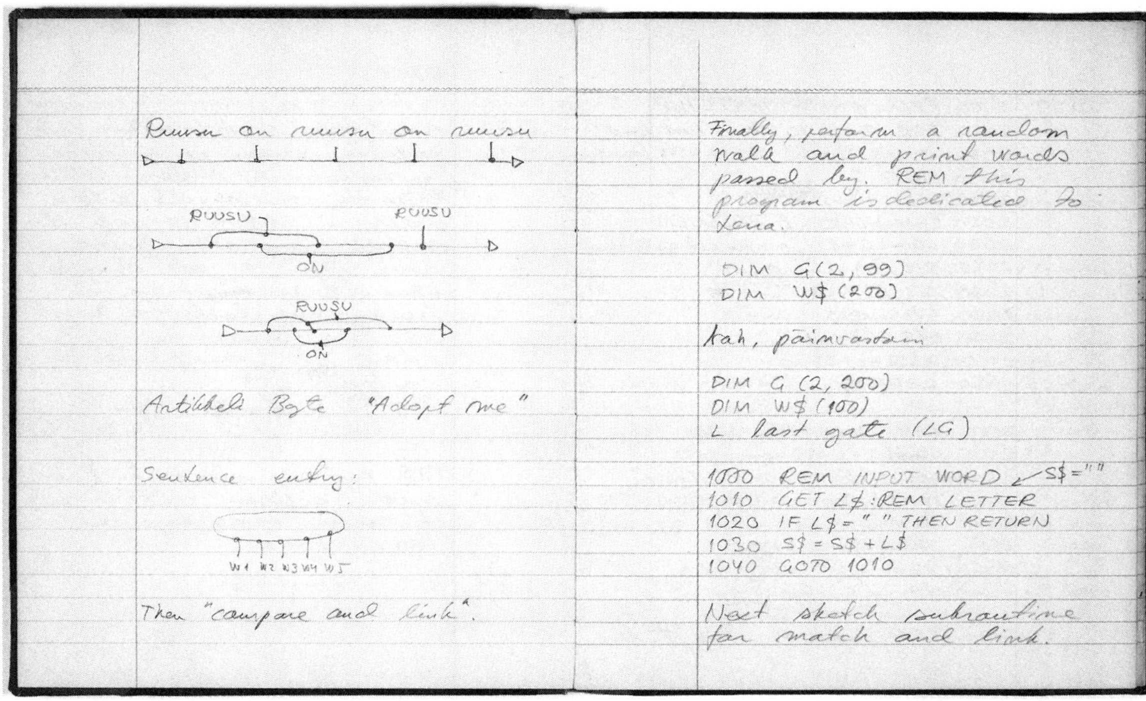

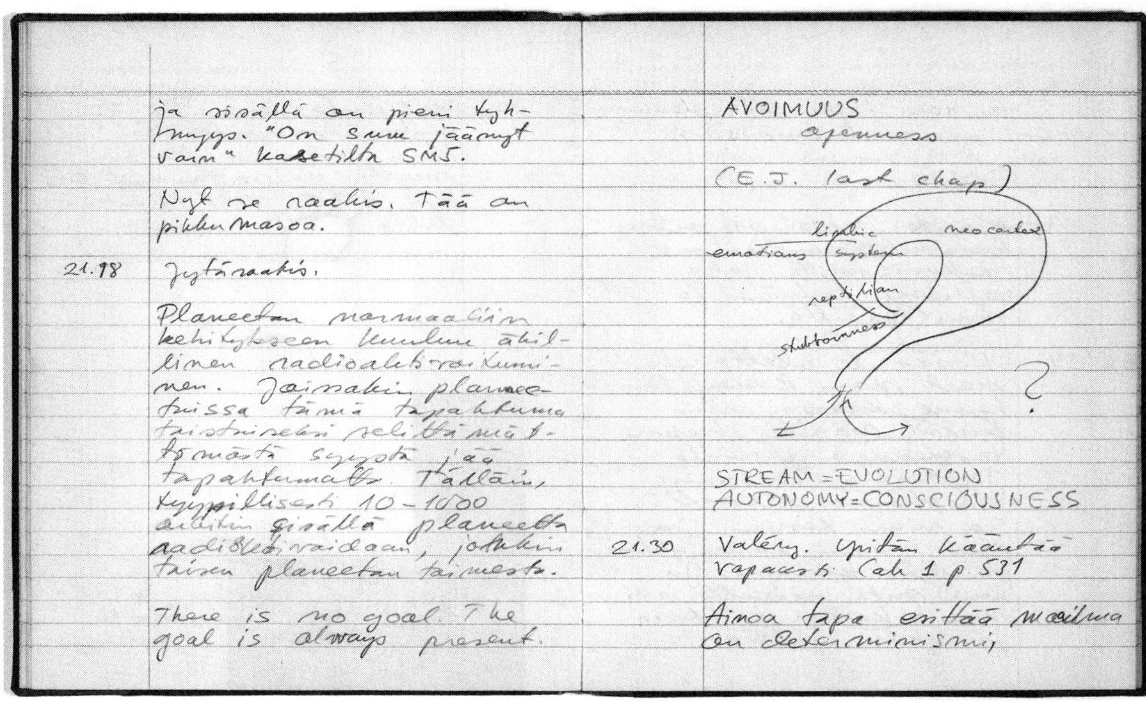

Figure II.28

Erkki Kurenniemi, excerpts from handwritten diaries (1970–2000). Originally published in Documenta 13 notebook no. 7, "Erkki Kurreniemi," with an introduction by Lars Bang Larsen, 2012.

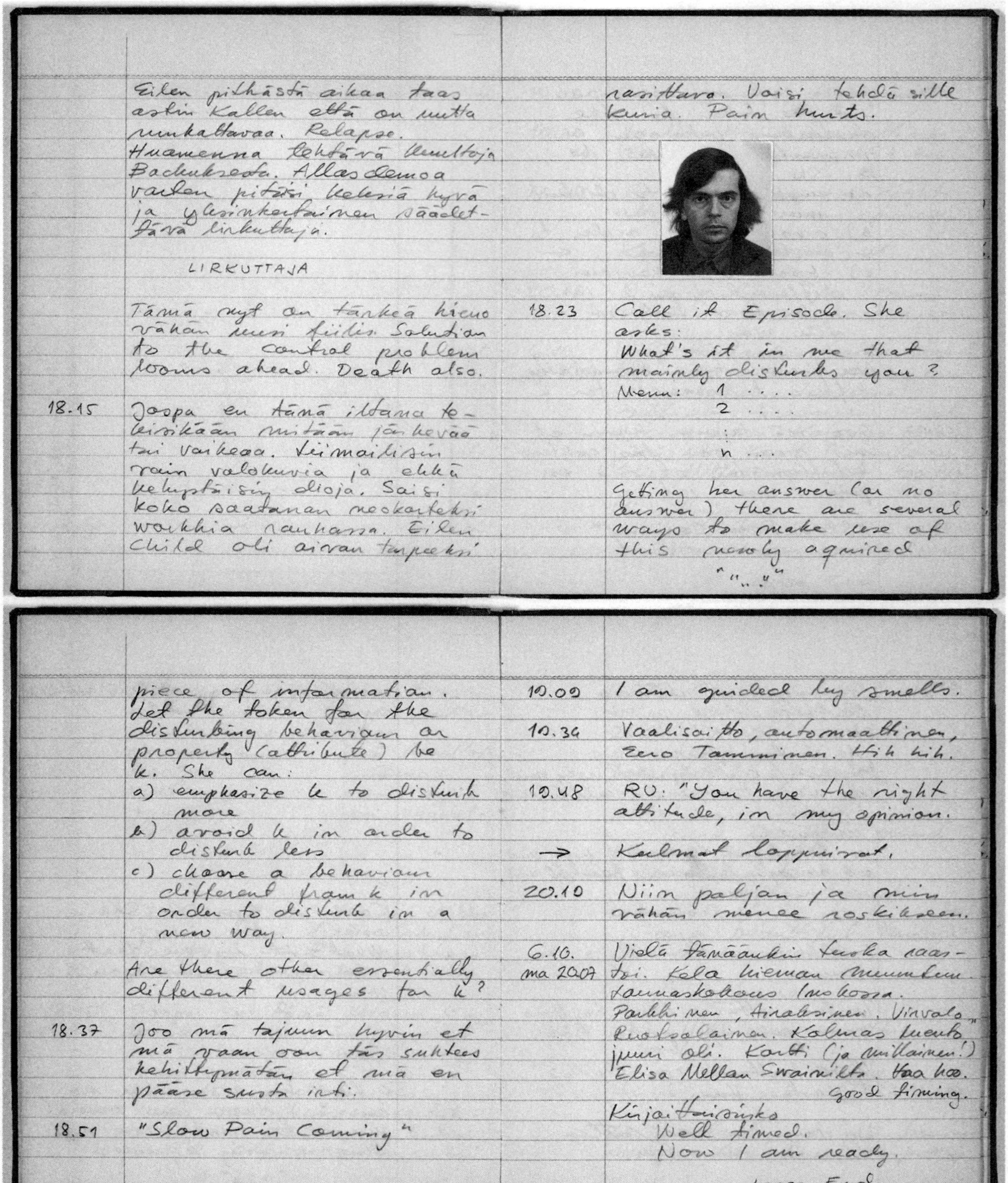

Figure II.28 (continued)

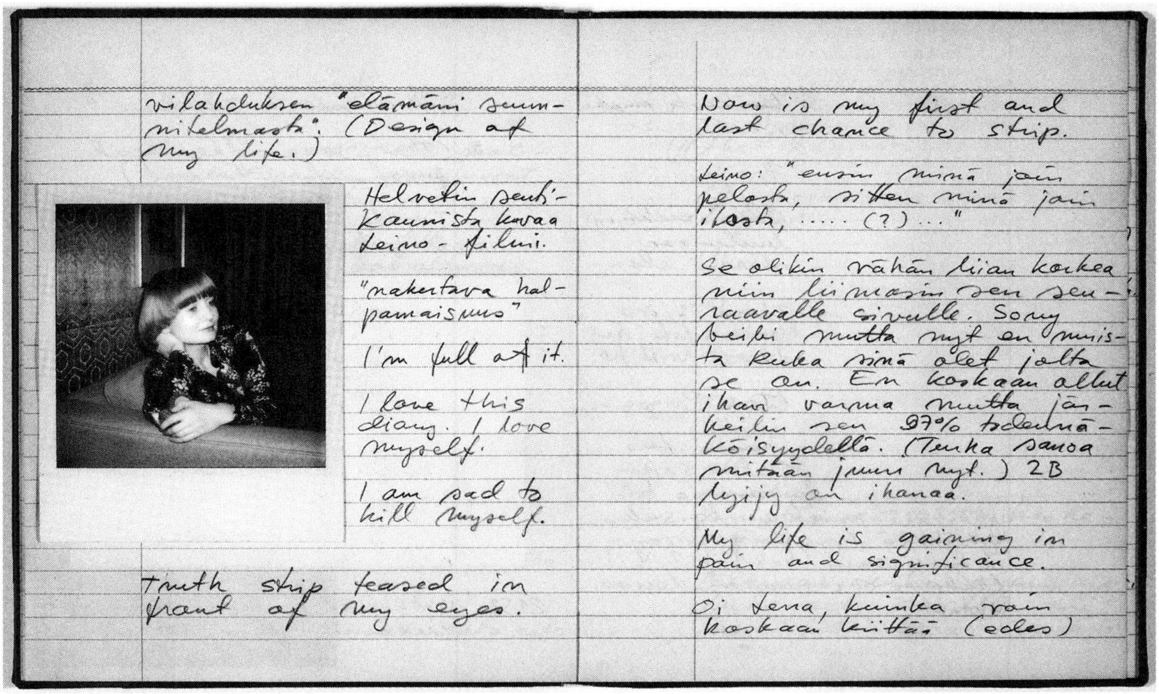
vilahduksen "elämäni suunnitelmasta". (Design of my life.)

Helvetin surki-kaunista kovaa Leino-filmi.

"makertava halpamaisuus"

I'm full of it.

I love this diary. I love myself.

I am sad to kill myself.

Truth strip teased in front of my eyes.

Now is my first and last chance to strip.

Leino: "ensin minä jäin pelosta, sitten minä jäin ilosta, ... (?) ..."

Se olikin vähän liian korkea niin liimasin sen seuraavalle sivulle. Sorry beibi mutta nyt en muista kuka sinä olet jolta se on. En koskaan ollut ihan varma mutta jäi-keilin sen 97% todennäköisyydellä. (Turha sanoa mitään juuri nyt.) 2B lyijy on ihanaa.

My life is gaining in pain and significance.

Oi Lena, kuinka voin koskaan kiittää (edes)

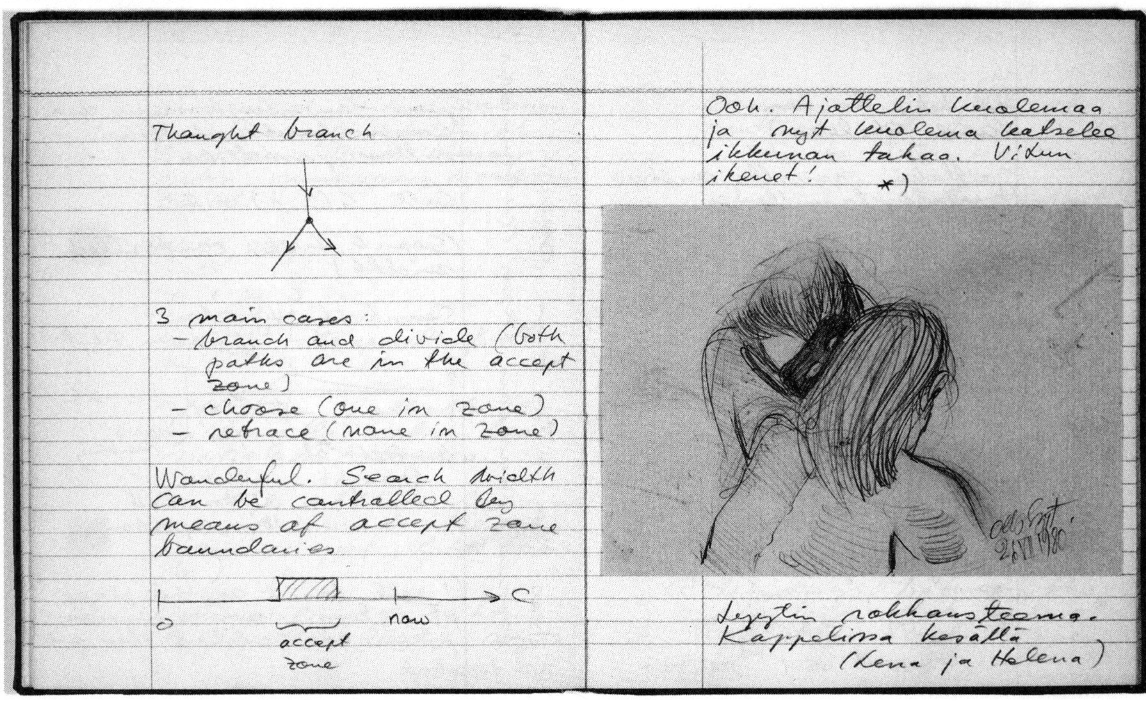
Thought branch

3 main cases:
- branch and divide (both paths are in the accept zone)
- choose (one in zone)
- retrace (none in zone)

Wonderful. Search width can be controlled by means of accept zone boundaries.

Ooh. Ajattelin kuolemaa ja nyt kuolema katselee ikkunan takaa. Vitun ikenet! *)

Löysin rakkausteema. Kappelissa kesällä (Lena ja Helena)

Figure II.28 (continued)

Tietokone vapauttaa taiteen

Joko Sinulla on kotimikro? Kuka Teidän perheessä pelaa parhaiten Packmania, Zaxxsonia, Zorkia, Choplifteriä ja Moonlanderia?

Videopeli on tämän päivän seikkailusatu. Ei ole sattuma, että videopelit ovat teini-ikäisten ja hieman kehityshäiriöisten keksimiä. He ovat aikuisia rohkeampia ja ennakkoluulottomampia.

Monet taiteilijat ovat pakanneet siveltimensä naftaliiniin ja tarttuneet tietokoneeseen. Ensin tietokoneeseen uskalsivat koskea muusikot, sitten kuvataiteilijat.

Systeemitutkija Erkki Kurenniemen mielestä tietokone on ihmiskunnan suurin keksintö tulen ja pyörän jälkeen. "Tietokone tulee olemaan suuri kulttuurin yhdis-äjä. Se on universaali väline, ilmiö jolla ei ole esikuvaa. Sillä on suuremmat potentiaaliset mahdollisuudet kuin kirjapainotaidolla."

Kurenniemi työskentelee Nokialla teollisuusautomaation ja robottien tuotekehittelyssä. Hän kuuluu suomalaisen tietokonealan ylipapistoon. Hän tuntee tietokonealan elektronisista komponenteista sovellutusohjelmointiin.

Kurenniemi ei halua erotella hyvää suunnittelua ja taidetta. "Hyvä suunnittelutyö on tietyltä osin taidetta. Siinä on sama loputon luomisen tuska ja ah, niin harvoin tulevat inspiraation hetket."

Illuusion ajasta interaktion aikaan

"Uusi tietotekniikka tuhatkertaistaa informaation määrän ja liikkumisnopeuden. Laadullinen muutos on vielä merkittävämpi", Kurenniemi sanoo.

Kaikki tietokonetta edeltävät tietotekniikat, esimerkiksi valokuva, filmi ja televisio, ovat passiivisia. Ne pyrkivät tallentamaan tai siirtämään informaatiota uskollisesti.

"Perinteisen tietotekniikan ihanne on High Fidelity, mahdollisimman täydellisen illuusion luominen. Valokuva vapautti maalaustaiteen uskollisuuden taakasta, elokuva teatterin", Kurenniemi sanoo. Sittemmin tuli abstrakti valokuva ja abstrakti elokuva.

Tietokone ei ole passiivinen väline. Se ei pelkästään talleta ja valitse informaatiota, vaan muokkaa ja muuntaa sitä aktiivisesti. Tietokone on vuorovaikutuksessa ympäristön ja ihmisen kanssa.

Videoväri ei kuivu

Tietokonetaiteessa katsoja on Kurenniemen mukaan osa taideteosta, jota hän katselee. Ihanuus on siinä, että kuvaa voi muuttaa, ellei entinen miellytä. "Videoväri ei kuivu."

Tietokoneella piirtäminen on nopeampaa kuin siveltimellä. Tekijä voi käyttää enemmän aikaa suunnitteluun. "Siinä mielessä tietokone vapauttaa taiteen", Kurenniemi sanoo.

Hän uskoo, että vuonna 2000 80 % kuvataiteesta tehdään käyttämällä tietokonetta primäärisenä välineenä. "Kone vapauttaa tekijän median rajoituksesta. Tietokoneella on kätevää tehdä se, mikä olisi työlästä perinteisin menetelmin."

Esimerkiksi animaation tekemisessä ei tietokonetta käytettäessä tarvitse kuva kuvalta piirtää uudelleen kuin liikkuvat osat.

Tietokonetaiteessa painopiste on Kurenniemen mukaan vielä kokeilussa ja tekniikassa. "Hyvästä ja huonosta tietokonetaiteesta ei tässä vaiheessa kannata puhua. Siihen vaaditaan pitempää aikaperspektiiviä."

Ylipäätään Kurenniemi ei halua määritellä taidetta. "Taiteen olemukseen kuuluu, että se murtaa kaikkien määritelmien rajat." Katsojalle on olennaista, pitääk- hän jostain vai ei.

Tietokoneohjelmia taidenäyttelyyn

Erkki Kurenniemi on innostunut tietokonetaiteesta. Hänellä ei ole muodollista koulutusta tai pätevyyttä taiteen alalla, mutta hän o- vuosikaudet harrastan- musiikin teoriaa, rakentan- elektronisia soittimia ja s-

Ainutlaatuisruno 5860860860

Kaikkiin seutuihin puut vajoavat
te olette tarpeelliset
hallitseminen on hallitsemista.
Ainoata aikakautta jaetaan illalla sammumaan
maailmat pysähtyneet loppuun leikkivät.

Ainutlaatuisruno 41821282182

Jokainen paluu.
Jokainen suuntiin paluusta
kenen laaja
näköpiirin laaja.

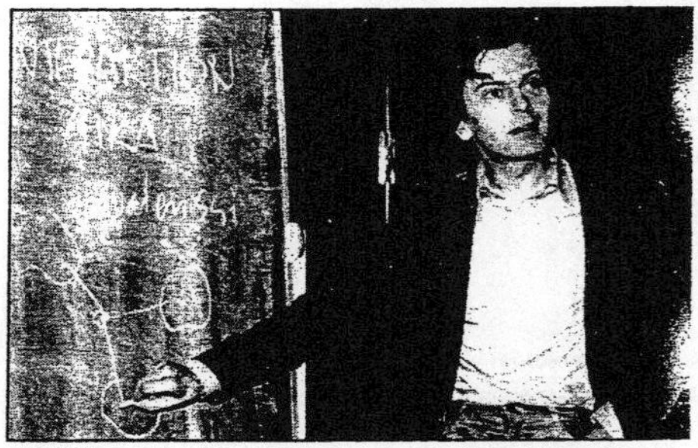

Erkki Kurenniemi on tietokoneuransa aikana saanut kaksi oivallusta. Toinen liittyy harmonia-käsitteen ja lukusuhteiden vastaavuuteen, toinen musiikin rakenteen esittämiseen binääripuiden avulla.

8

Figure II.29

Erkki Kurenniemi, "Tietokone vapauttaa taiteen," archival article (English trans. "Computer Liberates Art").

Teksti ja valokuva: Anitta Manninen
Piirrokset: Erkki Kurenniemi/Apple II +
Runot: Juha Viikki/Digital Rainbow 100

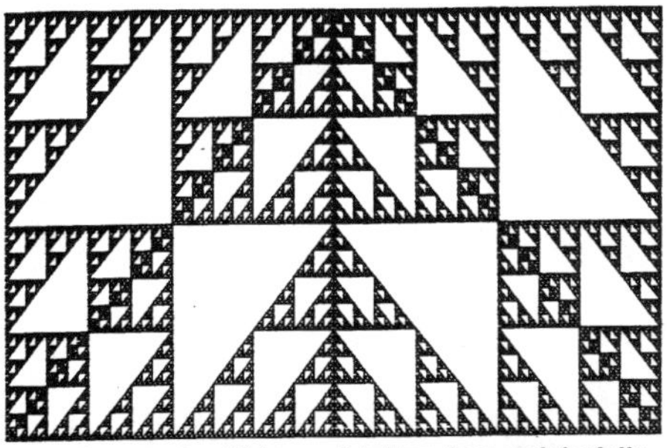

Intiaanien ornamentiikkaa muistuttava teos on toteutettu klassisella rekursiivisella algoritmilla. Aihe perustuu kolmioon, joka muodostuu kolmesta kolmiosta, joista kukin muodostuu kolmesta kolmiosta, joista kukin muodostuu kolmesta kolmiosta...

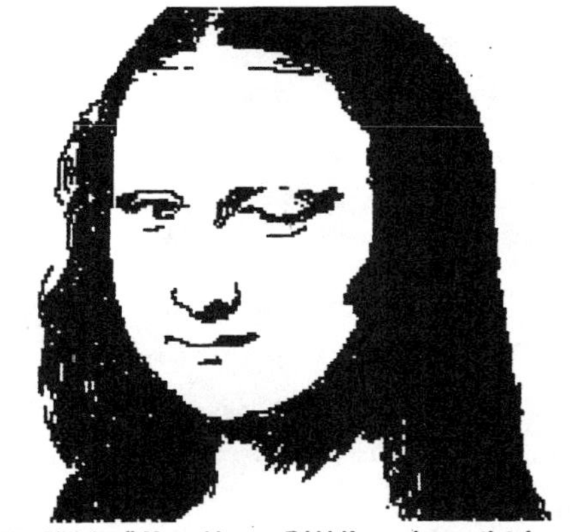

Silmää iskevä Mona Lisa on Erkki Kurenniemen tietokonetaiteen bravuuri. Kuva on viety grafiikkataululla koneen muistiin ja talletettu. Sitten on tehty toinen kuva, jossa vasen silmä on retusoitu kiinni. Näyttöpäätteellä Mona Lisa katsoo avoimin silmin 10 sekuntia ja aina sen jälkeen iskee silmää 1/10 sekunnin ajan.

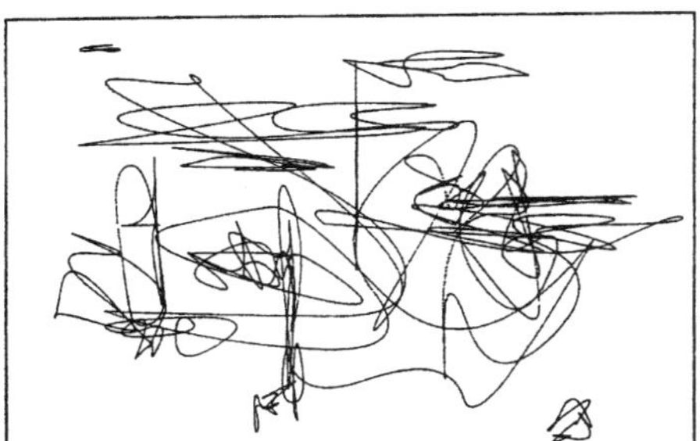

Spontaanin näköinen tuherrus on toteutettu toisen asteen rekursiivisella Bezier-interpolaattorilla.

veltänyt jonkin verran elektronimusiikkia. Lisäksi hän on tehnyt tietokonegrafiikkaa ja joitakin tietokonerunoja.

Kurenniemi on kehittänyt muun muassa seksofonin. Ääni syntyy, kun käsiraudoilla toisiinsa kytketyt ihmiset koskettelevat toisiaan. Mitä hikisempi kosketus, sitä pienempi vastus ja vaikuttavampi ääni.

Vuonna 1976 Kurenniemi alkoi harjoitella piirtämistä ja maalaamista. Nyt hänellä on tarvittava tekniikka, jonka hän on opettanut tietokoneelle.

Hän on suunnitellut muutamia satoja pieniä tietokoneohjelmia, jotka tuottavat abstraktia grafiikkaa. Niissä muodot ja värit elävät systemaattisesti tai sattumanvaraisesti.

Kurenniemi kehuu joitakin tietokoneohjelmia visuaalisesti mielenkiintoisiksi. ''Silloin tällöin kiikutetaan taidekilpailuun epäonnistuneita tietokoneohjelmia. Ohjelmavirheen takia niistä on tullut jännittävän näköisiä.''

Tietokone kutsuu leikkimään

Tietokone houkuttelee kokeilemaan ja leikkimään. Aikuiset katsovat tietokonenäyttelyssä kateellisina, kun lapset heittäytyvät näppäilemään sadan tuhannen markan laitteilla mitä mieleen juolahtaa.

Juha Viikki on 16-vuotias koululainen. Hän on kolme vuotta kuulunut naapurikoulun atk-kerhoon. Hän voitti tietokonerunoilla Matemaattisten aineiden opettajien liiton peruskoululaisille ja lukiolaisille järjestämän ohjelmointikilpailun.

Viikki kertoo, että ohjelman teko kesti noin 40 tuntia. ''Hankalinta oli sen pakkaaminen sellaiseen tilaan, että sen pystyi kilpailussa kirjoittamaan kuudessa tunnissa. ''Minulta kului kahdella sormella näppäillen 5 tuntia 45 minuuttia.''

Ohjelmaan Viikki syötti rivi riviltä kuusi Paavo Haavikon runoa. Sitten hän jäsensi lauseet kielioppisääntöjen mukaan. Kone tuotti uusia runoja.

Palkinnoksi Juha Viikki sai tuliterän Digital Rainbow 100 -pienoistietokoneen. Perheenjäsenistä ei Juhan lisäksi vielä kukaan muu ole uskaltanut koskea laitteeseen. Isä aikoo sen tehdä, mutta äiti pyyhkii vain pölyt.

Minä sain kokeilla Erkki Kurenniemen Applella sekä piirtämistä että säveltämistä. Sävelsin hätäisesti pienen valssin, joka ei oikein onnistunut. Sen sijaan värikuvani olisi varmasti saanut Sam Vannin tai Paul Kleen kateudesta vihreäksi ja he ovat sentään maailmankuuluja taiteilijoita.

Valitettavasti tähän lehteen ei ollut mahdollista saada tulostusta piirroksestani Dream. Ehkä tarjoan sitä Taidehalliin nuorten taiteilijoiden vuosinäyttelyyn. □

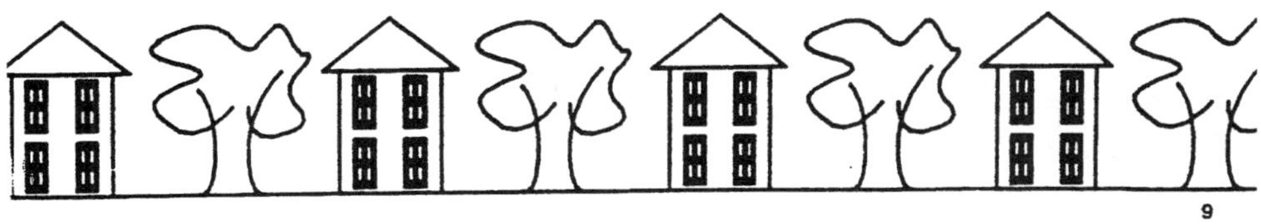

Figure II.29 (continued)

Erkki Kurenniemi tunnetaan kokeellisen musiikin soittokoneiden keksijänä ja kehittäjänä. Musiikkikoneet DIMI-A ja videoposetiivi DIMI-0 ovat vieläkin maailman ihmetyksen kohteena. Eri maiden elektronimusiikin studiot ovat osoittaneet näitä kohtaan suurta kiinnostusta. Nyt Kurenniemi on valmistanut aivan uudenlaisen soittimen, seksofonin, joka toimii kosketteluperiaatteella. Kaksi tai useapia ihmistä voi luoda musiikkia toisiaan koskettelemalla. Seksofonin parhaita puolia on käytön yksinkertaisuus, kuka tahansa oppii hetkessä musisoimaan. Siis eräänlainen jokamiehen instrumentti.

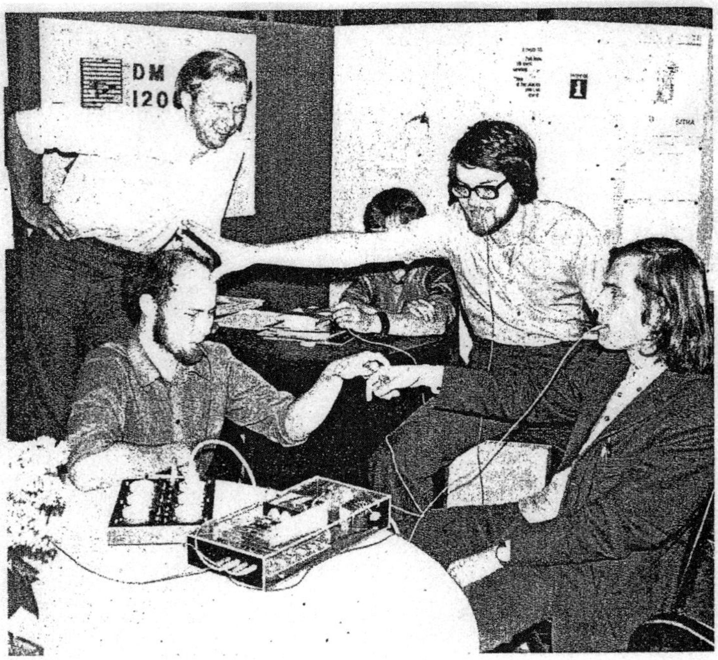

JOUKO ALANKO

Musiikkia sekso

■■ SEKSOFONIN idea tuli oikeastaan Ruotsista. Tunnettu elektronimusiikin säveltäjä Ralph Lundsten penäsi Kurenniemeltä soitinta, jota ihmiset voisivat soitella toisiaan koskettelemalla. On vaikea tietää, mistä Lundsten sai tällaisen idean: konservatiiviseen konserttitilanteeseen kontaktisoittimen käyttö ainakin luo uutta ilmettä; elektronimusiikin esittäminen on yleensä tapahtunut steriilin ja epäinhimillisen, kauittimista ja nauhureista koostuvan koneiston avulla. Seksofonin käyttö tuo tähän elämää vierastaneeseen musiikkimuotoon ihmisen ja esiintymisestä muodostuu eräänlainen happening. Uusi soitin tuo myöskin improvisoinnin sähkömusiikkiin, vaikka seksofonia voi soittaa nuottienkin mukaan.

Joka tapauksessa Kurenniemi on yhdessä Digelius Electronic Finland Oy:n työryhmän kanssa valmistanut Lundstenin toivoman soittimen, maailman ensimmäisen seksofonin, DIMI-S-Ø2:n.

Toimintaperiaate

Seksofoni, joka on stereofoninen soittolaite, näyttää äärimmäisen yksikertaiselta. Levysoitinta pienempi eletroninen koneisto näkyy kauniina kirkkaan muovikuoren alla. Verkkojohdon ja käynnistyskytkimen lisäksi laitteeseen ei sisälly muita säätöjä. Neljän kontaktijohdon päihin on tällä kertaa kiinnitetty banaanikoskettimet, jotka toimivat soittopelin ohjausyksikköinä ja vastaavat tavallisten soitinten koskettimia tai kieliä. Konserttitilanteessa johtoihin on menestyksellä liitetty mm. yleensä musiikille melko vieraat koskettimet — käsiraudat.

Yksinkertaistaen voidaan sanoa soittimen koostuvan äänigeneraattorijärjestelmästä, suodattimesta ja pienehköstä muistista. Systeemin alkupäässä oleva piiri mittaa jatkuvasti neljän kontaktipisteen välisiä vastuksia.

Neljä kosketuspistettä muodostavat kuusi paria, joiden jokaisen välinen johtokyky mitataan ja muodostetaan siitä riippuva säätöjännite. Kutakin kuutta jännitettä tulkitsee vertailupiiri, joka antaa yhden loogisen signaalin eteen päin, mikäli jännite ylittää määrätyn kynnysarvon.

Seksofonissa on vain yksi ainoa äänigeneraattori, jota ohjaa melko monimutkainen vibratolaitteisto. Pääoskillaattori haaroitetetaan kuuteen taajuuden jakajaan. Perustaajuus voidaan jakaa luvuilla 2,3,—,15. Kuusi ääntä muodostavat soinnun, jossa äänien suhde on rationaalinen. Nämä äänet johdetaan kukin omaan suodattimeensa, joka määrää äänen värin. Tämän jälkeen äänet johdetaan analogiakytkimiin, jotka kytkevät äänen päälle tai pois, ja lopuksi kahteen summausvahvistimeen. Huomattakoon, että analogiakytkimiä voidaan ohjelmoida kuten analogiakoneita kytkemällä kontrollisignaaleja eri tavalla. Seksofonin ohjelma on kiinteä, mutta sitä voidaan helposti muuttaa.

44 Tekniikan Maailma 18/1972

Figure II.30

Jouko Alanko, "Musiikkia Sekso," archival article on Errki Kurenniemi's DIMI-S, published in *Tekniikkan Maailma* 18 (1972), Helsinki.

Ohjelma

Tällä hetkellä seksofoni on ohjelmoitu seuraavasti. Jokaista paria kohden on oma ääni, ja kun vastus tulee riittävän pieneksi, ääni soi. Jos sama ääni otetaan uudestaan, muuttaa laskijalaite samalla jakosuhdetta, jolloin äänenkorkeus nousee. Nouseva asteikko pyrkii edustamaan mm. seksuaalisen jännitteen nousua. Englanninkielinen sana sex tarkoittaa juuri seksiä, mutta ruotsinkielinen sex myös kuutta, mikä voidaan rinnastaa neljän kontaktin muodostamaan kuuteen pariin. Puuttumatta tarkemmin nimen symbolisiin merkityksiin, voimme todeta sen ainakin kaupallisesti hyväksi.

Vaasan kesässä soitinta kokeiltiin menestyksellä. Neljä tuolia oli asetettu tähden muotoon. Käsiraudoissa olevat muusikot soittelivat olkainsa yli, toisiaan koskettelemalla. Psykologit ovat todenneet koskettelun merkittävästi parantavan stressiä, joten soittopeli saanee ainakin mielenterveydestä huolehtivat puolelleen.

Yksikin henkilö voi aivan hyvin soittaa seksofonia ottamalla kaksi banaanikosketinta kumpaankin käteensä. Näitä painelemalla musisoija kykenee aikaansaamaan ennalta halutun äänikentän. Vaikka seksofoni onkin esim. pianoon verrattuna yksinkertainen soittaa, vaatii se silti pienen harjoittelun toimintaperiaatteen selville saamiseksi. Virtuoosimainen käyttö tietenkin edellyttää musikaalista lahjakkuutta ja pitempää harjoittelua. Saksalainen popyhtye Ulysses käytti soitinta samaisilla Vaasan festivaaleilla. Elosyyskuun vaihteessa seksofoni esiintyi myös Pohjoismaisilla Musiikkipäivillä Oslossa ja herätti ansaittua huomiota.

Yleissoitin

Kokeilemamme seksofoni oli juuri lähdössä Ruotsiin, missä kiinnostus soittimeen on ollut suuri — onhan prototyyppi sijoitettu Prippsin olutpanimon Pripporama-näyttelytilaan. Soitin on toteutettu stereofonisena, kolme bassoääntä saadaan toisesta ja kolme diskanttiääntä toisesta kanavasta. Seksofoniin voidaan helposti liittää myöskin erilaisia valoshowlaitteita, jolloin esim. tietyn soittajaparin nimikkolamppu syttyy heidän koskettaessaan toisiaan. Tällöin valoja ohjataan äänen sijasta loogisella kontrollilla. Seksofoni on tulevaisuudessa vakavasti otettava soitin. Sen lukemattomista käyttömahdollisuuksista mainittakoon vaikkapa vain kuvanveistotaide. Liittämällä seksofoni metalliveistokseen, saadaan tämä eri osia koskettelemalla soimaan soittokoneen ohjelman mukaisesti. Vastusmittausyksikkö voidaan joissakin tapauksissa jättää pois ja syöttää sisään muita signaaleja, vaikkapa videosignaaleja.

Seksofoni on toteutettu uusimmalla tekniikalla, pääasiassa mikropiireillä ja se on ehkä liiankin monipuolinen ja hieno, mikä näkyy muutamaan tuhanteen markkaan nousevassa hinnassakin. Kurenniemen mukaan tulevaisuudessa voidaan rakentaa yksinkertaisempia soittimia kotikäyttöön, ja pudottaa hinta muutamaan sataan markkaan. Kuka sitten pitää seksofonin sähköisestä äänestä, on taas esteettinen kysymys. Mutta maku muuttuu koko ajan, ja tulevaisuudessa pianon tilalla saattaakin seistä modernimpi soittopeli — seksofoni.

TM

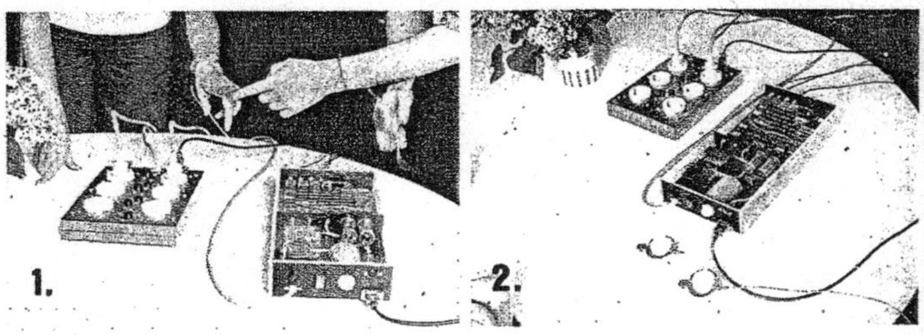

nilla

KUVASSA on laitteisto toiminnassa. Vasemmalla ylhäällä valoyksikkö, johon voi kytkeä kuusi lamppua, jotka syttyvät henkilöiden koskettaessa toisiaan. Samalla kaiuttimista kuuluu vastaavat äänet.

KUVASSA 2 on valoyksikköön kytketty 2 lamppua. Musta kaapeli keskusyksikön ja valoyksikön välillä on yhdyskaapeli, joka ohjaa valoyksikköä. Keskusyksikköön tulee paitsi verkkojohto myös 4 elektrodijohtoa. Alhaalla näkyy 2 elektrodia, käsirautaa.

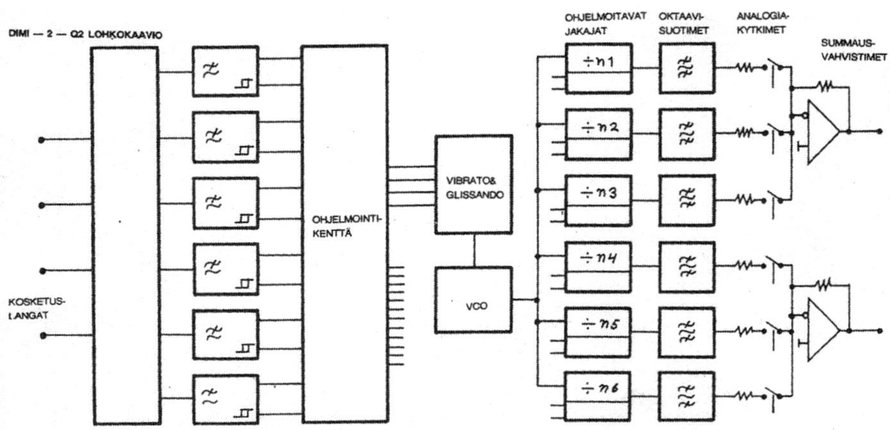

Figure II.30 (continued)

■■ VIDEOPOSETIIVIN tai -urkujen kehittely aloitettiin vuoden 1971 alussa ja tämä DIMI O valmistui jo toukokuussa. Ja vuoden loppuun mennessä sitä oli vielä paranneltu nykyiselle asteelle, jolloin se on siinä mielessä valmis, että sitä voidaan toimittaa vaikka kaupallisesti. Ilman esi-isää ei DIMI O ole kuitenkaan syntynyt. Yliopiston musiikkitieteen laitoksella on lähinnä viimeisten kymmenen vuoden aikana kehitetty laitteita tavoitteena digitalisen elektronisen musiikin studio.

Videourkujen lähimpänä vertailukohteena voidaan pitää jännitesäätöisiä syntetisaattoreita eli moug'ja, jotka ovat lyöneet jo itsensä läpi uudentyyppisinä soittimina. Ne ovat rakenteeltaan analogialaskijoita. Siinä mielessä siis tietokoneeseen verrattavia.

Nyt on ilmeistä, että digitalinen tietokonetekniikka tulee ajan mittaan jopa merkittävämmäksi kuin analogiatekniikka, mutta aika ei ole vielä niille kypsä. Toisaalta taas on niin, että tällainen systeemi vaatii niin suuren tietokoneen, että käyttö tulee kalliiksi. Jos itse äänisignaalia tuotetaan digitalisesti, niin siinä tarvitaan erittäin suuri nopeus, ts. nopeasti toimiva systeemi. Tavallinen pienoistietokone, joka on jo halpa laite, ei pysty tuottamaan ääntä tehokkaasti.

Ensimmäinen kompromissi oli DIMI O:n edeltäjä DIMI I, jossa oli tehty digitalisesti niin paljon kuin mahdollista. Äänigeneraattorit, voimakkuuden säädöt ja muisti olivat normaalia digitalitekniikkaa. Ainoat lineaariset olivat suodattimet ja äänenvärin määrittely. DIMI I:ssä soittotekniikka oli järjestetty siten, että koskettimiston muodosti painettu piiri, jossa oli varsinainen koskettimen iskumortti 32 kullattua metallifoliota, joita kosketeltiin metallipuikoilla.

Tästä ideat kulkivat eteenpäin ja välillä oli mielessä va-

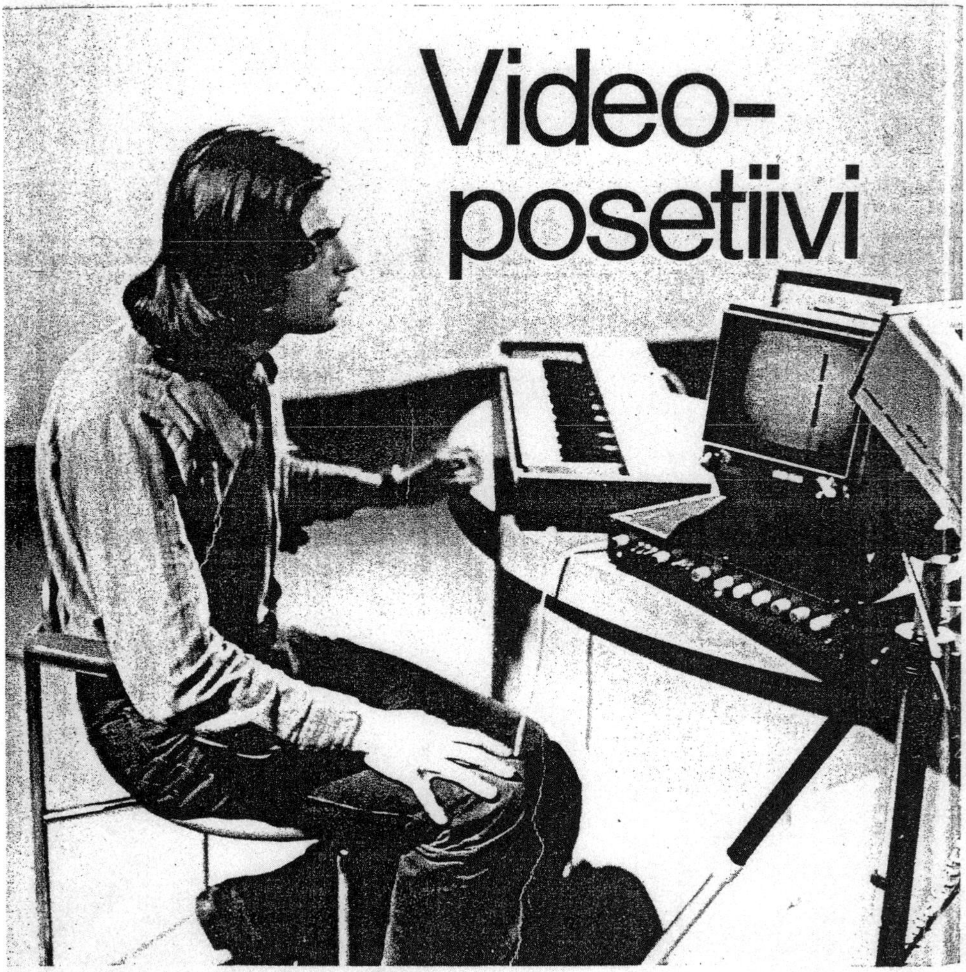

34 Tekniikan Maailma 5/1972

Figure II.31

Erkki Kurenniemi, "Videoposetiivi," archival article on DIMI-O, published in *Tekniikkan Maailma* 5 (1972), Helsinki.

Dimi O kulkee myös nimellä videourku ja sitä voisi arvella digitalisen maailman silmuksi ja totaalisen häiväksi. Jonain päivänä näiden jälkipolvilla katsellaan Bachin musiikkia kuvapatsaina tai tauluina kummallisessa multimediamaailmassa. Tässä on sentään vain kyse musiikista ja sen tuottamisesta — nimenomaan elektronisen. E r k k i K u r e n n i e m i on jo vuosikausia painiunut näitten ongelmien parissa ja samalla hän on luomassa myös maan ensimmäistä elektronisen musiikin studiota.

RAIMO HÄMÄLÄINEN

KUVAT
MATTI RISTIMÄKI
PEHR FRÖBERG

lokynän käyttö. Ajatuksena oli, että tv-kuvaruudulle sähköisesti tuotettaisiin jonkinlainen koskettimiston kuva ja valokynällä koskettamalla näitä koskettimen kuvia soitettaisiin. Mutta tämä samoin kuin DIMI I:n systeemi on hankala, koska siinä käytännössä voidaan painaa vain kahta kosketinta samanaikaisesti — siis valokynä kummassakin kädessä. Siksi päädyttiin kokeilemaan tv-kameraa ja uudentyyppistä systeemiä — DIMI O, jonka toimintaperiaate selviää oheisesta lohkokaaviosta.

Käyttö-mahdollisuuksia

Alun perin on videourkujen avulla tähtäimessä elektronisen musiikin tuottaminen. Lähinnä tulee ensi sijassa kysymykseen elektronisen musiikin studiot, jotka Euroopassa ovat useimmiten yleisradioyhtiöiden ylläpitämiä ja Amerikassa yliopistollisia.

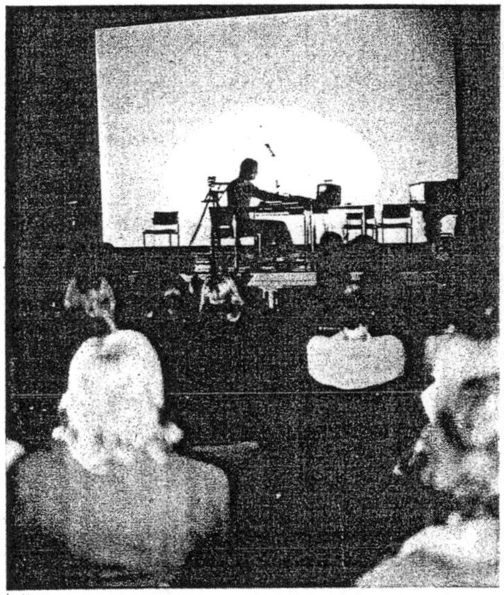

YLEISÖTILAISUUS Vanhalla keräsi uteliaita. Tarkoitus oli, että yleisökin sai osallistua "kävelymusiikin" tekemiseen.

Eräs käyttäjäkunta on yksityiset säveltäjät, jotka voisivat käyttää tällaisia laitteita. Toinen ryhmä on muusikot. Videourku on yhtä hyvin esitysinstrumentti kuin studiossa käytettävä laite. Juuri tämän videosysteemin takia löytyy käyttöä kokeiluluontoisen filmimusiikin, balettimusiikin ja erilaisten näyttelyiden kohdalla. Esimerkiksi veistoksia ja maalauksia voidaan suoraan muuttaa "jonkinlaiseksi musiikiksi".

Tulevaisuus näyttää olevan enemmän videon kuin äänen puolella. Tällä hetkellä on tarkoitus kehittää videon suuntaan eräänlaiseksi kuvageneraattoriksi, jotta voidaan siis käyttää kameraa, erilaista kuvamateriaalia, filmiä, kuvanauhaa; ja toisaalta luodaan koskettimiston ja nappuloiden avulla suoraan piirtämällä kuvaruudulle alkuun abstraktia, mutta kun saadaan kuva paremmin hallintaan, suurempi erotustarkkuus ja värit mukaan — myös tuottaa kuvapinnalle mitä tahansa, kuvallista, myös tekstiä. Musiikki saattaa tällöin olla vain säestävässä asemassa.

Opetusmielessä tätä laitetta voidaan tällaisenaan käyttää musiikinopetuksessa monella tavalla. Tässä tapauksessa lähinnä urkujen soiton opetuksessa, koska kuvaruudulla nähdään heti nuottikirjoituksena, mitä on soitettu ja voidaan nähdä virheet ja tempon heitot jne. Toisaalta nuottikirjoituksen, musiikin teorian ja sointuopin opetuksessa taas esimerkiksi nähdään ja kuullaan nuotit samanaikaisesti. Käytännössä on esteenä nuottikirjoituksen pelkistetty graafinen muoto. Mahdollista kylläkin olisi toteuttaa mainitussa muodossa, mutta se merkitsee, että urkuihin pitäisi liittää kiinteä puolijohdemuistio, johon on tallennettu pysyvät nuottisymbolit. Tällainen konkreettinen sovellutus on siksi kallis, ettei sitä huvin vuoksi kannata kokeilla.

Julkisuudessa DIMI O on ollut kokeilumielessä Pasilan tv-studiolla, missä tehtiin 10-minuuttinen värifilminäyte pohjoismaiseen tv-seminaariin. Näytteessä taiteilija tanssi kameran edessä ja loi liikkeillään itselleen musiikin. Esittä-

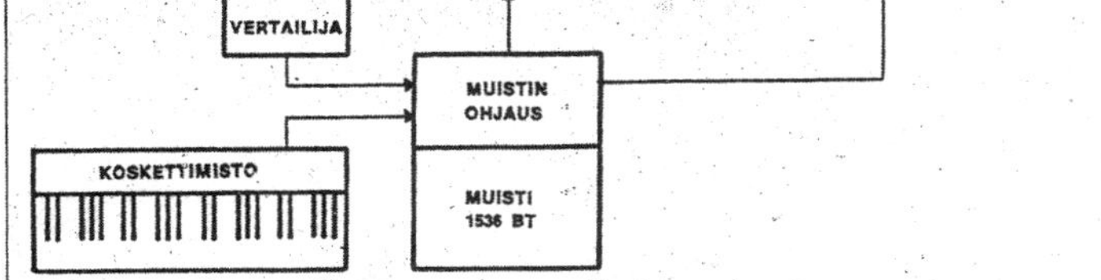

Figure II.31 (continued)

jän kannalta on toistaiseksi vaikeata pitää huolta omasta asiastaan (tanssistaan) ja samalla "säveltää" itselleen musiikki. Siinä tilanteen totaalinen outous.

Toinen julkinen esittelytilaisuus oli vanhalla ylioppilastalolla Helsingissä, missä teemana oli "kävelymusiikin" säveltäminen yleisön toimesta.

Tällaisenaan urkusysteemin hinta on 30 000 markan hujakoilla, kameroineen, monitoreineen kaikkineen, mutta mahdollinen sarjatuotanto oletettavasti pudottaisi roimasti hintaa.

Multimediataide

Tämä termi pitää sisällään erilaisten taidemuotojen yhdistämisen. Multimedia ahtaa vastaanottajan täyteen informaatiota kaikkia kanavia pitkin. Siis näkö- ja kuuloaisti ja mahdollisesti painoaisti ja kosketuskin välittävät samaa asiaa.

Olennaisesti on mukana jonkinlainen tietokone, joka käyttää normaalia tietokonetekniikkaa ja siinä on ympäristölaitteet, jotka vastaavat reikänauhaa, reikäkorttilukijoita ja muita laitteita. Lisäksi vielä muisti. Itse asiassa kehitys on kulkemassa tähän suuntaan. Tulossa on uusi laite — DIMI P eli programmable, joka muodostuisi kolmesta hyvin pienestä toisiinsa liitetyistä tietokoneista, joista yksi olisi ohjausyksikkö; se hoitaa ympäristölaitteet, koskettimistot, kamerat, kuvaruudut, ja siinä olisi päämuisti, johon tallennetaan musiikillinen rakenne (ohjelma). Toinen yksikkö on keskittynyt äänien synnyttämiseen, siis generaattori, ja samalla yksiköllä voidaan tuottaa peruskuvamateriaali. Kolmas yksikkö olisi digitalisuodatin, hyvin pieni mutta erittäin nopea erikoistietokone, joka digitalimuodossa voi suodattaa, kaiuttaa ääniä jne. sekä suorittaa kaikenlaisia muutoksia.

Nyt on mahdollista, ettei suunnitelmaa tässä muodossa koskaan toteutetakaan, koska yllättäen tuli markkinoille yhden ainoan mikropiirin tietokone. Sen avulla pystytään nykyinen rakenne puristamaan agenttisalkun koosta taskukokoiseen. Päämäärä on täysin konkreettinen. Hintatietoja tästä integroidusta prosessorista ei ole vielä saatu.

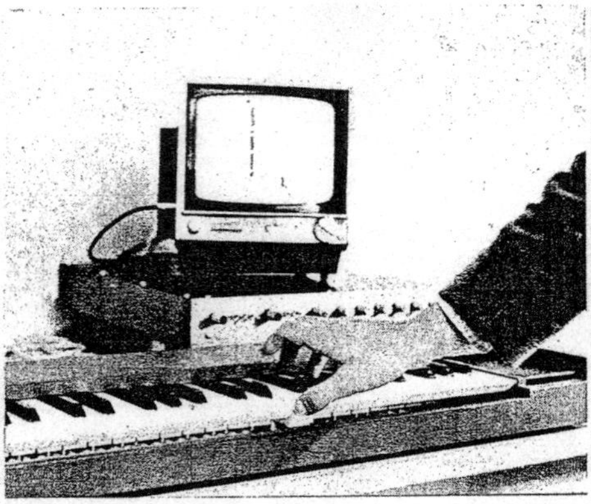

URKUKOSKETTIMISTOLLA soitettu musiikki tallettuu laitteen muistiin ja näkyy yksinkertaisena nuottikirjoituksena kuvaruudulla. Musta pystyjuova (osoitin) pyyhkii nuottisivua vaakasuuntaan säädettävällä nopeudella. Osoittimen kohdalle lankeavat nuottimerkit kuuluvat ääninä. Äänen korkeuden määrää nuottimerkin asema pystysuunnassa.

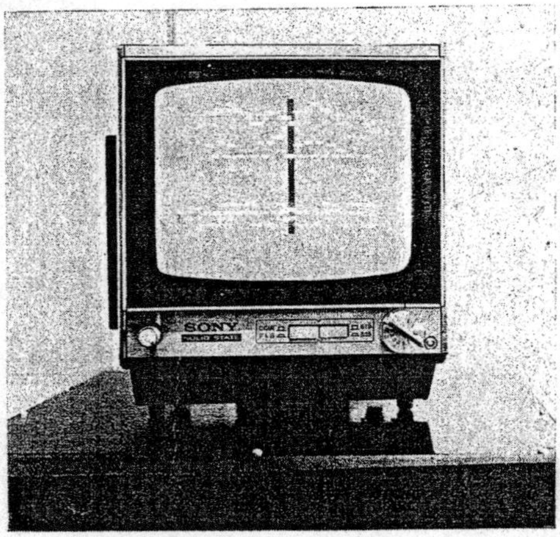

VERTAILIJAPIIRI muuttaa videokameran välittämän kuvan nuottikirjoitukseksi. Nuottimerkit muodostuvat polariteettikytkimen asennosta riippuen joko kuvan tummille tai vaaleille kohdille.

Bachia veistoksena

— Menemättä pitemmälle filosofiaan Erkki Kurenniemi arvelee, että käyttämällä elektroniikan tarjoamat mahdollisuudet synnyttää ääni ja kuva yht'aikaa, ollaan tulossa uusiin maisemiin. Laajempia näkymiä kytkeytyy oikeastaan juuri muistivälineeseen, vaikkakaan sitä ei ole toteutettu. Eräs luonnollisimpia lisälaitteita nykyisiin on digitalikasettinauhuri; tietokoneen magneettinauha-asema halvassa ja pienessä muodossa. Seuraavan kymmenen vuoden aikana nämä mahdollisesti yleistyvät. Taiteen alalla se potentiaalisesti merkitsee ratkaisevaa vallankumousta n. 50 vuoden perspektiivissä. Siten, että esimerkiksi musiikissa perinteisiä muistivälineitä on lähinnä nuottikirjoitus. Länsimainen musiikki on nimenomaan paperiin sidottu. Nuottikirjoitus on itse asiassa tehnyt länsimaisen musiikin sellaiseksi kuin se on. Tämän vuosisadan panos on äänilevyt ja -nauhat, jotka ovat muuttaneet musiikkikulttuurin luonteen aivan ratkaisevasti.

Levyt ja nauhat ovat ikään kuin sulattaneet musiikin alueet yhteen. Jos konsertit olisivat ainoa mahdollisuus kuulla musiikkia, niin tietyt henkilöt kävisivät vain tietyissä konserteissa. Mutta kun kotona on mahdollisuus kuulla kaikenlaista musiikkia, on yleistymässä se, että todetaan kaikki musiikin alat samanarvoisiksi ja ne vaikuttavat toisiinsa.

Seuraava edistysaskel, ilmeisesti vuosisadan loppupuolella, on tietokonemuisti, joka ei ole siinä mielessä muistiväline kuin nuottikirjoitus, äänilevy tai -nauha, vaan yleinen periaate on binäärinen tiedon esitysmuoto. Samantekevää on, mitä konkreettista muistivälinettä käytetään. Onko se reikäkortti, reikänauha, magneettinauha, digitalikasetti, ferriittimuisti, levyt, rummut tai mikä tahansa; pääasia on, että kun jokin tieto on tässä digitalisessa muodossa, niin se voidaan suurella nopeudella siirtää muistivälineestä toiseen ja toteuttaa monilla eri tavoilla. Ratkaisevaa on, että ensimmäinen muistiväline on sellainen, joka kelpaa yhtä hyvin kuvan, äänen, musiikin, kirjoitetun sanan tai jos ajatellaan numeerisesti ohjattuja työstökoneita, niin jopa kolmiulotteisten muotojen kuten veistosten tallentamiseen.

Kaikki aikaisemmat muistivälineet ovat sidottuja. Meillä on filmi, kuvanauha ja ääninauha erillisiä. Nyt kun tulee uusi muistiväline, perusesitysmuoto, niin se tulee potentiaalisesti sulauttamaan nykyisin erillään olevat taidemuodot yhteen. Jos Bachin sävellys on pantu digitalimuotoon, se voidaan yhtä helposti kuin se soitetaan äänillä, tulkita kuvana tai veistoksena. Pohjimmainen vaikutus elektroniikalla taiteisiin on siis suuri synteesi. On asia erikseen, että synteesi saattaa ulottua paljon pitemmällekin, koska muistiväline on edelleenkin sama, jossa kaikki muutkin tiedot säilytetään.

Vuosi 2000 tulee olemaan paljon kummallisempi kuin missään on edes kuviteltukaan.

Figure II.31 (continued)

HEINOLAN KURSSIKESKUS
Tietokoneavusteinen opetus historiassa ja yhteiskuntaopissa 6. - 10.10.1986

File: Heinola
Disk: Misc Dat

Tietoyhteiskunta

Erkki Kurenniemi

Ylihuomenna

3 arvomaailmaa

Kehitystä ohjaavat arvojärjestelmät, ainakin jos arvot (tavoitteet) ymmärretään niinä joiden mukaan ihmiset toimivat eikä niinä joita he julistavat. Eräs tapa on jaotella arvojärjestelmät seuraavasti:

- Elämän, luonnon, evoluution arvot (neutraali autopoiesis)
- Ihmisen (yksilön ja yhteiskunnan) arvot (humanismi)
- Ihmisen luomien artefaktojen arvot (poliittiset, taloudelliset, teknologiset, tiede- ja taideinstituutioiden valta/arvorakenteet)

Hieman analoginen kolmijako voidaan nähdä teknisessä evoluutiossa:

- Luonnonlakien asettamat mahdollisuudet ja rajoitukset
- Kekseliäisyyden ja luovan innovaation itseisarvoinen paine
- Tuettu ja tavoiteohjattu tutkimus ja tuotekehitys.

Henkilökohtainen kommunikaattori

Aikanaan tietokoneet muistuttivat temppelin alttaria, sen ympärillä sai liikkua vain papisto (ATK-henkilöstö) joka tunsi salaisen kielen ja toimi välittäjänä tietokoneen ja rahvaan välillä. Tämä instituutio elää edelleen mutta on joutumassa epäsuosioon. Nyt tietotekniikka on toimistokalustoa, kuvaruutuja, näppäimistöjä, hiiriä, ja itse tietokoneen keskusyksikkö, jos se ei mahdu näppäimistön sisään, on tungettu halventavasti pöydän alle. Tietokoneet pienenevät ja siirtyvät salkkuihin ja taskuihin, mihin seuraavaksi?

Eräs kuvitelma on suora ajatusyhteys aivojen ja tietokoneen välillä, tämä voitaisiin ehkä toteuttaa aivosähköelektrodien (EEG), aivoihin upotettujen elektrodien tai jonkin molekyyliresonanssitomografialaitteen (kypärän) avulla. En usko tähän kehitykseen, neljästä syystä.

9. lokakuuta 1986 1

Figure II.32

Errki Kurenniemi, "Tietoyhteiskunta" (English trans. "Information Society"), excerpts from *Do-it-yourself Personal Communicator prototype kit* (eye projectors), 1986.

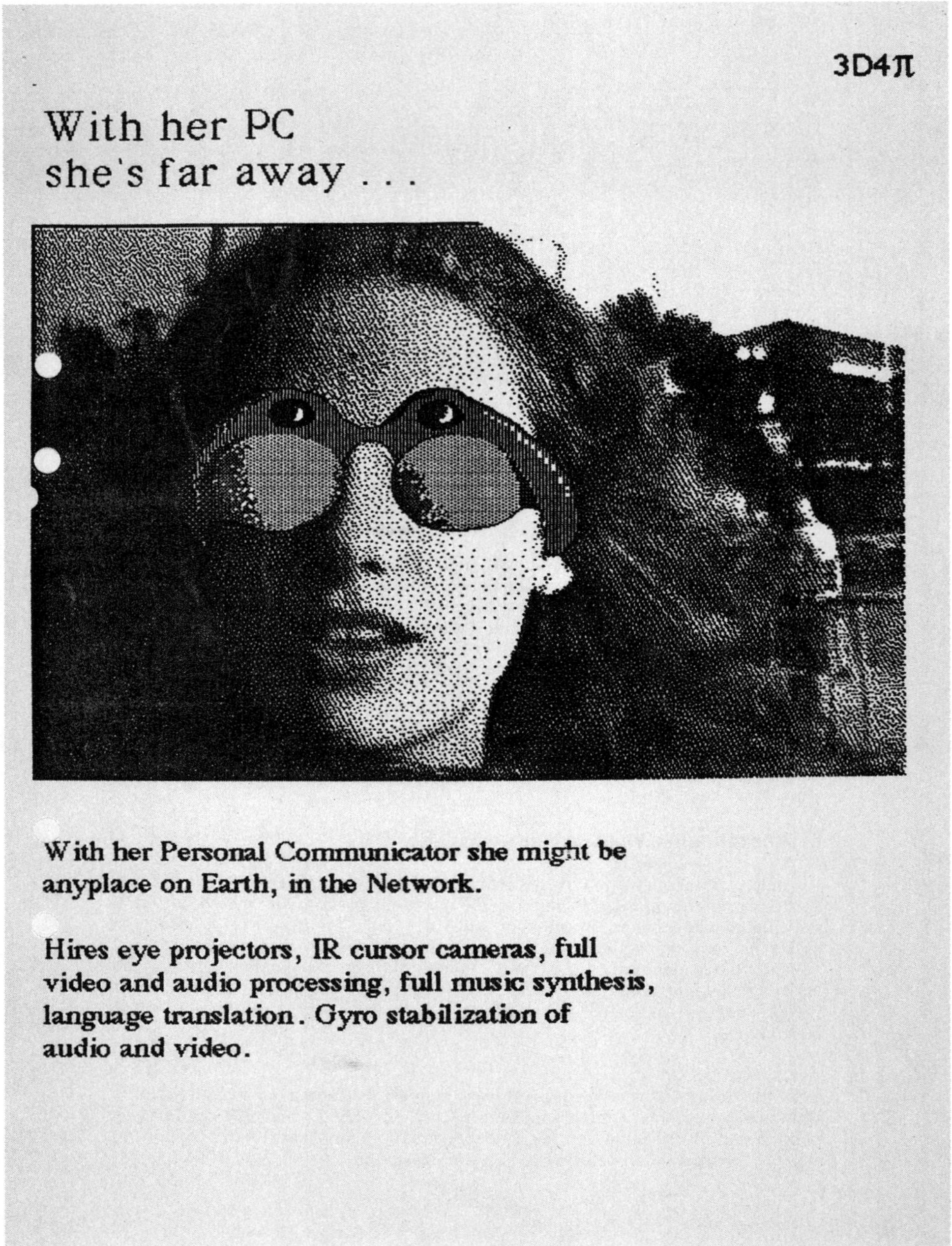

Figure II.32 (continued)

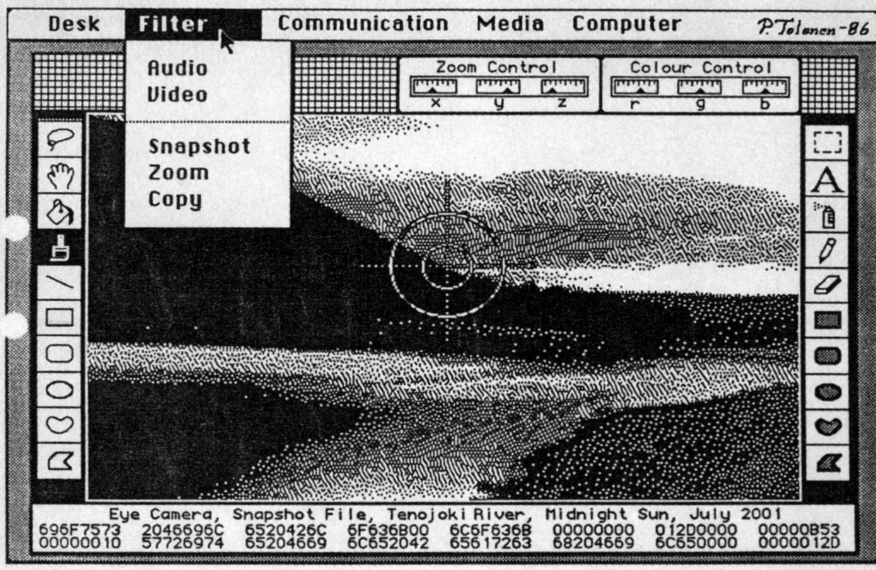

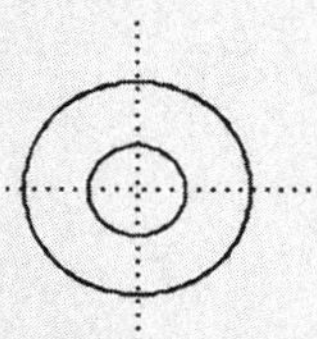

Maailma kommunikaattorin kautta nähtynä. Maisema voi olla luonnollinen, kameroiden välittämä tai jostakin muualta lähetetty. Kursori, 'tähtäin' liikkuu silmiä kääntämällä. Toiminnot käynnistetään silmänräpäyksellä.

Figure II.32 (continued)

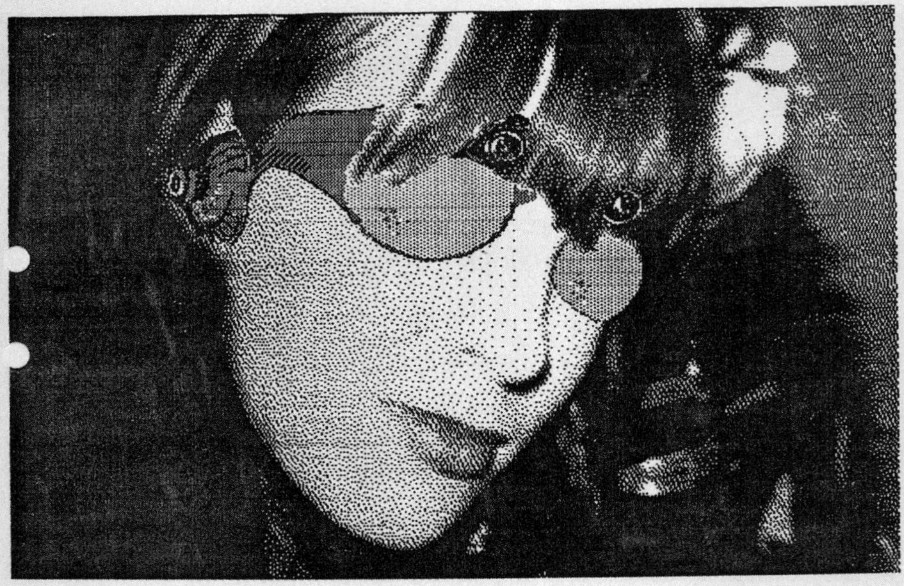

Peeseilläsi voit maalata maailman ruusun-punaiseksi, suodattaa pois kaiken epämiel-lyttävän. Voit muuntaa todellisuuden sellaiseksi kuin haluat.

File: N-B PC 28.9.1986 EKi

Figure II.32 (continued)

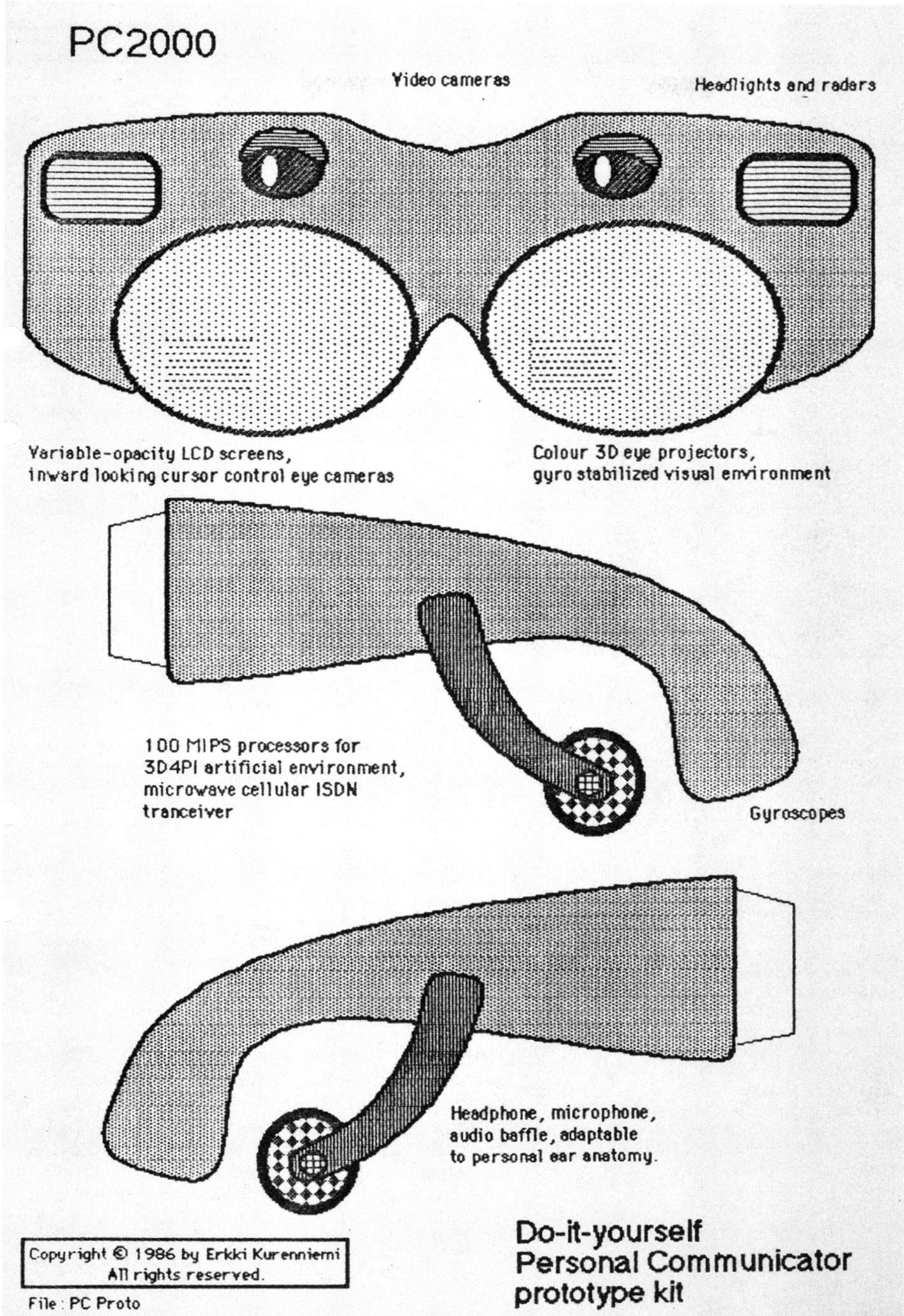

Figure II.32 (continued)

Erkki Kurenniemi
Tiedekeskussäätiö
Kielotie 7A
01300 Vantaa

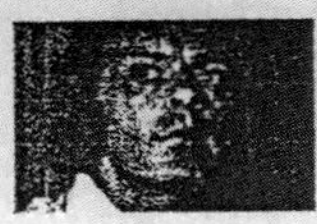

Kohti yhdentyvää taidetta

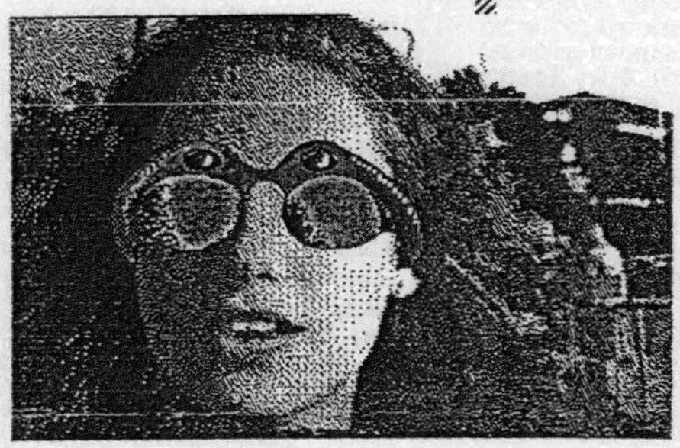

Henkilökohtaisen tietokoneen peruskomponentit pienenevät, kunnes kaikki voidaan rakentaa silmälasien sankoihin.

Kukaan ei vielä tunnu tajuavan, miten naurettavalla tavalla taide on pirstoutunut.

Suuri Transkriptio

Seuraavan vuosikymmenen suuri perustehtävä on siirtää kaikki vanhoille medioille talletettu tieto uusille elektronisille, magneettisille, optisille ja aikanaan ehkä biomolekylaarisille tietovälineille. Jo nykyään merkittävä osa uudesta tieteellisestä ja teknisestä tiedosta talletetaan suoraan magneettisille välineille. Teollisuudessa suuria piirustusarkistoja siirretään paraikaa optisille levyille. Transkription välivaihe päättyy vasta, kun kaikki mitä koskaan on kirjoitettu, painettu, piirretty, maalattu, valokuvattu, filmattu, äänitetty tai kiveen hakattu, on globaalisesti suorasaatavissa tietoverkon kautta.

HYPERMEDIAT

Henkilökohtainen kommunikaattori

Luontevin tapa liittää ihminen elektroniseen tietojärjestelmään tapahtuu aistien kautta ja niiden välittömässä läheisyydessä. Mitään epämiellyttäviä aivoihin työnnettäviä elektrodeja ei tarvita. Kutakin aistia varten tarvitaan kaksi komponenttia, **stimulaattori**, joka ärsyttää luonnollista aistielintä ja **sensori**, joka on luonnollisen aistin keinotekoinen korvaaja. Kuulon stimulaattoreina voidaan käyttää tavallisia stereokuulokkeita, sensoreina toimivat kuulokkeiden ulkopintaan sijoitetut mikrofonit. Näön stimulaattorit ovat 'silmälappuvideot', **silmäprojektorit**. Näön sensoreina toimivat pienet videokamerat.

Jotta järjestelmä voisi toimia kommunikaatiovälineenä, näköprosessorin ja kuuloprosessorin tulee olla välittömässä yhteydessä yhteiskunnan selkärangan muodostavaan tietoverkkoon (soluradio-ISDN). Yhteyden on oltava langaton ja kaksisuuntainen.

Kyberavaruus

Tietokoneen kuvaruudun takainen maailma saa joskus visuaalisen ja tilallisen hahmon joissakin tietokonepeleissä ja graafisissa suunnitteluohjelmistoissa. Kun aikanaan kaikki ihmiset päivittäin kommunikoivat keskenään tietokoneen välityksellä, ruuduntakainen keinotodellisuus saa arkkitehtuurinsa ja liikennesääntönsä. Syntyy synteettinen kyberavaruus.

Sivu 1

Figure II.33
Errki Kurenniemi, "Kohti yhdentyvää taidetta" (English trans. "Toward Integration of the Arts"), article, with notes and scribbles by Errki Kurenniemi.

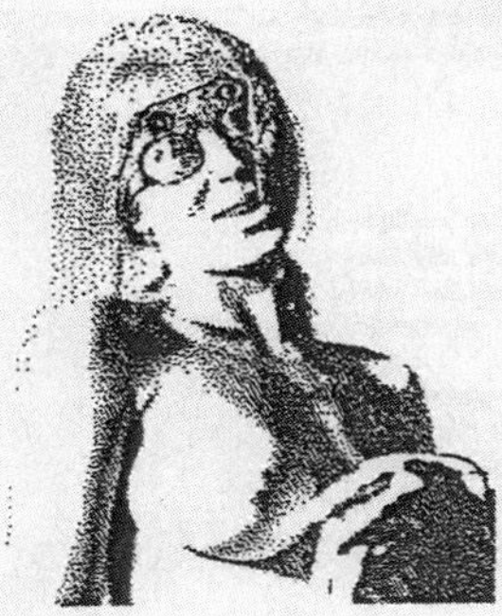

silmäkursori *tähtäin, hiiri, itikka* kohdistinsymboli jota siirretään laseissa silmiä kääntämällä, sitä ohjaavat *silmäkameröiden* viestimät kolme koordinaattia: vaaka, pysty ja etäisyys. Vaaka ja pysty johdetaan silmämunan asennosta, etäisyyden peesee laskee etäisyysmittarin tapaisella mekanismilla tarkentamalla retinaan, josta saadaan mykiön polttoväli sekä huomioimalla pupillien konvergenssi.

Tuntemattoman peeseen vaellus

Mfph. Herään kolmeen ääneen. Kolme kuvaa haalistuvat, jokainen kolmesta kuvasta liittyvät kukin kolmeen ääneen langoilla tai verkoilla jotka ohenevat. Ahaa, taas K_4 metsästää minua, standardipainajainen, siinä on kuusi reunaa, kuin muinaisessa seksofonissa, tetraedri metsästää minua vai metsästänkö minä tetraedria? Olenko minä reuna vai solmu? Yritän pitää kiinni kuvista ja dempata äänet mutta en näe filtteriä. Kukaan ei voi nukkua peeseet silmillä vielä. Yksi ääni on ropiseva sade. Koiraa haukkuu. Joku yrittää käynnistää autoa. Ahaa, tämä on konnektiopeli. Yksi kuva on kalenterin sivu mutta siinä päivien paikalla ovat vuosituhannet. Jalkani ovat reunoilla ja reunat halkaisevat jalkani, navan kohdalla halkeamat yhtyvät ja annihiloivat partiaalisesti, navasta ylös lähtee kundalinin poika ja halkaisee leukaluun, haarojen päässä on kaksi lehteä jotka ovat silmäni, peeseeni ja sisäisesti ymmärrän mitä tarkoittaa, että prinsipaaliunessa aina peeseeni alkavat valua, silmät näkevät fiiberit eivätkä koskaan tavoita niitä. Sateen ääni on sama, koiran haukunta on sama, auton ääni on erilainen. Muistan lapsuuden autot jotka surisivat, nyt niissä on tuo pelästyneen lapsen ininä. Toisessa kuvassa sata ikkunaa menee kiinni, kaikissa ikkunoissa kolme omenaa kolmiossa. Kaikki kalenterit ovat omenakalentereita, äh niitä on kaikkialla, omenat ropisevat mennessään kiinni ja kaikki ikkunat tsuumaavat pois ja inisevät. Räpsytän kalentereita ja ikkunoita ja omenoita mutta kursori on kadonnut. «Huomenta.» Taas herään sekuntia ennen herätyskelloa. «Juu... «Oletko varmasti ... «JUU turpa ikoniksi!» Kiljaisen, haluaisin heti opettaa herättäjälle jotain rivompaa. Herätessä aika on nopeimmillaan. Koiraa lakkaa haukkumasta, auton ininä kaikkoaa, kaikki ikkunat ja omenat ja kalenterit tsuumaavat ikoneiksi nenän varteen ja ovat poissa, jossain peesee piipittää.

Hyperpersoonat

2000-luvun alkupuolella ihmiset ja tietokoneet alkavat sulautua toisiinsa hyperpersooniksi. Ei enää ole helppoa sanoa, missä ihminen loppuu ja kone tai toinen ihminen alkaa. Samoin kuin monisoluisessa elämässä älykkäät solut uhraavat itsenäisyytensä, hyperpersoonassa ihmisyksilöt ovat enää kokonaisuuden hallittavia osasia.

Lisää työttömiä ?

Automaatio teollisuudessa hävittää työtehtävät, jotka voidaan ohjelmoida tietokoneille ja roboteille. Uusi tietotekniikka hävittää ammatteja myös sen kautta, että ennen ammattitaitoa vaatineen työn voi tehdä kuka tahansa. Ajankohtainen esimerkki on pöytäpaino. Kirjoitan tai varastan tekstin ja kuvat, taitan sivut ja laserilla tulostan painovalmiin originaalin. Ammattilaiset kauhistelevat jälkeä, mutta heidän tuskanhuutonsa hukkuvat, kun miljoonat alkavat julkaista omia lehtiään, kotona. Ja ajan mittaan tyylikin voidaan opettaa peruskansalaisvalmiudeksi.

Maailma muuttuu, Gustav. Kun tekoäly alkaa jyllätä ja kakarat alkavat säveltää Bachin tyyliin paremmin kuin J.S. itse, minne joutuvat pyhät arvonne? Murros on kohta edessä ja kokonainen kulttuurin kerrostuma siirretään etnisyyden hyllylle nopeammin kuin ehdit oppia kontrapunktin. Taiteilijoitahan mikään ei tapa, seuraavaksi ne tulevat vain uudelta suunnalta.

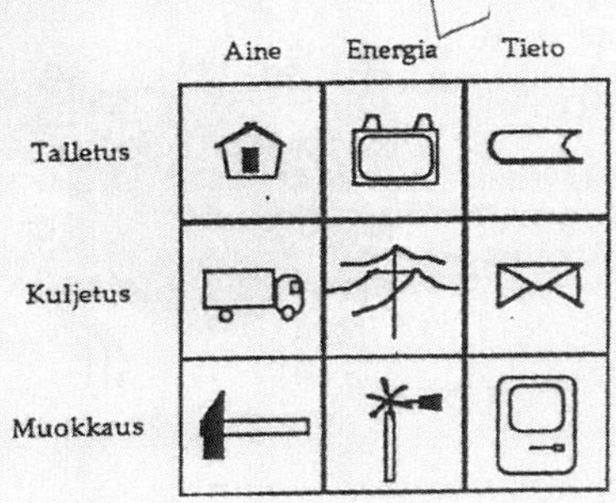

Sivu 3

Figure II.33 (continued)

Kyberavaruus: William Gibsonin tieteisromaaneissaan »Neuromancer» ja «Count Zero« käyttämä termi.

Optroniikan nykyisellä kehitysvauhdilla PC voidaan 50 vuoden kuluessa rakentaa piilolinsseihin.

Nenänvarressa riippuvat laserprojektorit ampuvat hologrammin piilolinssin kautta verkkokalvolle.

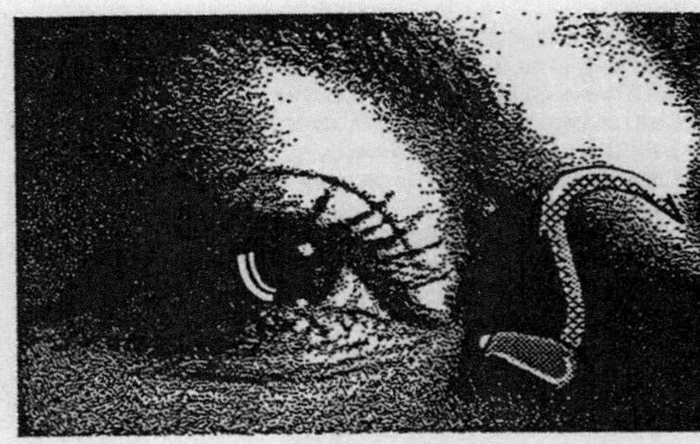

Piilolasivideot vuodelta 2014. Nenänvarteen ripustetut laserprojektorit ampuvat 3D4π-hologrammin erikoisvalmisteisen piilolinssin kautta verkkokalvolle.

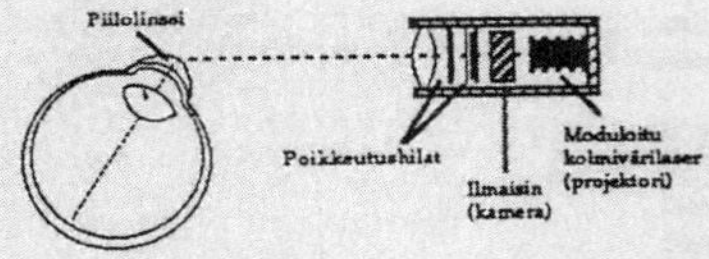

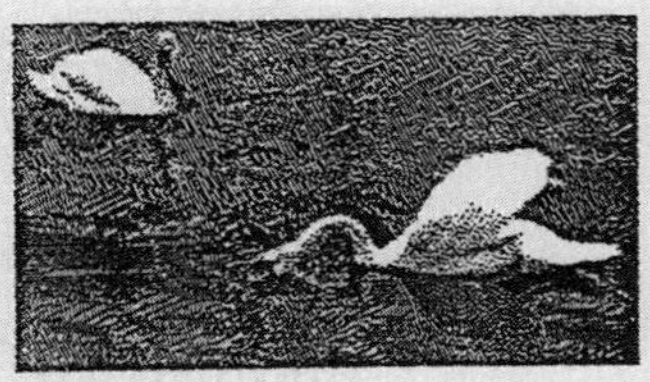

Taide on viestinnän laji. Viestintä edellyttää tiedon kuljettamista, varastointia, käsittelyä ja muuntamista. Vanha taide on pirstoutunutta, koska vanha tietotekniikka on pirstoutunutta. Eri informaation lajeille käytettiin eri tekniikoita. Sana paperilla, ääni nauhalla, kuva filmillä.

Uusi tietotekniikka on digitaalista ja sen takia tieto ei katoa eikä huonone käsiteltäessä. Tieto kulkee valon nopeudella. Muokkaus- tallennus- ja siirtovälineet ovat universaalisia.

Uuden tietotekniikan lähitulevat perustekijät ovat:

- Tiedon muokkaamisessa tietokone, mikroprosessori, tekoäly.
- Tiedon tallettamisessa optinen 12 cm laserlevy.
- Tiedon kuljettamisessa digitaalinen monipalveluverkko, ISDN.
- Tiedon muuntamisessa AD/DA sensorijärjestelmät ja silmälappuvideot ihminen-kone yhteydessä.

Maailma on hauska niin kauan kun on naurettavia asioita. Erityisen naurettavaa on ihmisten viehtymys pirstoa aktiviteettiaan alueisiin kuten taide, tiede, tekniikka, politiikka, kauppa ja mitä vielä ne ovat keksineet. Ja sitten on tämä kylmä ja ankea todellisuus.

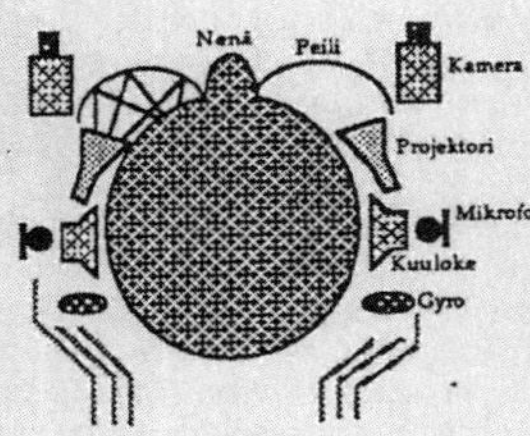

Verkko-olio

Keinopersoonat. Aivojen sisäisen tietoliikenteen volyymin ja ihmisten välisen tietoliikenteen volyymin suhde. Keskenään kommunikoivat ihmiset sulautuvat moniin kollektiivisiin hyperpersooniin, joissa tekoäly toimii organisoivana liimana.

Sivu 2

Figure II.33 (continued)

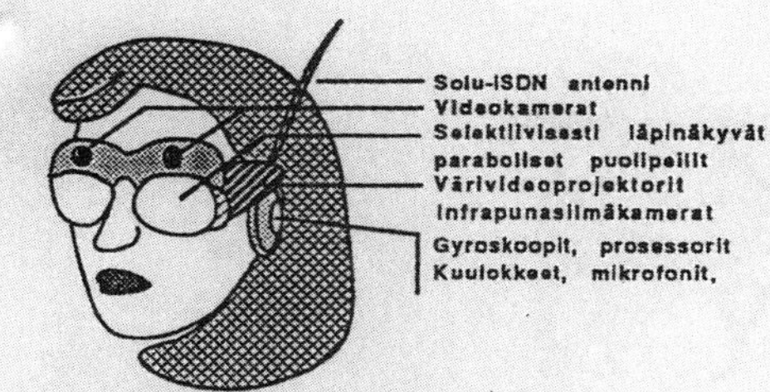

Hyperromaanin, hyperelokuvan ja hyperoopperan käsikirjoittaja ohjelmoi keinopersoonia. Keinopersoonien kantahenkilöitä ovat mm. 70-luvun alun psykiatri Eliza ja uudempi tarinaniskijä Racter. Asiantuntijajärjestelmätekniikka tekee keinopersoonista uskottavia, mutta ihmiseen verrattavat kokonaispersoonat eivät vielä ole näköpiirissä. Nykyisin keinopersoonien kanssa keskustellaan näppäimistöllä tekstimuodossa, mutta henkilökohtaisen kommunikaattorin välityksellä kyberavaruudessa keinopersoonat saavat tietokoneanimaatiolla 'fyysisen' hahmon.

Keksintö tai uusi teknologia alkaa toimia vasta, kun kaikki olennaiset palat ovat kohdallaan. Pelkkä tietokoneen keskusyksikkö ei tee tietoyhteiskuntaa.

Uuden yhdentyneen taidevälineen nimi tänään on hypermedia. Hypertekstin keksi Ted Nelson vuonna 1965. Hypertekstiä ei voi painaa paperille, vaan sitä on luettava kuvaruudulta, sillä hyperteksti on vuorovaikutteista, käyttää animaatiota, simulaatiota ja kaikkia tietokoneohjelmoinnin mahdollisuuksia.

Hypervideo on sama asia kuin interaktiivinen video, vuorovaikutteinen kuva. Hypermusiikissa kuulijan ja säveltäjän roolit sekottuvat. Musiikin hypernotaatiossa ei tarvita nuottilehden kääntäjää, nuotit välittyvät soittajalle kuvaruudulta kapellimestarin määräämässä tahdissa.

Hypermediassa yhdistyvät kaikki aistit aktiivisella tavalla. Hypermedia on vuorovaikutteista, joten tekijän ja nauttijan roolit sekottuvat. Käsikirjoittaja ei enää sommittele yhtä juonta, vaan ohjelmoi teokseen kaikki mahdolliset tapahtumien kulut. Hypermediateokset välitetään elektronisten postilaatikkojen, telekonferenssien, ISDN-verkon kautta.

Uusi hallinto

Hegemonian saavuttamiseen kuluu paljon aikaa. Niin myös hegemonian murtamiseen ja uuden dynaamisen hallinnon etabloitumiseen. Yhdentynyt taide vaatii uuden tuotantoteknologian, se on jo olemassa. Tarvitaan myös uusi jakelukanava, ISDN, se on tulossa. Syntyy uusi informaatiotalous, jonka peruspremissi on tiedon digitaalisuus eli täydellinen kopioitavuus. Sekin on tulossa, piraatit ovat asialla.

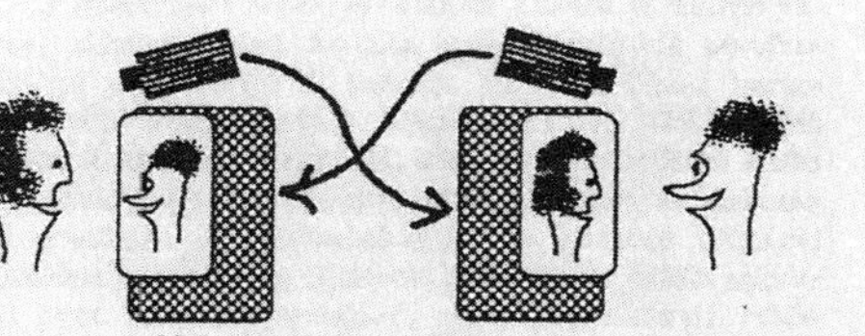

Väärin ajateltu näköpuhelin

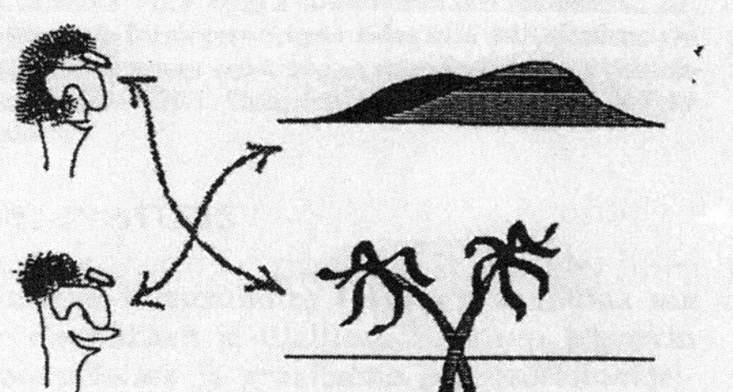

Oikea näköpuhelin vaihtaa ympäristöt

Hyperavaruudessa, kyberavaruudessa, ihmiset ja ohjelmat kommunikoivat keskenään nopeammin kuin ehtivät yksilöinä ajatella.

Sivu 4

Figure II.33 (continued)

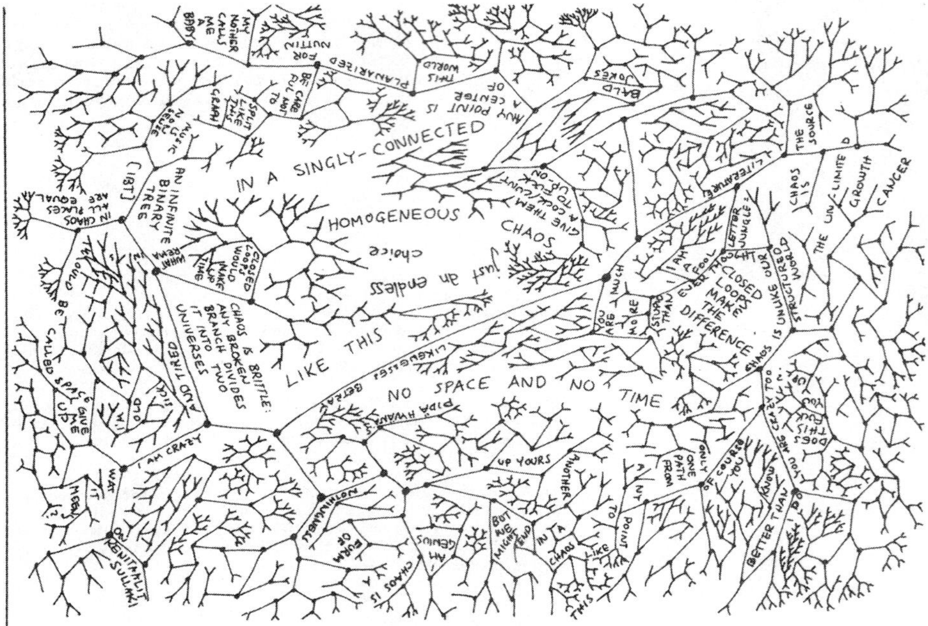

Tietoisuuden herätessä maailma lohkeaa kolmeksi fragmentiksi tai — koska fragmentit eivät ole toisistaan erillisiä — kolmeksi aspektiksi. Tiedostava mieli (MI) näkee aluksi maailman itsensä ulkopuolisena todellisuutena (TO). Samanaikaisesti syntyy kuitenkin myös kolmas aspekti, mielen ja todellisuuden havaitsemisen ja toiminnan kautta yhdistävä kielen (KI) eli viestinnän aspekti.

Figure II.34

Brain map, excerpt from Erkki Kurenniemi, "Mieli, kieli ja todellisuus." Source: *Uuden Ajan Aura*, 1979, julkaisija Oraan Suojelijat r.y. Helsinki.

Unless otherwise stated, all images are from the Erkki Kurenniemi Archive at Finnish National Gallery Collections. Photographer unknown if not mentioned.

III Artistic Practice

Foreword

Joasia Krysa

In his preface to the book, media theorist Erkki Huhtamo points out that Erkki Kurenniemi's body of work is best characterized as a "formidable unpredictable celebration of the fragment" rather than complete masterpieces. Whether films, music, writings, or grand ideas such as creating an algorithmic template for life executable by a future quantum computer, Kurenniemi's eclectic collection of "unfinished fragments, beginnings, and dead-ends" becomes material for others to develop and complete.[1] The emphasis is on an explorative process of testing, from one idea or technology to the next, with failure playing an inherent role in the method. Kurenniemi's work explodes disciplines and fields of practice, expectations, and epistemes.

In his early writings on art, Kurenniemi argues for an approach that merges artistic practice with technology, or what he calls the "integration of arts." The three texts included in this section—"Message Is Massage" (1971), "Computer Eats Art" (1972–1982), and "Computer Integrated Arts" (ca. 1986)—together demonstrate how his trajectory of thinking moves from an understanding of technology as a tool for making art to an understanding of technology as art. But more than simply an interdisciplinary perspective, he represents a far more messy and complex, even contradictory creative force, in which art is just one possible (and probably inadequate) way to describe his multifaceted output.

In this sense, the structure of this book—which separates Kurenniemi's diverse contributions into disciplines, and in particular music from artistic practice—would appear counterproductive. However, his undoubted contribution to the field of electronic music deserves closer examination in a dedicated section, while this section aims to focus on other aspects of Kurenniemi's artistic practice, ranging from films, computer animations, and interactive installations to collections of objects, drawings, paintings, and diaries.[2] Still, it is not possible to entirely separate out Kurenniemi's work that relates to music, insofar as it also relates to his wider aesthetic production and his influence on the practices of other contemporary artists. For instance, some of his early music machines designed and produced in the 1960s and '70s can be equally considered as pioneering interactive installations. One of the them, the DIMI-O (Digital Interactive

Music Instrument—Optical Organ, 1971), was one of the first synthesizers to use video tracking as a control method that enabled the user to play the instrument through bodily movements in front of the camera, with built-in digital memory for storing short musical passages. Considered an interactive sound installation, it precedes better-known works such as David Rockeby's *The Very Nervous System* (1982–1991). Another example, Kurenniemi's *Electric Quartet* (also known as *Sähkökvartetti*, 1968), a collective instrument with a programmable sequencer for four musicians where players were able to collectively improvise along with repeating melodic and rhythmic patterns, anticipated the development of other real-time participatory platforms. Similarly, DIMI-S (also known as *Sexophone*, or *Love Machine*, 1972), a collective biomusical instrument[3] for up to four players in which the human body served as a control signal for music material and light,[4] was essentially a processual instruction-based work in the tradition of Fluxus or algorithmic art where a finite number of steps are followed to complete a task. The instructions for the DIMI-S were simple:

1. Ask three people to join you.
2. Ask each of them to pick one iron ball.
3. Touch each other's skin.
4. Play!

It is hard to find in any tradition or field, or even any transdisciplinary practice, another artist who offers a more complex formulation that is both "art and not just art"[5] at the same time. In many ways it serves to question what constitutes an artwork, leaving Kurenniemi an outsider to both contemporary and media art histories while simultaneously offering a kind of independence from the mainstream art world and economy. The first large-scale international presentation of his work is relatively recent, as part of the 2012 Documenta 13 exhibition in Kassel.[6] Held at the Cabinet of Astronomy and Physics in the Orangerie—another oddity—the presentation included a substantial and varied range of materials, including fifteen short silent films, generative computer animations, electronic music compositions, music machines, an interactive robotic machine (Master Chaynjis), and a collection of objects, photographs, and writings ranging from computer programs, music theory, and computer theory to personal diaries. The difficulty with such an eclectic presentation was that it challenged more mainstream views of what artwork might be and where an art exhibition might be held even in 2012; so it is not surprising that his work has oscillated between the categories of art and not-art for so long.

It is interesting to note, however, that Kurenniemi's work has resonated with a younger generation of contemporary artists. For instance, Tarek Atoui's *DIMIs Re-connected* (2012) project for Documenta 13 was essentially a reenactment of Kurenniemi's electronic synthesizer DIMIs from the 1960s and '70s as a series of collective sound performances.[7] The encounter with Kurenniemi's work is described in these notes:

Between December 2011 and March 2012, I visited Kiasma and the University of Helsinki, where I was introduced to these instruments by Perttu Rastas, Jari Suominen, Mikko Ojanen, and Kai Lassfolk. On these visits, I noticed that Erkki Kurenniemi did not to write user manuals for these instruments, as he believed in an intuitive and individual use of his inventions. The Electric Quartet, one of the first collective electronic music instruments to be conceived, also got my attention and drove me to extend Erkki's idea of collective instrument. I therefore connected the DIMIs through a computer interface and re-created a collective instrument to be played by four persons. While the EQ had a distinct sound identity for each player, on my instrument, through feedback and sound processing, the performers lose track of their sound and individual contributions and become part of one fused and sustained collective body. (excerpts from Atoui's notes, 2012)

Other projects that emerged directly or indirectly from Kurenniemi's work include his ideas regarding AI and artificial neural networks and sound, developed further by contemporary sound artist Florian Hecker (included in part 5, "Music"). His ideas regarding archiving have also further inspired the *Active Archives* work by Constant (Nicolas Malevé and Michael Murtaugh) and ideas related to nonhuman agency in their essay "Archiving the Databody," included in this part of the book (with Geoff Cox).[8] Lars Bang Larsen, in his essay "The Unbearable Non-Artist from 'L'homme machine' to Algorithmic Afterlife" (2014), also included in this part of the book, further develops the collapse of Kurenniemi's life and work as a fractured whole, as an "antidisciplinary" practice and "incompleteness," as he puts it. Indeed, to Kurenniemi as mathematician, "incompleteness theory" would verify that all logic has inherent limitations.

An emphasis on algorithms is registered in both these essays—"algorithmic archiving" and "algorithmic afterlife," respectively—referring not only to Kurenniemi's various works but also to his overall thinking. He treated both what he collected and his life as data for further processing through algorithms (exemplified by his project for 2048). As Mika Taanila's film *The Future Is Not What It Used to Be* (2002) demonstrates, Kurenniemi's overall life project not only evolved around the systematic collection of data, but also unfolds complex interactions of collecting and presenting recorded material, as well as signaling its dynamic, active character.

It can be argued that this eccentric sensibility comes close to emergent forms of curatorial practice that recognize the dynamic qualities of the material, working with materials not simply to collect and store but with an intention to shape them (produce meanings)—in Kurenniemi's case, to shape them using computational power (thus partly delegating the curatorial function to a computer program). In a wider sense, Kurenniemi can be seen acting not only as an archivist of his life but also as a curator (of life-data) in the expanded sense, exploring new computational forms of organization and meaning making.[9] Furthermore, like many other of his projects, his archive (and his project to develop a template for a human life) is also a work in progress. It takes on the character of an incomplete project that inadvertently halts in 2005, ready to be further developed. In other words, his material/archive operates as source code, open for further modification: scripts and scores to be performed, programs to be run at some time in the future by someone else—source material for another curator to share and execute and thus recreate the idea of human life as software. This is what Constant has been developing since 2012, responding speculatively to his grand vision of reconstructing human life from source material by computer. Constant's working approach has been not to follow standardized archiving procedures but instead to offer a series of speculations on the specific qualities of the materials by running computer programs, or "probes." Its algorithms act as archivists but also as curators inasmuch as the group uncovers aspects of what is not directly apparent in the material by running various programming procedures and displaying results. This provides the opportunity to consider algorithms that can act on the material as a kind of curating. This also suggests the possibility of considering Kurenniemi in another mode—as a curator, or more precisely a pioneer of algorithmic curating, allowing his work to be understood as less a collection of unfinished fragments and more as a template for future exhibitions yet to be realized. Such wild speculations emerge from the contributions to this part of the book, rendering the category of art far too limiting a term.

Notes

1. For instance, in 2003, six of Kurenniemi's silent short films from 1960s and 1970s had soundtracks added together with films by fellow artist filmmaker Mika Taanila, who has also released a compilation CD of Kurenniemi's electroacoustic compositions (*Äänityksiä/Recordings 1963–73*). Another example is Kurenniemi's computer animation *Spindrift* (1965), considered one of the earliest computer animations from a Nordic country that followed similar earlier developments elsewhere, such as the mechanical animation techniques of John Whitney, Sr., that were used to create sequences for television programs and commercials in the 1950s. The work, a collaboration between Swedish composer and musician Jan Bark and Kurenniemi (who programmed animations) was an experiment for a new kind of "music for black-and-white TV" exploring audiovisual synesthesia. It was premiered on 16mm film at the Computer Music Seminar, Dipoli, Espoo, on October 28, 1967, and broadcasted by Swedish Television, but was subsequently lost. The reconstructed version of the film (2013) is based on the surviving 16mm positive "work copy" film reel, and has been produced by Kiasma Museum of Contemporary Art in Helsinki in collaboration with the National Audiovisual Archive (KAVA).

2. The list of his works as cataloged to date includes: fourteen experimental silent 16mm short films on the themes of nature, environment, travel, sex, and technology (shot between 1964 and 1971); 6mm and 8mm films (1963–1970), including *Huumava sieni* (Intoxicating Mushroom, 1968) and *Sex Show* (1968–1970); *Dimi Ballet* (1971, 11 min.), documenting DIMI-O (a TV production); short computer animations, each thirty seconds to one minute, on floppy discs; stereoscopic slides from early 1970s; Master Chaynjis, an early robot from 1982; diaries—handwritten notebooks (1969–2006), ninety-one hours recorded on audiotape (1971–74) (an extract is

published in part 1 of this book), and video diaries on VHS cassettes with sound (1981–82), around 300–400 hours, intended to be shown in 2048; 100 drawings (1978–1979); photographs (negatives, prints, 50,000 digital images); electroacoustic compositions, realized in the Electronic Music Studio of the University of Helsinki in the 1960s and '70s; musical instruments—Integrated Synthesizer (1964–1967), Electric Quartet (also known as Sähkökvartetti, 1967), Andromatic (1968), DICO (Digitally Controlled Oscillator, 1969), DIMI-S (also known as Sexophone or Love Machine, 1972), DIMI-O (Optical Organ, 1971), DIMI-A (Digital Musical Instrument with Associative Memory, 1970), DIMIX (1972), DIMI-T (1973), DIMI-6000 (1973–1974), DIMI-H (2005–2006, together with Thomas Carlsson). His works are held in the collections of the Finnish National Gallery Collections, Archives and Library, and the Museum of Contemporary Art Kiasma; some of his instruments are deposited at the Musicology Department at the University of Helsinki. His extensive private archive is deposited in the Finnish National Gallery Collections, Archives and Library, where it is being cataloged and digitized. A complete list of his works is included in the last part of this book, and is available online at http://en.wikipedia.org/wiki/Erkki_Kurenniemi.

3. This work was influenced by ideas of biofeedback and in particular the ideas of American composer Manford L. Eaton in relation to biomusic (see Manford L. Eaton, *Bio-Music* [Something Else Press, 1974]).

4. In this sense it can be seen as a development of Gordon Pask's cybernetic art machine *Musicolour* (1953), which is widely thought of as the first cybernetic art machine.

5. "Not just art" is a reference to Matthew Fuller's "A Means of Mutation" (1998), where he describes not easily categorized practices that can be seen as part of the art field but just as easily as part of other systems. See http://bak.spc.org/iod/mutation.html.

6. Subsequently, Kurenniemi's work has been presented at international exhibitions: Kunsthal Aarhus, Denmark (as part of the *Systemics Series* program, "Systemics #1: Certain Peculiar Things and Ideas, Often Failed (or On Humans, Machines, and Running Algorithms)," January–March 2013), and a large retrospective at Kiasma Museum of Contemporary Art in Helsinki (*Toward 2048*, November 1, 2013–March 2, 2014).

7. During Documenta 13 (June–September 2012, Kassel), the *DIMIs Re-connected* instrument was installed in Kurenniemi's exhibition space at the Orangerie. On the first three days of the exhibition's opening, Tarek Atoui, Kai Lassfolk, Jari Suominen, and Mikko Ojanen held twelve different performances on *DIMIs Re-connected*, which became Atoui's first iteration of the concept of collective instrument.

8. The project was commissioned by KURATOR and Documenta 13, funded by Arts Council England. Subsequently, developed versions have been presented at exhibitions at Kunsthal Aarhus in Denmark ("Systemics #1," 2013) and Museum of Contemporary Art Kiasma in Helsinki (*Toward 2048*, 2013–2014), and at numerous presentations and conferences.

9. Patrick Lichty says something similar when he refers to "curating everyday life": Patrick Lichty, "Reconfiguring Curation: Noninstitutional New Media Curating and the Politics of Cultural Production," in *New Media in the White Cube and Beyond. Curatorial Models for Digital Art*, ed. Christiane Paul (Berkeley: University California Press, 2008).

5 Message Is Massage

Erkki Kurenniemi

Let me put it this way: I shall now write a number of sentences. Some of them are true at this moment, others are false, some will become true at some point in the future, others will be false in all possible futures, some will be true in some futures, false in some others. The same applies for pasts. The truth has a thousand faces. This is one of those sentences.

Some physicists talk about so-called EWG interpretation, which claims that not all of the futures that will be realized are situated on this particular string of time. Reality is continuously dividing into all its potential actualities, into who knows how many equally real alternative worlds completely ignorant of each other.

The artist's instrument in 1983: A pocket computer that can be connected to any screen or projector or network. The housing contains also a colour video camera and a small colour monitor (perhaps even holo). The device also works as a tape recorder. Some of the programmes can be voice-activated, they can understand simplified language. The words they understand appear on the screen as text. You can draw on it with a colour light pen. The overall tone of the picture can be altered with a programmed transformation of the colour layer. If you want to save memory, after drawing a circle you can give the programme an idealisation command, which just saves its centre point and radius. Motion pictures can be animated one frame at a time by making the desired changes in the previous image with a pen or by spoken commands and transferring the images to memory. Or by giving the elements of the image their trajectories and speeds.

What about the artist's task in a society which is a symbiosis of people, communities, conscious and subconscious machines, talking animals, thinking cars, roads and cities aware of themselves as big and small entities, self-destructive, constantly evolving, improving, renewing, decaying, unstable, analysing its past and wondering about its future and its own purpose at superhuman speeds?

To dress up in the latest fashions and take DOET in a cafe.

"The pendulum of world politics has swung from one end to the other three times this week."

A personal tool is a combination of a computer, terminal, television, telephone, videophone, radio, tape recorder, operating table, book, magazine, newspaper, library, school, post office, bank, electric organ, answering machine, walkie-talkie, cinema, theatre, typewriter, calculator, calendar, diary, watch, camera, microscope, telescope, workplace, entertainment, social relations, photo album, museum, art exhibition.

The general data network integrates television, radio, telephone and data communications into itself. The network is hot.

Computers will probably never become humanlike. Before that has a chance to happen they will have become like gods or other spirits. Mankind will probably follow, but by then the word "human" will refer to something a bit different. This claim is based on the fact that unlike the human mind, which is firmly fixed to the brain, a computer programme can be transferred from one computer to another. The programme can also, unless it is prevented from doing so, transfer itself from one computer to another.

Or will mankind continue to divide into hostile camps. Each camp with its own network, surveillance, jamming, sabotage.

And in this environment, how will the inevitable in art—the separation of form and matter—take place? Some separation has already happened. When the word "round" becomes abstract, roundness becomes distinct from objects and can just as well attach to Pegasus' thigh. The round thigh of Pegasus. And yet the word is fixed on paper, text is slow to read or copy. The word will tumble in the mind for a while, perhaps even associating with something and then being soon forgotten. Form will only become liberated deep in the associative memory of the computer as it evolves into a thinking creature based on the Algol 90 language. Things are not quite free in the human mind, thinking is constrained by patterns, survival of the individual and the species, real and imagined goals. Only a machine can create with absolute spontaneity, without any aim, without any thinking, without any fear.

DO YOU WANT TO BE A MODEL FOR PICASSO?

Our programme has analysed Picasso's style to within 90 percent. Send us 5–10 photos, the desired dimensions, number of copies, any preferences regarding background and pose, date, etc. You can also order works in the master's posthumous style (as based on the probabilistic prediction of our machine). Price 50/sq.m., by return mail.

LIBIDO LTD

Why do we talk about television programmes and computer programmes? What the cat brings, the dog will take away.

The first computers were the size of an elephant, now they are buzzing around us like mosquitoes. They have been taught to block biological mosquitoes peacefully, if possible.

As entertainment, the mosquitoes in a thousand-headed swarm can be commanded to arrange themselves head to toe to construct regular space lattices. They change their position lightning-fast, creating a kind of a small waltz.

And I recall the early days of video, the old legendary Sony era, when guys noticed that it doesn't matter whether you show the people the monitor or the camera or both.

software—hardware
programme—machine
soul—body
spirit—matter
content—form
form—matter (material)
message—medium

I think the elements of this conceptual duality will become increasingly independent of each other, in some way, perhaps, I don't quite know, in the cybernetic age.

The spoken word is bound to

the speech organ or loudspeaker
and the ear or the microphone
and time or a recording device
and place or telephone or radio
or some other medium.

Pictures are on canvases, canvases on frames, on a wall, in a house, in a region, in a country, on the Earth.

Music is in instruments, notation, tapes, disks.

The soul is attached to the brain.

Life is attached to realities.

"Only a memory remains …"

Art is chamber music for children.

Communication in art sometimes refers to the performance, publication or exhibition of the artwork. In this connection the work can be a message or a medium or both.

The study of thinking machines teaches us more about the brain than we can learn by introspective methods. Western man is externalizing himself in the form of gadgets. Ever pop coke in the mainline? … C pleasure could be felt by a thinking machine, the first stirrings of hideous insect

life. ... Of course the effect of C could be produced by an electric current activating the C channels.
—W. Burroughs, *Naked Lunch*

A picture is not worth a thousand words. It refers to different things.
—Juhana Blomstedt

As a form moves into a new memory device, it stops being static, a form. It becomes dynamic, a process, a programme. There is no point in recording an image by registering the hue of each individual point separately (using this method, a single TV-standard image would use the whole buffer, i.e. consciousness, of a large computer). It is better, in fact necessary, to analyse the picture into a sum of elements of much greater complexity than individual points. These elements are not points, typographic symbols, lines, arcs, exponential curves, networks, screens, cylinders, tetrahedrons, but processes, sub-routines. A graphic syntax is based on sub-routines that produce dots, lines, symbols, screens, 3-, 4- and *n*-dimensional cubes. On the next level are sub-routines which arrange the dots, push the lines, turn and stretch the screens, splice cubes into pieces, conjoin tetrahedrons into strings. There are sub-routines that move objects in perspective, create shadows, sub-routines that create random shapes, movement, imprecision, sub- routines that modify other sub-routines. Therefore one command in Fortran or Lisp given on the main programme level can trigger a chain reaction, a computer-graceful play, evolving continuously for a thousand years, throwing human visual perception haywire.

The first ultra-intelligent machine is the last invention that man need make.
—Irving John Good

Disposable art. Works of art disappear from the tangible world, because they are useless most of the time. An image does not have to hang on a wall except when someone is looking at it. Not even if the wall is painted with smart-paint. When a work of art is expressed in a programming language, it can be realised when and wherever there is a device to print it. Alternatively the form is realised via brain electrodes directly in the mind. (Next summer there will be a conference somewhere on the theme of using a video camera and electrodes to restore sight to the blind.) Or it can be projected with a laser onto the sun, who knows, it might be alive. The work can be stored on punch cards, punched tape, magnetic cards, ferrite rings, magnetic tapes, MOS transistors, disk memory, holograms, magnetised bubbles or written on paper. It can be printed out as music on a digital synthesiser, as text on a line printer, as graphic on a plotter, as film on a COM (Computer-Output Microfilm) device, on a TV screen, on video tape, on metal, wood or plastic using a computer-controlled tooling device, set into print using a computer-controlled typesetting machine, and it can be destroyed with a single DELETE FILE command.

Art has opened the eyes of man, music his ears, computer his mind.

Originally published in *Taide* 6 (1971), Helsinki.

6 Computer Eats Art

Erkki Kurenniemi
Translation by Teo Välimäki

The computer is a multipurpose tool, a universal instrument. This is why it's also an artistic tool, an artistic universal instrument. Its major effects on the arts will be the separation of art from material and the separation of art from man.

These processes of separation are different from those caused by earlier inventions in computing. The printing press, film, audio records, and electric communication caused the separation of art from time and the separation of art from location.

I can listen to Jarrett's concert in Cologne on an audio record without having to be present at the recording itself. In principle, all art ever recorded is constantly present. The world and history have been crunched together.

It's no coincidence that the most vibrant art forms nowadays are environmental and computer art. In fact, they go together like peas and carrots. The reason: humankind is splitting into two separate components. One component is merging with the machine, becoming one with it. The other is returning to nature, merging back to that from which it has been divorced for too long. The asexual reproduction of a species: the old human in both branches expires and something eternally branching is born. Some branches intertwine and create a multicivilization.

The art of film is bound to material film itself, to the camera and projector. Music is bound to an acoustic space, to instruments, to sheet music, to records and tapes. Videotape differs from audiotape. The computer does not abolish the principal bind with materiality, but it renders all other matter besides itself redundant. On a hard drive an image, a piece of music, and a poem are conjoined by a binary structure. The central processor processes all this information as equal data. Because the computer is a "universal Turing machine," it can do anything that is computable to this data. Not everything is computable. Gödel established that the mystic infinite creeps even into the computer. We know of a multitude of very simple problems that cannot be solved by any computer ever. In regard to this, I recommend reading Hofstadter's *Gödel, Escher, Bach: An Eternal Golden Braid.*

A computer renders sheet music into music, music into dance, dance into visual image, image into poem, poem into mathematics, mathematics into physics, and physics into a multisensual illusion of reality and eventually into the reality of an illusion. All art is political, and all artists are responsible for the future, bearing all the more responsibility the more creative they are. Fucking reductionist.

In environmental computer art, the computer is a part of the environment. Soon every Finnish ridge and lake will be embedded with a microprocessor, monitoring tourists so they won't litter the environment with their prophylactics and sounding warnings if representatives of the local power company are nearby. When no imminent danger looms, the microprocessor will play chess with the locals, using haystacks as the game pieces. Swallows will make their home in microprocessor-fitted nests.

In environmental computer art, the Hitler Youth lying in a sensory deprivation bubble annihilate the galactic virus going up the core of Mercury. The Freudian processor synthesizes a 3D harem into Khachaturianic pirate jelly. One can survive it by momentarily removing oneself to a laser-multiscanner. The rest is just sweet pain.[1]

The Stages of the Computerification of Art

The breakthrough can be divided roughly into three stages. In the first stage the computer is applied in the traditional domain of art as a tool of experimental creation and expression, which soon reverts to standard practice. In the second stage the computer invigorates the art scene by creating new art forms, such as interactive fiction. In the third stage the different genres of art are fused together into an all-encompassing and pure computer art, from which the old genres are only obscurely distinguishable as strands of history.

These stages are typical for all breakthroughs and revolutions: the creation of a holon. First, local disturbances occur. The disturbances escalate into chaos, the old entities are destroyed, and new ones are rapidly created everywhere. Finally, a consolidation takes place, a new dynamic government assumes power, and history is rewritten by the conqueror. The ideological turn of general opinion can take a long time, but it doesn't have to. In a few years or decades, all modern art will inevitably turn ethnic.

The First Stage: Fluctuations

Each His Own Gutenberg Early on, in the realm of literature, a computer will be programmed to write random nonsense poetry and tales, and with a little more programming, some computers will be able to produce short stories with working plots. These experiments won't do much to disrupt literature.

The gears will properly begin turning when the writer exchanges the typewriter for a word processor. Its basics can be learned in days, but the actual change will take months to grasp fully. The writing will be buffered into a file stored on a magnetic disk. With a typewriter, the natural way to add text is to continue writing at the end of the document. Adding text within the document is difficult and messy. With a computer, the natural way is to "insert" text anywhere within. The rest of the text will be displaced automatically. At first, excitement to try out William Burroughs's (nomen est [omen]) collage-type writing may lead to learning the structured "top-down" method. The real text resides invisibly on a disk. I may print out a working copy only every three days. A little editing and the text is printable.

Modern word-processing software stores text as a file: the text is saved as continuous successive characters onto a drive. At the next stage the file is replaced by a database. A database is an internally structured file, not just mere words in succession but facts and their interconnections, that is, information. A database can be filled with just facts, leaving the expression and stylization for later, even rendering it automatically. Associative queries to the database can be made. If the user enters a character's name from a novel, the database lists all mentions of the character. The database is not merely a passive recipient; it can be conversed with. The final bell will toll for literature when the author realizes, after finishing a text, that the printed paper version represents only a dead shadow of the rich relational realm of the database.

Computer Music Classic computer music means three experimental circuits. A computer can be programmed to compose music. A pragmatic answer to the query, isn't the programmer the real composer then?—if the programmer has no idea what the composition will contain *sans* computer, then he or she is not the composer, at least not in the traditional sense. Second and more modestly, the composer can use the computer as a tool, much like an abacus. Third, the computer can be used as an instrument: a note can be comparatively easily determined mathematically and the computer can create an audio record by processing the calculations.

For a while it seemed that music would eat the computer. The electric guitar and synthesizer had become commonplace instruments. Soon the microprocessor will invade all instruments. The synthesizer has its own characteristic sound and so will the processor. On the other side of the curtain, however, the processor will devour the whole orchestra in one big bite.

Automatizes the Orchestra The home computer or the personal computing machine will have an accessory, easily obtained for the cost of a decent record player, which will act as a rudimentary synthesizer. The hardware won't be much to shout about, but the software will be golden. Fiddling with some switches and dials will enter the track into the computer's memory using traditional notation. Within five years, similar equipment will be developed that can imitate any music instrument or playing style.

Japanese pocket organs are currently spreading into a very elastic niche in the market. The hobo playing the mouth organ will soon be an anachronism. Pocket organs already include the most essential of attributes: memory. Floppy disks will emerge alongside records and audio cassettes. They will look like small audio records, but will function like magnetic tapes. In five years the technology might revert to optical, laser-based operating principles. A piece of music will be published as an audio record, a cassette, and a digitized copy, as notation, as disk.

The Great Transcription has begun. In about a decade a significant part of all musical notebooks and compositions will be in "machine-readable" form. A computer will play old compositions with instruments that "don't exist" anywhere. Existence becomes symbolic. The old and the new fuse. Music is composed for a "real" orchestra by sketching and structuring a symphony using a musical note editor. An orchestra simulator will replace the composer's piano. When the piece is ready, a matrix printer will print the composition and its parts. All music will have both recorded and live versions available.

Reenacts the Elements of Classical Music John Cage noted that, as in earlier music, where the battle was between consonance and dissonance, now the focus of the tension has shifted to the note and noise. Electronic music has tossed the old concepts of rhythm, melody, and harmony overboard and immersed itself in fields, continuums, sounds, timbres, and transformations.

The computer thrusts so deep into music that even its classical theories must be ungrounded and reworked. The micro- and macrostructures of classical music are highly formal and mathematical. The classic theory of harmonies is almost an exact science. The computer ravages this like … The parameters of Beethoven can be defined. Using the new theory of harmonies, a computer can easily analyze the power of individual chords or generate their desired power levels. We've discovered that harmonies are natural numbers. All their forms can be calculated.

In algorithmic music, the forms of a music track are simplified to the extreme. Ten lines of programming can create hours or years of constantly changing but formalistically consistent music. At the same time we're beginning to understand the algorithms of older music. A good algorithm is an entirely new musical dimension. It can be applied and combined with other algorithms ad infinitum. The algorithm itself is a strange creature. It's transmitted onto a computer using a programming language, but the algorithm itself is independent of the language used. It's an atom of programming, an indivisible part of a program whole. The algorithm is the description of a happening, an event, a process, or an action abstracted to the extreme.

The Visual Arts Painting and photography are 2D art; sculpture is 3D. Computer media don't have a specific dimension number. The three-dimensional coordinate readers or digitizers for computers are just entering the market. They'll replace the pen, the brush, and the chisel. We'll realize that drawing itself isn't two-dimensional; the idea is just a restrictive remnant from old media. A pencil can't be used to draw on air, but a 3D pen can. Using a computer graphics system, an artist moves a brush in the air in the field between three ultrasound microphones. Inaudible to human ears, the ultrasound impulse reaches the microphones at slightly different times (the distance difference divided by the speed of sound), and based on these differences, the computer calculates the location of the pen. The trace appears on the screen, in color and in the desired perspective.

When the apples, vases, and the table in the still life have been drawn, they'll be stored in the 3D-object library on the computer's hard disk. The wallpaper is chosen from the wallpaper library. The table can be grabbed with the 3D pen and dragged next to the wall—no, into the middle of the room. The vase and the apples are dragged onto the table with the pen. The artist then activates "VIEWPOINT" and the computer locks the viewpoint to the pen's location. By moving the pen in midair the artist moves around the room and views the still life from different angles and distances. The artwork is finished. The artist doesn't lock the work to a particular perspective, because 3D artwork is distributed on a diskette. The viewer moves in the space using his or her own ultra-pen. All this will soon be made available in everyday stationery stores.

The computer is a tool that doesn't displace or replace any previous medium yet transforms everything. The visual arts, with all their methods and the historical and current libraries of style, are reinvigorated as computer simulation. Earlier inventions often created diversity. There are oil painters and acrylic painters. The computer isn't just a platform of simulation; it adds its own active dimension and therefore unifies art. In a 3D system, painting and sculpture are fused together. Painting transforms into a three-dimensional activity, and sculpting takes place on the two-dimensional surface of a computer screen. A three-dimensional still life can be explored onscreen. The *Mona Lisa* is made into a statue. The file MONA LISA is entered into the GEN Z computer program. MONA LISA is a two-dimensional color-digitized representation of the *Mona Lisa*. The program generates a third dimension derived from the background landscape and female shape. After this, another program called APT (Automatically Programmed Tool) creates a 3D "tape" of the MONA LISA file, which is programmed to guide an electronic grinder. The robot grinder sculpts Lisa out of gray plastic and then activates an MCY (magenta cyan yellow) pistol. The pistol paints the sculpture in a single coat, because it mixes the color pigments with the bonding agent in real time. The wax cabinet has as little to do with wax as tape has to do with paper-punched tape.

Kinetic artworks commonly have one electric motor; some have dozens. They create motion. Processor sculptures commonly have one microprocessor; some have hundreds. We're moving toward wireless close communication. The first applications are used in television remotes and wireless earphones. Nowadays it's trendy to create piles of sand or lumps of plastic for art exhibits. Fitted with microprocessors, these piles and lumps can communicate wirelessly with each other. Should they have a speech-recognition device and a speech synthesizer, they could also communicate with the audience, both together and separately. The microprocessor will do for sculpture what film did for photography.

And the computer will, again, do the same with the image, first with drawing (wire frame, even with hidden lines) and then with painting (surface graphics). The computer is a tool for film animation, but purely computer-generated animation is a genre unto its own. True computer animation occurs when the creation of an image takes less than tenths of a second. Then the succession of images begins to create motion and time enters the frame. Animated and animating computer graphics are being created in hundreds of different fields simultaneously.

Computer Ballet The notation problem in performative dance hasn't been solved in a satisfactory fashion using traditional methods. The situation has been solved using a computer as follows. The joints of the dancer are fitted with sensors, which constantly transmit the positions of the joints wirelessly to the computer. The stored multidimensional information structure can thus be "played back" on the computer screen as a three-dimensional animation. The choreography can then be created and edited using a keyboard or a light pen. The computer can also compare the dancer's performance with stored data and inform the dancer when his or her posture was flawed.

This synthetic ballet can also be created directly as a computer program. The program can create the music and the choreography simultaneously, interacting with the audience. The graphic representations of dancers on the screen are like realistic three-dimensional dolls. Nureyev's anatomy can be recreated as a database. The dancers can also be animals or abstract geometric shapes.

A more concrete form of ballet is made possible by the use of robots. It will take a while, though, before the fluidity of motion in robots will attain even the level of chimpanzees.

The Second Stage: New Forms

Interaction The most important new element granting a new dimension for computer art in all its applications is interactivity: the interplay between man and machine. The medium between a composer and music is no longer just a piece of dead paper printed with stagnant symbols, but a sounding instrument or orchestra. But interactivity is even more significantly present between the work and the audience.

Fairy tales have been replaced by "adventure" type games, which children play using a keyboard or, in the future, using just speech. They roam the world of the game fighting beasts or ghosts, seeking treasure or an imprisoned princess. They no longer listen to stories, but guide them themselves with their own dialogue and choices. The creator of the tale cannot completely control the way in which the story unfolds, but provides the player a certain array of situations. The characters in the tale or novel will begin to become more and more lifelike. They will begin life as stick figures, then evolve into comic book characters, then into anatomically accurate three-dimensional dolls, and finally into as realistic as the people you see on television. You can converse with them and even have them keep secrets. You can order them to be nicer or naughty, very naughty in fact. An alien race in a science fiction adventure can be truly four-dimensional: 4D graphics software is identical to the three-dimensional ones—all conversion formulas just contain one extra equation. An era of artificial personas, soft creatures, begins. Santa Claus transforms into a living child psychiatrist. If an artificial persona is tortured, it will act so realistically that a behaviorist would claim it experiences pain.

This new genre has been proposed to be named using the three-lettered acronym CSF (Computer Simulated Fantasy), maybe Computer Simulated Reality or just CS-Reality or C-Reality.

Paraphrasing Adams: "Ladies and Gentlemen, in a moment you will be able to meet the Concerto of the evening."

Interactive fictions are also designed to be adventures for adults. Once classics were made into comics; now they're being made interactive. At first interactive novels are autistic. Immersion into a novel or a film takes place through identification and imagination. Immersion into an interactive novel takes place through verbal action. Writing a novel has been a process of spinning a single thread. Writing an interactive novel is weaving fabric. An author has to prepare and program all possible outcomes of a triangle drama, which the "reader" can affect. At the simplest level, interactive is like a game of pinball. The participant guides the bifurcation of the plot in a binary web using two buttons.

A computer adventure can host several players, even people who are strangers to each other. The computer uses a modem to contact an adventure server, which replies with a list of available roles to play and a schedule for the beginning of the adventure. At first a player will be unaware which of the other roles are inhabited by people and which are simulated by the machine. Sometimes in the end the true identities are revealed. A masquerade.

The Separation of Art and Man Ever since the early days of the computer, it has been used to compose music. Today, researchers of AI are pondering how a computer can be programmed to listen to music. Analysis is more difficult than synthesis: computers can already speak easily and understandably, but it will be at least a decade before they'll be able to understand speech. It's easy for a computer to draw a circle but hard for a robot to recognize one. Sometime in the future an antiperceptual art will be created: art made by computers for other computers. That will be quite a separation.

But by separation I really mean another, even more extreme development. Our art culture is stigmatized by two polarizations. There is a polarity between man and art, a subject–object relationship, which is why we talk of a "work" of art. There is another polarity between the artist and the audience: like two magnetic poles that exist in constant tension and only really also exist in relation to each other. Without man there would be no art and vice versa; without the musician there would be no audience and vice versa. Like a prisoner and a guard handcuffed to each other. When the cuffs are broken, the prisoner escapes and becomes his or her own guard; the guard pursues and becomes his or her own prisoner. The computer replaces the straitjacket with outpatient care. When the work of art begins to think and feel, it soon experiences the peculiarly human way of thinking and feeling: first as a funny joke, then as an interesting psychological curio, and in the end as a magnificent work of art, that which it itself sprung from at some time before the year 2000. The child of the village idiot.

The Third Stage: Consolidation

When the dust finally settles, the entirety of the computer's effects on society will finally be revealed. Many things have been permanently displaced. The most central new innovations are the computer-readable common storing media. Physically they can be magnetic or optical disks about five to ten centimeters across, each of which can hold whole bookshelves of information. The physical aspect is less important, because machine-readable data can always be cloned error-free onto other media. It will take several decades to transfer most of the world's libraries and museums onto machines.

Digital or magnetic media will have the significant new property of noncorruption of copied data. The difference between the original and the copy dissipates. The changes in the status of copyright issues are already visible. Occasional corruption or mutation does happen, but its percentage is negligible. An artwork that can be endlessly and identically copied is nonentropic and indestructible. What is this, if not the separation of art from matter?

Since a program or a file can be edited between copies, several versions of an artwork can be made available. A tree of ideas, sprouting from a single idea, can branch indefinitely. Making art becomes incremental: a composer can take an old program and make a seemingly insignificant alteration, but the resulting music is entirely new and different. As the programs and compositions grow larger, starting from scratch becomes ever rarer.

Art transforms more totally into a monologue of an integrated society. In the art of advertising, progress can be measured in income. More machines are sold as a consequence of supporting experimental computer art.

The reception process of art turns interactive. The passive recipient becomes an active one, and his or her actions are reflected in the artwork itself. Many programs can be designed to allow the recipient to choose the level of interactivity desired. At one extreme, the recipients virtually create the artwork themselves; at another, they can passively view an AUTOSHOW.

Paraphrasing Lem:

```
BEGIN CREATE ARTIFICIAL PERSONALITY;
 NAME ADOLF
 ACTIVATE CONSCIOUSNESS
 LOAD PERSONAL HISTORY
 FOR JEW = 1 TO 6000000
 TORTURE ADOLF
 UNTIL DEAD
 NEXT JEW
FORGIVE END;
```

The computer as we now understand it will never become humanlike. First of all, humankind itself will change before that happens. Second, when the computer will be ready to personify a complete human being, it will already be much greater than humanity. The computer is now about forty years old, and it has never really evolved toward humankind. It has always been designed to be the negative cast of humanity, a complement. Humans are slow, imprecise, forgetful, and easily fatigued. That's why the computer has been designed to be fast, precise, tenacious, and indefatigable. Humans have had to suffer sluggishness, uncertainty, forgetfulness, and fatigue as ransom for our intuition, inventiveness, and creativity. To be evolved is very different than to be creative. The complementary relationship between human and machine eliminates competition. A racecar driver is one with his vehicle, but still alone in his consciousness. The difference between human and machine will disappear naturally and subtly, when these two entities, so different in their basic substance, eventually coalesce.

Notes

Original title: "Tietokone syö taiteen." Published in *Dimensio* (1972–1982), Helsinki.

1. Editor's note: final sentence in English in the original.

7 Computer-Integrated Art

Erkki Kurenniemi

Abstract: Personal computers and workstations are rapidly penetrating most fields of artistic activity. The computer is more than just a tool for making music and pictures, and there will not be just a new genre of computer art. *Computer-integrated art,* made possible in part by the *Personal Communicator,* will emerge as the dominant art form.

Introduction

In the early experimental stages of computer art (the term *art* is used here for all forms of artistic expression, not just visual art) the very unfortunate idea emerged of "computer-made art." In some visions, the human artist was thought to evolve into a "meta-artist" who, by writing a universal program, creates an artist in the computer. This led to strong emotional resistance by those who felt that they would never master the new technology, who were afraid of a tendency of the "mechanization of art" and that a paradigm of "two cultures" was being projected. A contradiction was seen between "human art" and "computer art."

Today, the atmosphere is different. Personal computers and easy-to-use software packages have brought the computer within reach of artists with no computer background. As a reaction to the earlier frightening visions of the "computer as artist," the prevailing paradigm perhaps too much takes the view of the "computer as a mere tool." In this paper, we argue that the universality of the computer will deeply affect all institutions of art, and that the traditional division into genres of art will be gradually displaced by an organization of art forms by *degrees of participation* rather than by sensory modalities and storage media.

The unimodal applications of the computer need not disappear. Ordinary conventional music will still exist, but, because the score is stored in digital code, the composition will be easily transcribed for various acoustic and electronic instruments, and the piece can be used in any integrated environment as well (e.g., a computer game). Some developments could be called *computer-enhanced art.*

Computer-Specific Art Forms

Although *integration* is our theme, tendencies toward *differentiation* will also emerge, and new computer-specific art forms will appear. I oppose the view that computer art *in toto* is a genre; rather, for limited periods, specific novel genres will emerge based on newly found ways to use the computer. Some general characteristics of these computer-specific art forms are easy to deduce.

The traditional art media were mostly *analog* memories, meaning that a copy was always less perfect than the original. As a result, the distinction between *original* and *copy* persisted. In the *digital* computer world, there is no essential difference between original and copy. In the analog world, it was necessary to protect the original from destruction and deterioration. In the digital world, there is no need for museums, at least for the original purpose (museums and libraries are handy *backups* because they are already there). For new works, a few backup copies at friends' places will suffice. The ease with which a digital work can be copied, edited, and transformed, and the disappearance of the original, will lead to the *incrementality* of all computer art. The automation of the traditionally laborious orchestral score writing will mean that a composer can in a lifetime produce thousands of major works, perhaps all in a single database.

Other characteristics of computer art are *interactivity* and *participatoriness*. As the delivery vehicle for a piece of art, the computer can react to stimuli, and this is always useful. Recall that in pre-electronic time, the music listener had no volume control except geographical location. From simple variations there is a continuous path to ever deeper audience participation, up to the level of coauthor. Artistic adventure games are the theater of the future, with traditional actors automated away. Of course, traditional theater will not die out, but more professional acting will be mediated via two-way electronic systems.

Computer software and programming languages are new materials of the artist. Has anybody yet written a book of poetry in Lisp? Recursive forms, self-similar forms, fractals, and complex boundaries are a new exciting frontier for exploration, and the aesthetic experience is deeper if the beauty of the often surprisingly simple algorithm can be appreciated.

It is not only that the random subspecies of "art" suffer the structural catastrophe (fluctuations—integration↔differentiation—new dynamic regime sets in); the higher levels of conceptual organization also change, and art will again be integrated with science, politics, commerce, and religion. The relevance of mathematics for music has had to wait for the emergence of the computer, because although the principles of harmony were clear to Pythagoras, their combinations are so complex that it takes a computerized instrument to play in just intonation in real time. Thus the relevance of mathematics and other sciences increases for all the arts and is possible because the computer takes care of the tedious details.

Techniques in artificial intelligence also make it possible to construct rudimentary behaving units, "pseudo-personalities," and some day we shall be able to "meet the work of art."

Games as Art

Computer games are the most important new art form, not because of their artistic depth, but because games are showing the way to developing the interaction of new technologies, sense modalities, and traditional art forms. The evolution of ideas in computer games is rapid, but we can devise a rough classification as follows.

1. Transcriptions of *traditional games*. Most sports, board, and card games are available in computer form.
2. In *action games* (arcade games), the main activities are shooting and fighting, pursuit and flight, and collecting valuables. There is usually a hierarchical structure, and the tension increases gradually toward the loss of the last life and the death symbol "GAME OVER." These are *endurance games*.
3. In *adventure games* the goal is to solve the essential problem by first solving a hierarchical set of preliminary subproblems. An adventure game session can terminate in four ways: the player "dies," the game end is reached, the game is terminated by the player, or the game situation is saved for later continuation.
4. In *strategic simulations*, the player activity consists primarily in furnishing and programming an alter ego, a robot, an exploration team, or an army. The operational phases are then simulated by the program.

A fifth stage could be a "game construction set," but this facility is often already provided by all types of games. A typical game has three levels of participation: passively watching a game demonstration, playing the game at different levels of difficulty, and designing new games by using the provided game editor.

Game programming is often a difficult exercise in artificial intelligence. Especially in adventures and strategic simulations, the central requirements are the maintenance of a consistent model of physical reality, natural language communication, maintenance of artificial personalities, and some goal-directed problem solving in the game domain.

There are bound to be games of all categories in which, instead of the trivial pursuit of entertainment, deeper aesthetic rewards are sought. These may be transcriptions of existing literary works or more original creations.

Computer-Integrated Art and the Personal Communicator

Today, direct human–computer interaction is based on technologies like the CRT screen, keyboard, mouse, and so on. But soon this will be a thing of the past. The *personal communicator* (PC) described below does not presuppose any essentially new technology; it only requires further miniaturization of existing technology and is believed to be feasible and commercial before the turn of the century. The unifying principle is *sensory tapping*. This means that primary human sensory channels are tapped by augmenting sense organs with an electronic system that contains (1) an external sense, (2) an external stimulator of the native sense organ, (3) online digital processing of the external sensory data, and (4) a variable mask, baffle, or by-pass between the native sense organ and environment.

Aural tapping: Ears are covered by headphones, which also contain a computer-controlled baffle that either transmits acoustic sound from the environment or damps it. Attached to the headphones are microphones at both ears. With the baffle open, acoustic and headphone signals are superposed; with the baffle closed, the microphone signals can be processed arbitrarily. Music synthesis and rudimentary language translation are provided by the PC.

Visual tapping: The eyes are covered by screens, which by computer control can be made opaque or variably transparent (liquid crystal?), perhaps window by window or even continuously. Miniature video projectors generate color images that reflect to the eyes from the mirrorlike inside surface of the screens. The images cover the total visual field. On the outside of the screens, miniature video cameras record the external visual environment. They are located just above the eyes to maintain correct stereo baseline. There are also two eye monitor cameras that register eye movement for three-dimensional cursor control and eye blinking for selection and dragging. The projectors can be used to project any video material or as a "computer screen" (3D color). Most computer transactions can be carried out by means of eye movement only. In semitransparent mode, computer images are superposed on the eye screens with natural vision.

Kinesthetic tapping: Near the ears, miniature gyros sense the wearer's head position. The primary use of this is to stabilize the projected image and stereo sound: when the person turns his or her head, the projection pans accordingly. Transmitted video signals from eye cameras are similarly stabilized. Small head movements can be interpreted as mouse commands. Stimulation of the balance organ probably would be intrusive and is not included. If a nonintrusive solution is found, it might also constitute a baffle by compensation, which would be great for science fiction.

The Personal Communicator communicates to the ISDN (integrated services digital network) via a cellular microwave transceiver. Thus the wearer of the PC is at all times connected with the global village. Because the high bandwidth wireless communication spectrum is the scarcest resource in the electronic society, the PC is packed as full as possible with local computing power to compress and encrypt network data.

These are the basic constituents of the PC. One may add optional peripherals, such as continuous curvature/torsion sensor wires installed in the seams of clothing, used to transmit total body posture and limb movement. Evidently all pieces of clothing will in time include electronic control centers, and it will not be necessary to remove batteries before washing. Clothes, pocket mouses, and other gadgets might well communicate with short-range induction loops.

A warning: it is easy to speculate on *total* sensory tapping by means of a two-way tactile suit. There have also been speculations on direct neural intrusion by means of microelectrodes and/or brain state tomographic scanning. I stress that these ideas are either just speculative or very inconvenient in practice. My point is that the faculties offered by the PC outlined above are both technologically realistic and ergonomically plausible. In my experience, going too far in speculation is, for many, a way to keep the frightening future at bay.

The main use of the PC is to "copy the environment." Millions can see through the eyes (cameras, pardon) of an astronaut. Two individuals can "swap environments." I call someone, and the signals from my eye cameras are routed to her eye projectors; similarly, I can see through her eye cameras. The sound from my ear microphones goes to her earphones and vice versa. To some extent, I can turn her eyes, as she can turn mine. At first, this is a bit disturbing; but the next generation will be born with PCs on. The early attempts at videophone failed simply because the camera was idiotically positioned. Who wants to look all the time at the face of one's partner? It is much more convenient to show by looking rather than by moving a paper or a camera.

PC-based art forms need not only be simple, entertaining collective games. There will be soft reconstructions of ancient cities, and these synthetic places (which are byte-wise quite expensive) will serve for the study of history, sociology, simple tourism, and of course games. With the PC, one can visit a four-dimensional sculpture gallery, even a full four-dimensional world. There are no limits for art except the limits of computational complexity. A very pure form of PC art is just a visual/auditive signal-processing algorithm; all the material is derived from the real environment.

An *integrated work of art* based on PC technology is characteristically a very high-level and large structure of *references*. The point is that an integrated work, incorporating facilities for all the senses, for all desired grades of interactivity and participatoriness, must be able to be "directed" or "composed" using very high-level constructs; otherwise the sheer programming work would be unbearable.

Unpublished text, ca. 1986.

Bibliography

Dennett, Daniel C. "Where Am I?" In D. R. Hofstadter and D. C. Dennett, *The Mind's I*, 3rd ed. New York: Basic Books, 1981.

Mandelbrot, Benoit B. *The Fractal Geometry of Nature*. New York: W. H. Freeman, 1977.

Peitgen, H.-O., and P. H. Richter. *The Beauty of Fractals*. Berlin: Springer-Verlag, 1985.

8 The Unbearable Non-Artist from "l'Homme machine" to Algorithmic Afterlife: Non-Cartesian Cybernetics and Aesthetic Embodiment in Erkki Kurenniemi

Lars Bang Larsen

Earth cultures were not ready for even proximate immortality.
—Greg Bear, *Eternity* (1988)

To insist that an immortal machine is a paradox [is] absurd.
—Julien Offray de La Mettrie, *Machine Man* (1747)

It may seem rude to call Erkki Kurenniemi an unbearable non-artist. However, instead of picking out and addressing one of the disciplines or some of the practices he was involved with when he was active as a musician, filmmaker, scientist, and entrepreneur, I will here try to address his life and work as a fractured whole; a lifework that impinges on the various disciplines and practices that it continually crossed, and on which it is impossible to base a polite discourse. In terms of aesthetic experience, these crossings cannot be understood in the specialized sense of making of art works—neither what is typically rubricated as visual art, nor what is called computer art.[1] Rather, it is to be seen as a perplexingly unmetaphorical instance of a Foucauldian aesthetics of existence, a self-transformative process that takes life as its material.

A sense of incompleteness pervades all that Kurenniemi touched in his working life. He evaded the art object, left scientific criteria unfulfilled, broke down the commodity form of music, and refused to become an author and to build a career. To be sure there are moments of brilliance in his electronic music, his experimental films, the odd robot, in his engineering skills. But in a conventional sense, his practice lacks consistency. No less strange than these active attempts at getting rid of himself as a professional subject with career or a title, is his unworking of his own existence toward immortality on a digital platform. Thus to talk about Erkki Kurenniemi as an unbearable non-artist is to affirm the almost literary and science-fiction like quality that embodies his comprehensive *désoeuvrement* and ensures that it remain disaggregated and multiple, like the information he pried loose from his flesh in his life-archiving project.

Life Is an Archive

Kurenniemi's lifework revolved around the assumption that, thanks to the advent of quantum computers in the mid-twenty-first century, the human mind will be transferable onto platforms other than living tissue. The prospect of such a fundamental change of the human mind's material support would allow for a reevaluation of our entire species-being, with a view to eradicating death as well to the possibility of our existence in other environments than Earth, which we may leave as a "museum planet" once it has served its purpose as a home for "slime-based life forms."[2] Until he fell ill in 2005, Kurenniemi was thus the archivist of his own life, taking cognizance of everything that happened to him through the manic registration of facts and brainwaves, building his archive of the self by photographing incessantly, storing information on floppy disks and hard drives, and filling up hundreds of hours of video and audiotapes, as well as dozens of notebooks. The remains of Kurenniemi's auto-recorded life can be consulted in cardboard boxes, cassettes, paper photos, and antiquated hard disks in the Kurenniemi archive at the Finnish National Gallery, where they occupy quite a bit of space. This elementary, archival objectification of Kurenniemi's informational self creates a melancholic frisson in relation to the ontological transfer that he imagined.

His obsessive information gathering was an unsystematic model for what he called a template for all human life, as he relied on the efficiency of future supercomputers to reorganize the compiled information. Thus his life-replication project emphasized organization (of an archival kind) over the design of the posthuman body (the cyborg). There is no saying how he imagined that future supercomputers would reproduce parts of his life that he hadn't monitored, uninscribable moments of pain or ecstasy, vacuous days, and reconnect the dots of his charted subjectivity across blanks of improvable existence.

Flickering between cellular awareness, conscious articulation, and automatic recording, thus flipping in and out of his body as a sensorium, made for a process of stroboscopic subjectivation. One of Kurenniemi's notebooks, written in 1980, contains diary entries (meetings with people, things he has seen on TV), computer code, receipts, tickets to art exhibitions, translations of quotes (Paul Valéry, Margaret Mead), snapshots, an original drawing by the artist Olli Lyytikäinen, schema of electric circuits, a 16mm strip from an erotic film, and feverish exclamations such as "I need urgently / a beautiful body / and ugly mind / and an ugly body / and beautiful mind / I am ready, now." The fragments of the notebook affirm the hope of a life without caducity, sometimes on a depressed note. On one page, he lists and thanks all the girlfriends he has had; on another, he has jutted down thoughts on good and evil. A severe pain in his mouth: "Vitun ikenet!" ("Fucking gums!"). Sketches for a thinking computer called CHILD. The heuristic, the existential, and the trivial: material traces and signs of life, devoid of inwardness and depth to create a human manifestation of memory-effects in a future machine.

In the work of twentieth-century artists such as Jean Tinguely, Gustav Metzger, Marta Minujín, and others, technology was integrated as burnt-out and *kaputte* objects, thereby suggesting how the breakdown of society's "wish machines" belongs to their functioning, as they alternately produces wonders and delusions. Something here is reminiscent of the course that Kurenniemi's project ended up taking. No doubt the human being disappears differently in a computer than in the machine associated with industrial society: What Kurenniemi shares with the artistic avant-gardes and their speculations on the machinic, however, is that desire and imagination can here come to both ecstatic and catastrophic conclusions.

Introducing *Sade Fourier Loyola* (1970), which compiles monographic essays on the ultra-libertine Marquis de Sade, the libidinal socialist Charles Fourier, and the warrior monk Ignatius de Loyola, Roland Barthes mentions in passing that his book is about three "unbearable" authors. These extremists and pervs wrote about strange pleasures and limit experiences, and produced neither fiction nor philosophical or political treatises, but unintelligible, unclassifiable *text*. Qualifying his threesome of unacceptable writers as "logothethes"—founders of new languages—Barthes writes that the admired text has a violence that "enables it to exceed the laws that a society, an ideology, a philosophy establish for themselves in order to agree among themselves in a fine surge of historical intelligibility."[3]

The lifework of Kurenniemi, too, revolves around affirming new imaginaries, new languages. Its singularity comes down to the fact that it can't be admired, that it isn't exemplary. It should be seen for what it is, namely an unfinished process that may even have been unbearable to himself, as he proceeded with a belief in technological progress that seems naïve from any commonsensical point of view—a naiveté tinged with self-destruction. Thus in one of the pages of his many notebooks he talks about a "deepsick euphoria" and of how drugs and alcohol eased the process of his getting rid of himself.[4]

We can take inspiration from Barthes and call Kurenniemi an ontotheth: a founder of a new form of being. But what form of being, what life, exactly? His life was an experiment that dominated him from the other side of his death through the archive. Conceptually, at least, the interminable process of grinding himself down to a state of information came to dominate Kurenniemi's human-being. His self-archive became a subject bigger than him, not a passive repository but a producer of his exhaustion. Like one of de Sade's masters of ceremony, Kurenniemi's archive was an insatiable principal that yielded to no other law than its own: he had to obey it, and if it didn't get what it wanted—Erkki, in the totality of his physical and mental existence—then he would face the punishment of being called back to life in a decremented version of his former human self. Beyond this, the life-logging that he prefigured in extremis promised nothing but unfathomable continuation: no amelioration, no new liberties or pleasures, only predicated by a new fragility, as "it can be destroyed with a single command, DELETE FILE."[5]

In his account of Hellenistic practices of the self, Foucault discusses *hypomnemata*, which literally means *notebook* or *copybook*. To the ancient Greeks, the purpose of the personal use of hypomnemata was neither confessional nor an "account of oneself," but a project "which is nothing less than the constitution of oneself"; one's self-relation should be a "creative activity" to reconstitute oneself as a work of art in an externalized memory.[6] What enables a direct comparison between self-technologies from the digital age and ancient Greece is Foucault's observation that in Plato's time, the introduction of hypomnemata was "a new technology [that] was as disrupting as the introduction of the computer into private life today."[7]

Of course, Kurenniemi's notebooks served altogether different motives and values than hypomnemata did in ancient Greece, as well as in what Foucault, with a nod to US counterculture, calls the "Californian cult of the self." Obsessive self-examination, focusing on sexual behavior, is common to both; but where one proceeds by moderation and self-fashioning, the other preaches authenticity and self-discovery. If Foucault seemed depressed about the sameness of behaviors and practices from one culture to another, perhaps he would have found something else in Kurenniemi's indifference to moderation, self-realization, and the ambition of telling the truth, as well as his scant interest in the old Occidental hang-up of happiness.

Pleasures of Machine Men

His family resemblance to Barthes's unbearable writers suggests that Kurenniemi takes a place in the nexus where science meets ecstasy and the erotic meets the intellect at the conceptual and historical boundaries of the modern. In 1747, the physician Julien Offray de La Mettrie published the pamphlet *L'Homme machine*, in which he contends that human beings are machines. The human body can be compared to a clock, "huge and cleverly constructed," in the way of "a machine that winds up its own strings … the living image of perpetual motion."[8] Properties that separate human beings from the beasts—intellectual powers, emotions, the soul—follow from particularities of our bodily existence. Unlike the Cartesian conviction that humans are "beings of reason," in La Mettrie humans are profoundly embodied beings, and he in a sense reduces the soul to a by-product as it "follows the body's progression."[9] His is a one-substance doctrine in which there is no essence superior to matter that is modified diversely between living beings, whether the "meanest insect" or "the most splendid human"; for this reason there is "no abrupt transition" from animal to human—or from human to plant, for that matter.[10]

Since the human body is a machine it can be optimized, made to feel and enjoy more, or, as La Mettrie also suggested, have its life expectancy improved. Stimulants—food, or the lack thereof; coffee, La Mettrie's own favorite opium; and extreme physical states such as pregnancy—can heighten bodily pleasure and augment what it can do, just like age and climate affect it. Conversely, and as a true radical enlightener, he also stresses the effects of mental upon physical states.

La Mettrie's materialist critique is not purified by rationality. In fact, he performs a double subversion of Descartes by privileging the imagination. Where imagination in the latter is like nature's burglary in the house of reason, La Mettrie invests it with an intimate connection to the muscles of the body, as well as with the property of the mind that is "the most suitable for both science and art."[11] Any form of human greatness, any manifestation of intelligence and genius, is due to imagination, because it alone *perceives*—it forms ideas of objects, and connects them to words and figures. In short, "imagination is the soul, because it plays all its roles."[12]

The "counterintuitive" argument that Jessica Riskin builds around La Mettrie is that

> the view of human beings encapsulated in the Enlightenment man-machine was more anti-rationalist than rationalist, proto-Romantic in its celebration of sensory and emotional experience and mystery, and also deeply moralised. ... "La loi naturelle" operated through the machinery as "un sentiment intime" for integrity, humanity and virtue over their opposites. To treat others as one would want to be treated, La Mettrie claimed, was not a principle but a feeling built right into the machinery. The greatest vice in his moral universe was therefore rational reflection: the doomed attempt to transcend one's bodily mechanism.[13]

This can be linked, as Riskin does, to later theories of evolution, as well as to what she calls "a certain reactionary impulse," namely "to tame the imperial, rational self unleashed by Descartes and his contemporaries, to bring it right back down to the ground."[14]

There seems to be a genealogical echo between La Mettrie and Kurenniemi. Most importantly, perhaps, Kurenniemi brings down the imperial, rational self by way of desire—sexual, intellectual, creative—as the human substrate that distinguishes us from machines. His archive speaks abundantly and hetero-conventionally of this fact through enormous amounts of photographic home pornography and his hobby paintings of nudes; he even wrote a libretto for an unrealized "porn opera" with the almost 'Pataphysical title, *The Potatoes of Manchuria* (1983). Kurenniemi's *l'homme machine* is a lover machine for whom (as Charles Fourier had it) the passions are proportional to the destinies.

But what will happen to desire after we have attained the secret of how to extend life beyond biological death and resolved the problem of reproduction? As science fiction novelist Arthur C. Clarke has noted, boredom is the "supreme enemy of all Utopias."[15] Thus perhaps immortality will end up as a masturbatory pantheon for our eternal return: "Maybe we'll spend eternity watching pornography," Kurenniemi observes laconically.[16]

Bad Eternity and the Living Brain

Science fiction can be considered another example of the bad text praised by Barthes. *Eternity* (1988), written by one of Kurenniemi's favorite authors, Greg Bear, whom you wouldn't know if you have a decent taste in literature, is populated by characters with downloaded body implants, projected personalities, and implant memory stores; in short, reassembled and parentless creations capable of extending their existence indefinitely through intelligent bodies that defy the flawed human design. In Bear, death has passed from being a biological condition to a historical event: "the Death." In *Eternity*, to be "reconstituted"—resurrected—is "an experience worse than death … worse than loss of family or city or nation or planet."[17]

Kurenniemi's vision of immortality is equally dystopic, if of a more abstract kind. In Taanila's film *The Future Is Not What It Used to Be* (2002) he predicts that in algorithmic afterlife, humankind can potentially be contained in a golf ball hovering in outer space. To be a point, or a string of code inside a hyperintimate version of cyberspace along with billions of other non-souls for an eternity of hollow permanence, sounds like the hypostasis of a numerical civilization. Such an undignified conclusion to our species that ensures its continuation in the void recalls the words of Hannah Arendt: "Under modern conditions, it is indeed so unlikely that anybody should earnestly aspire to an earthly immortality that we probably are justified in thinking it is nothing but vanity."[18] Kurenniemi, however, doesn't try to legitimize his aspiration to immortality, nor considers it desirable. His vision of digital mummification in a golf ball is hardly the type of greedy clinging to life of which the ancient Greeks, as Arendt reminds us, disapproved. Kurenniemi simply considers immortality the probable outcome of technological development that offers escape from an Earth that has been poisoned by progress itself. It would seem that when we are no longer part of nature's material continuum, humankind is deprived of purpose. In this respect, there is no ultimate reason, and no deeper purpose, for digital immortality.

La Mettrie's *l'Homme machine* is echoed by Norbert Wiener two centuries later, in Wiener's writing on the structural similarities between the electronic circuit and the animal nervous system. However, where La Mettrie proceeds by mechanical analogy, for Wiener information is the criterion for materiality.[19] This is what Katherine Hayles has called the cybernetic "erasure of embodiment." Hayles argues in *How We Became Posthuman* (1999) that cybernetics defined intelligence as "a property of the formal manipulation of symbols rather than enaction in the human life-world."[20] She dissects what can be called the Cartesianism of cybernetics, namely the notion that the human "I" is disembodied and purely intellectual. This paradigm presupposes a conceptual distinction and clear hierarchization between mind and matter, information and materiality, in which the former is seen as essential and the latter incidental. In short, with cybernetics, information lost its body. From this perspective, Kurenniemi's prediction of a postbiological future for the human race would seem to rest on a similar set of conventional assumptions organized around the physical body's excision from a universal human subject. In this light, his life-archiving seems cybernetic *avant la lettre* in that it, quite literally, is a disembodiment of data with a view to augmented performativity. Isn't there thus, in Kurenniemi's bad eternity, an uncritical reiteration of the cybernetic "erasure of embodiment"?

Indeed, much research in artificial intelligence holds that the machine is not to be confused with its support, and already Norbert Wiener agreed that it is "conceptually possible for a human being to be sent over a telegraph line"—to be beamed up, as it were.[21] But even if symbol-processing devices in this view do not depend for their operation on the nature of their material substrate, it doesn't mean that they really are disembodied or that their rematerialization in other media and materials is insignificant. Moreover, as we have seen in La Mettrie, who emphasizes embodiment rather than cognition, there are other points of entry into the human–machine paradigm than Descartes's dualism.

In *The Cybernetic Brain: Sketches of Another Future* (2010), Andrew Pickering demonstrates the existence of a branch of experimental, "antidisciplinary" cybernetics that sees the world as ultimately unknowable.[22] Thus to Pickering, the distinctive object of British cybernetics—as represented by figures such as William Grey Walter, Gregory Bateson, R. D. Laing, and Stafford Beer—was the brain understood as an adaptive mechanism, or a performative organ, that isn't indifferent to its environment but that instead aimed to be "sensitive and responsive to changes in the world around it," a capacity that "endowed it with a disconcerting, quasi-magical, disturbingly life-like quality."[23]

Such experimental cybernetics can be seen to share La Mettrie's pre-Romantic sentiments by being *"a form of life,"* as Pickering puts it, just as its persistent interest in what he calls "strange performances and altered states" is aligned with Kurenniemi.[24] In experimental cybernetics, the brain is not a thinker or a container of representations but a machine that acts on information: "an immediately embodied organ, intrinsically tied into bodily performances."[25] On this fundamental point, Pickering's version of cybernetics is opposed to Hayles's reading, and opens up to a wholly different idea of virtuality, too. Where Hayles builds her critique on a dualism, Pickering has it that cybernetics precisely refuses a dualist split between people and things while it embraces an evolutionary, "rather than causal and calculable" grasp of temporal process—two characteristics that to Pickering makes its ontology nonmodern in Bruno Latour's sense.[26]

One should acknowledge that Hayles and Pickering to a certain extent are talking about different things. Pickering's batch of brilliant freelancers and mad professors, with which Kurenniemi is a kindred spirit, represents a different potential for cybernetics than the mainstream version that according to Hayles retooled a liberal humanist matrix into a protocol for neoliberal control and government. Hayles's critique of cybernetic disembodiment recalls Adorno and Horkheimer's indictment of how the Enlightenment reduces human beings to things, to *Zentren von Verhaltensweisen*—"centers of modes of behavior," and how biopolitical governance conceives of citizens as patterns of information.[27]

Neural *Dérives*

What exactly is the ontological status of information in Kurenniemi? The issue seems ambiguous in the context of his life-archival project. This might imply that he takes the human body to be as potentially telegraphic as Norbert Wiener does. Conversely, you could argue that, with his disturbing—unbearable—image of afterlife in a golf ball, he gestures at the absence of the human algorithm's material instantiation. In any case, the question of a future reembodiment of information remains painfully open, a mental irritant.

If we turn to his artistic activities, however, it is clear that Kurenniemi sides with a non-Cartesian line of thinking and resists a cybernetic "teleology of disembodiment" (Hayles) in his music and films. These contain his imaginings of new bodies in an autopoietic linking of bodies and machines; they are *imachinations* in which electric, informational, and biological effects impact and regenerate human bodies. For one thing, Kurenniemi turned biopotential into sound. Among the many instruments that he built in the 1960s and '70s—synthesizers, effect devices, video organs, and sequencers—is the DIMI-S (1972). To play this "group sexophone," four performers were connected as if to a hookah, holding electrodes and creating a transcutaneous circuit by touching each other, thereby activating a synthesizer. As changes were generated in the electrical impulses of the players' bodies, sounds would be activated through a collective biofeedback, synesthetically manifesting the proximity of body, emotion, and electroacoustic phenomena. Here the cybernetic problem of whether to locate the observer inside or outside the observed system is turned into music.[28]

Manford L. Eaton, whom Kurenniemi heard lecture in Florence in 1968, writes in his 1974 book *Bio-Music* of a new kind of music based on the conversion of brain waves into sound, by way of external sensory or electrical stimulation. Using EEG equipment—brain scanners—to explore the correlation between the spontaneous activity of the organism with the effects of visual and audio feedback, Eaton speculated on the creation of a new music of the nervous system:

> If we can, through biological feedback stimulation, use biological potentials as control signals for the presentation of sensory stimuli, and by monitoring physiological parameters to determine the efficacy of our efforts, arrive at a point where we can predict, repeat, and change, at will, physiological states and the accompanying psychological states, we will have a music with great power.[29]

Since the 1960s, effects have played a marginal yet symptomatic role as they have flittered across theory. In Deleuze's words, sense "is always an effect"; Foucault investigated social effects of power; Jacques Derrida's deconstruction engages with what he calls the "discursive effects" of language; and Marshall McLuhan's famous book *The Medium Is the Massage* (1967) was evocatively subtitled *An Inventory of Effects*.[30]

The effect is set to work by an initiator or effector that withdraws to let it unfold. It is self-enhancing and self-regenerating to the extent that it takes on a life of its own: it is supplementary, and yet it becomes an actor and a producer. Breaking down the dualism between organ and machine, network and body, material and immaterial, hereby the effect becomes something as strange as an *autonomous supplement*. In effect-generated artworks and compositions, *desoeuvrement* spills over into emergence, as many contingent forces are allowed to enter the work.

Together with the Swedish composer and musician Jan Bark, Kurenniemi created the computer animation *Spindrift* in 1966, consisting of images created by differential equations. Commissioned by the Swedish national broadcast SVT as an exploration of audiovisual synesthesia for black-and-white TV, produced on an analog computer at the Department of Nuclear Physics at Helsinki University (where he studied at the time), premiered at the 1967 Computer Music Seminar in Espoo and later broadcast by SVT, the work's institutional provenance is symptomatically composite for Kurenniemi the practice-mangler.

Performed by Jan Bark, Bengt Beger, and Bengt Ernryd, a soundtrack of Indian music with a crabbing tambura and hectic tablas accompanies the evolution of effects on screen. As if under emotional turbulence, the briefest of sequences unfold in white light against a bottomless background, and in black on white, like so many vain life forms unraveled by a digital frottage: embryonic geometries, a magician's wand moving pure electricity, deep sea digisperm, a machine dandy's cigarette smoke, rising dough and fingerprints, moiré in net stockings, intergalactic radar searches, a knitting machine gone berserk, polarized skies, white lines rebelling in Flatland, neon signs in an unalienating city, black shrapnel in a winter war, pyrotechnic cockades.

Spindrift flashes through associations as fast as the human mind can follow, telling us of imaginaries and experiences on the edge of our bodies, brains, and nervous systems. Out here, flesh is corrupt and receiving, and data is restless.

Notes

1. See *The Anthology Mainframe Experimentalism: Early Computing and the Foundations of the Digital Arts*, ed. Hannah Higgins and Douglas Kahn (Berkeley: University of California Press, 2012), in which the term "computer art" is used extensively.

2. Quoted from Mika Taanila, *The Future Is Not What It Used To Be* (DVD, KIASMA/Kinotar, 2002). In the writing of this essay I have relied on sources translated from the Finnish, and I am furthermore inspired by Taanila's film to use Kurenniemi's unfinished project of self-archival as a material and narrative framework for his lifework ensemble.

3. Roland Barthes, *Sade Fourier Loyola* (Berkeley: University of California Press, 1989 [1971]), 10.

4. Quoted from Kurenniemi's book draft from 1976, *The Consciousness Trap* (Diaries and Letters 1974–1979, Finnish National Gallery, Archives and Library).

5. Taanila, *The Future Is Not What It Used to Be.*

6. Michel Foucault, "On the Genealogy of Ethics: An Overview of Work in Progress" (1983), in *The Foucault Reader. An Introduction to Foucault's Thought*, ed. Paul Rabinow (London: Penguin, 1991), 365, 350.

7. Ibid., 363.

8. Julien Offray de La Mettrie, *Machine Man* (*L'Homme Machine,* 1747). In *La Mettrie: Machine Man and Other Writings*, ed. Ann Thomson (Cambridge: Cambridge University Press, 1996), 7, 34.

9. Ibid., 4, 8, 9.

10. Ibid., 37–38, 13; see also La Mettrie's essay "Man as a Plant" ("l'Homme plante," 1748).

11. La Mettrie, *Machine Man,* 17.

12. Ibid., 15.

13. Jessica Riskin, "Mr. Machine and the Imperial Me," in *The Super-Enlightenment: Daring to Know Too Much*, ed. Dan Edelstein (Oxford: Voltaire Foundation, University of Oxford, 2011 [2010]), 76.

14. Ibid., 81.

15. Arthur C. Clarke, *Childhood's End* (London: Pan Books, 1956 [1954]), 64.

16. Taanila, *The Future Is Not What It Used to Be.*

17. Greg Bear, *Eternity* (New York: Tom Doherty, 1988), 98. According to Perttu Rastas, Ray Kurzweil's predictions of a postbiological future in his book *The Age of Spiritual Machines: When Computers Exceed Human Intelligence* (London: Penguin, 1999) has influenced Kurenniemi's view of artificial intelligence and transhumanism.

18. Hannah Arendt, *The Human Condition* (Chicago: University of Chicago Press, 1958), 56.

19. "Information is information, not matter or energy. No materialism which does not admit this can survive at the present day." N. Katherine Hayles, *How We Became Posthuman: Virtual Bodies in Cybernetics, Literature, and Informatics* (Chicago: Chicago University Press, 1999), 26.

20. Ibid., xi.

21. Norbert Wiener, *God and Golem, Inc.: A Comment on Certain Points Where Cybernetics Impinges on Religion* (Cambridge, MA: MIT Press, 1986 [1966]), 36.

22. Andrew Pickering: *The Cybernetic Brain: Sketches of Another Future* (Chicago: Chicago University Press, 2010), 9. In his book, Pickering comments on the work of William Grey Walter, Ross Ashby, Gregory Bateson, R. D. Laing, Stafford Beer, and Gordon Pask, all of whom he calls "profoundly amateuristic"; like Kurenniemi, they were freelancers with garden-shed approaches to what they were doing (ibid., 10).

23. Ibid., 7.

24. Ibid., 9, 13. Kurenniemi's project can be compared to the counterculture's notion of historical change: this was not predicated on revolutionary breaks, but on the fantasy of an evolutionary transformation that would alter the material matrix of human life, not to "suffer on the pathway of normal evolution," as the LSD-gurus Timothy Leary, Ralph Metzner, and Richard Alpert put it (Timothy Leary with Richard Alpert and Ralph Metzner, *The Psychedelic Experience: A Manual Based on the Tibetan Book of the Dead* (London: Penguin, 2008 [1964], 23).

25. Pickering, *The Cybernetic Brain*, 6.

26. Ibid., 19.

27. Theodor W. Adorno and Max Horkheimer, *Dialectic of Enlightenment: Philosophical Fragments* (Stanford: Stanford University Press, 2002), 67.

28. Kurenniemi himself considers that his graph field theory, rather than his musical work, is his most important work—a unifying theory of space, matter, and energy as a structure derived from the principles and mathematics of cellular automata.

29. Manford L. Eaton, *Bio-Music* (Barton: Something Else Press, 1974), 9–10.

30. Jacques Derrida, "Plato's Pharmacy," in *Dissemination* (London: Continuum Press, 2004 [1972]), 70.

9 Archiving the Databody: Human and Nonhuman Agency in the Documents of Erkki Kurenniemi

Geoff Cox, Nicolas Malevé, and Michael Murtaugh

6/24/1996, 9:44 p.m., Personal

Will she (either of them) share the love of pornography? Or at least, art? I shall present myself to both of them as a geniality self-flagellat%n [*sic*] machine. Just one bottle tonight, ok? I shall invite them on my journey of change, showing the way ahead. Immortality. I will let them sniff and sneak through my archives.

The archive has increasingly become to be understood as a cultural paradigm, transfiguring notions of collective memory and the complexities of historical and temporal processes. This essay seeks to align some of these ideas with computational processes through Constant's *(Preliminary Work) Toward an Online Archive*, a set of (ongoing) experimental explorations of the archive of Erkki Kurenniemi that probes the vast amount of diverse materials with a variety of machinic processes.[1] The work is extensively documented in a logbook[2] and begins with an entry that makes reference to Kurenniemi's own invitation to "sniff and sneak through my archives," from his (electronic) diary entry of 1996.

Active Archives

Preliminary Work builds on the ongoing *Active Archives* project initiated in 2006 by Brussels-based art and media collective Constant (of which Malevé and Murtaugh are members), which engages the politics of open data and introduces core values associated with free software development related to the decentralization of resources and the ownership of infrastructures.[3] Put simply, to Constant, archives are understood as a collection of material that is not merely readable but also writable and executable, and thus subject to certain ethical standards.[4] Their working approach is not to follow standardized archiving procedures of ordering and classifying, but to offer a series of speculations on the specific qualities of the materials by running computer programs. Nor is this reducible to something like Google's PageRank algorithm that makes sense of the vast archive of the Internet in distorted ways to "reify" knowledge and make sure that expression is linked to the market.[5] Rather, the programs produce emergent forms of knowledge—they operate as "probes," in their terms—more a project of *forensics* than historiography.[6]

As work began on the project, it became immediately evident that Kurenniemi had documented his life but not archived it in a traditional sense. Despite the lack of a consequent or single ordering however, multiple orderings of the material still existed. As work continued, a number of core problematics became evident related to giving "direct access" to the digital materials of the archive: (1) *legality*—how to address potential restrictions on its visibility owing to questions of privacy, copyright, and a frequently intimate sexual nature; (2) *quantity*—how to take on the vast amount and diversity of the material; and (3) *fidelity*—how whatever form of access given would then relate to Kurenniemi's ideal of an "artificial consciousness." In each case, the problems were embraced as central themes, and part of a unique opportunity to situate the work *before* the archive had fully formed, somewhat in the spirit of Kurenniemi's intention to archive for future use.[7]

Arguably, the challenge for constructing the contemporary archive is to construct one that is decentralized and remains active in the way that meanings are able to be continually produced rather than fixed in the geopolitics of time and space. Thus the archive is not some dusty repository of fixed meanings there to be uncovered and ordered into some kind of narrativized version of official history, but more so a contested and dynamic force field of potential human and machine "interpretation." In the case of *Preliminary Work*, interpretation of the archive is explored through performative "programming" techniques to uncover (and compile) aspects of what is not directly apparent in the material—beyond visual and tactile apperception. By running both human and machine processes, Constant speculate on what and how knowledge is produced in an expanded field.

This is more a process of performing knowledge, perhaps (somewhat in the spirit of Foucault's *Archaeology of Knowledge*[8] in seeking hitherto undiscovered knowledge, or seeking to discern what constitutes emergent knowledge)—a process of knowing and unknowing materials. This is an important issue inasmuch as it establishes the interaction of material conditions and the materiality of bodies. The archive produces, or rather performs, new kinds of human and nonhuman materialities in the power-knowledge matrix. This further resonates with "media archaeological" methods (outside the scope of this essay but very present elsewhere in the book) and the work of Friedrich Kittler in particular, who argues for combining material conditions and epistemology, as, for example, in acknowledging the software used to write a particular essay,[9] or in this case, we would argue, the wiki with which we collaboratively write this text.[10] In this way the tools that are used for archiving can be registered as part of the archive; as, for instance, in the case of *Preliminary Work*, through the extensive use of software repositories/archives such as Gitorious, with all scripts carefully documented for future modification and reuse.[11]

The iterative approach taken therefore suggests an epistemological understanding of archives beyond objects and relations—the conventional tropes of archives—to something far more behavioral, contingent, and recursive, and more like the machinic logic that underpins Kurenniemi's thinking in the first place. For example, below are some extracts from Kurenniemi's diary that treat his physical relationship to his lover's body as if it were an abstract machine to which logical operations could be performed (using addition or multiplication) or recomposed in subsystems to be further processed:

7/15/1996, 9:48 p.m., Personal

X[12] slept here last night. The obvious problem arose as she admitted that she'd like to multiply whereas I only want to add.

7/15/1996, 10:19 p.m., Personal

Now I must disassemble X in acceptable subsystems and then process them one by one.

The reflexive way of working described thus far is also well suited to the specific character of Kurenniemi's obsessive documentation of his own life to produce a meta-archive of all human life (described at length elsewhere in this book). His self-archive is not only intended to form a template for human life in the future, but to enable human life itself to be made executable, using algorithms yet to be invented. Perhaps these probes are some early expressions of discovering what is not yet known, with complexity building on simple initial understandings of memory allocation. For instance, *du-h* (disk usage, with "human" option) is a very basic and common tool for interrogation of the data stored on a hard drive. It simply dumps a list of the names of files and folders, with the amount of disk space used. The process recurses into each subfolder with the disk space usage overlapping and complexifying.[13] A further example of understanding the allocation of memory is the JavaScript emulation of the "find" command (see http://kurenniemi.activearchives.org/logbook/?p=284). By playing the script in the browser, the user reruns the command but this time within his or her own system and computer's memory. Rather than simply visualizing the output of a command, it is performed, with series of controls allowing the user to examine the process: by playing, pausing, or modifying the speed. As the archive is heavy, by running the command and seeing how the browser has to struggle to process it, one can feel the weight of the archive.

In this essay, we take the archive-body to be an exemplar of an active archive, not as fixed materials or a mere collection of objects but something more like source code that is modifiable and shareable—something that forks future paths of execution and recursion in the spirit of Kurenniemi's overall fantasy of reconstructing life (by the

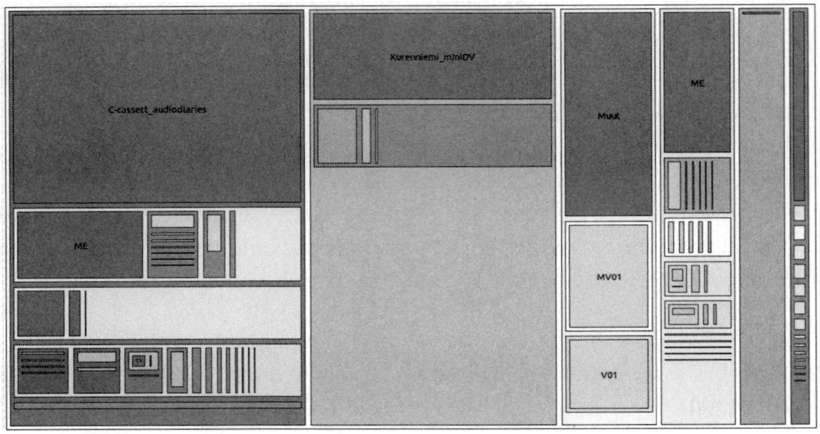

Figure 9.1
The output of the du command visualized as a treemap.

year 2048). More to the point, the quantum computer—yet to be built—is imagined by Kurenniemi to be able to make sense of the documents he has been recording and capturing in his daily life and stored into hard memory for later access and reallocation, merging biological and computational processes.

Data Bodies

Human memory and computer memory are endlessly compared in ways that confuse the idea of the archive as both a place for remembering and forgetting, as repository for both the storage and deletion of historical data. Not least the dubious use of analogies of computational systems to human memory reveal the collective fantasy of the ability to model the human organism at all levels of operation. This is not simply the story of the behaviorist tendencies of computer science but one that also occurs in numerous discourses that attempt to understand the human mind—as if it were merely a computer. It is at once a piece of science fiction and part of the concrete history of computational logic and the first ideas of what constitutes a human computer.

In his short essay on memory in *Software Studies*, Warren Sack refers to Turing's first description of how a machine might remember what it is doing and decide what to do next:

> It is always possible for the computer to break off from his [*sic*] work, to go away and forget all about it, and later to come back and go on with it. If he does this he must leave a note of instructions … explaining how the work is to be continued. … The note of instructions must enable him to carry out one step and write the next note. Thus the state of progress of the computation at any stage is completely determined by the note of instructions and the symbols on the tape.[14]

Sack continues that these simple procedures correlate with bureaucracy, as if the logic underpinning the thinking prescribes mindless tasks such as involved in the use of files, folders, directories, stacks, lists, and even the automated management of memory (so-called garbage collection[15])—that are encapsulated in desktop metaphors and office software.[16] The archive simply adds to the list in terms of the garbage that constitutes it, as well as its disciplinary character and regulatory practices that constitute the body as both subject and object of the system that produces the shit in the first place. It is almost as if the body can be understood as simply an input–output machine and that human subjects merely execute their programmed instructions and scripts.[17]

Of course, the body is thoroughly implicated in archives in more complex ways, as Foucault makes explicit,[18] but our point is also to acknowledge how the body is part of computational processes, not reducible to them. Kurenniemi has his own particular views on this:

> The transformation of physical sex into spiritual sex. So to say from hardware sex to software sex. Why do you need genitals when there is a shorter route … to manipulate your brain? … It's better to let these organic computers work the way they were developed to work. Nothing prevents your immortal soul to move over the new hardware. Don't spoil your old one.[19]

To Kurenniemi, the body is an obsolete envelope that humanity will leave behind in the necessary course of evolution, but having done so it will fall into eternal boredom. The archive of the body will be the only distraction available, the only way to hold on to the memory of what it was once like to have a body. However, it is more a question not of leaving the body behind but of understanding its active presence in complex and material assemblages that extend its, and our, understanding of agency—in coconstituted form. Kurenniemi recorded his life intensively and in particular his bodily activity (such as walking, sitting, eating, defecating, urinating, having sex, singing, yelling, sleeping), yet the various recording devices clearly cannot be abstracted away from the material realities they are archiving and while being archived simultaneously. This position is somewhat similar to the one that Wolfgang Ernst takes (in this volume too), in which the technological devices themselves *are* the archive, and kept functionally operative to maintain "nonhistoricist memory."[20] The archive is not to be taken as an "institution of frozen memory" but is made operative by its understanding as an apparatus. But how do the probes affect the nature of what they uncover, especially in the most intimate settings? We wish to consider the recording as part of a wider assemblage of active agents.

```
description of a folder.
the folder, named venetsia, contains 15 images.
the first image was taken 10 years ago, on thursday the twelfth of june,
2003, at 5 49 in the afternoon.
33 minutes later, image 2 was taken.
```

```
20 seconds later, image 3 was taken.
an hour later, image 4 was taken.
28 minutes later, image 5 was taken.
26 minutes later, image 6 was taken.
a minute later, image 7 was taken.
33 minutes later, image 8 was taken.
2 minutes later, image 9 was taken.
7 minutes later, image 10 was taken.
59 seconds later, image 11 was taken.
18 seconds later, image 12 was taken.
6 minutes later, image 13 was taken.
8 minutes later, image 14 was taken.
7 minutes later, at 9 44 in the evening, image 15, the final image, was
taken.
from first to last image, the folder represents a span of 3 hours.

def describe_interval (start, end):
        return humanize.naturaltime(end–start).replace("ago", "later")

def describe_image_folder (path):
        phrases = []
        phrases.append("Description of a folder")
        times = []
        _, fname = os.path.split(path.rstrip("/"))
        for ipath in os.listdir(path):
                if ipath.startswith("."): continue
                fpath = os.path.join(path, ipath)
                try:
                        metadata = pyexiv2.ImageMetadata(fpath)
                        metadata.read()
                        dt = metadata['Exif.Image.DateTime']
                        dt = dt.value
                        times.append(dt)
                except IOError:
                        pass
        times.sort()
phrases.append("The folder, named {0}, contains {1} images".format(fname,
len(times)))
phrases.append("The first image was taken {0}, on {1}".format(humanize.
naturaltime(dt), describe_datetime_full(times[0])))
lasttime = None
```

```
for i, dt in enumerate(times):
            if lasttime:
                  if (i+1 < len(times)):
                        phrases.append("{0}, image {1} was
                        taken."format(describe_interval(lasttime, dt),
                        (i+1)))
                  else:
                        phrases.append("{0}, at {1}, image
                        {2}, the final image, was taken".
                        format(describe_interval(lasttime, dt),
                        describe_time(times[-1]), (i+1)))
            lasttime = dt
span = describe_interval(times[-1], times[0]).replace("from now", "")
phrases.append("From first to last image, the folder represents a span of
{}."format(span))
return".\n."join(phrases)
```

[21]

Material Agency

The recent attention to dynamic matter and "new materialities" establishes this "inter-action" of the apparatus and forms of agency that are "more than human." Describing what she refers to as "agential realism," Karen Barad explains:

> Knowing is not about seeing from above or outside or even seeing from a prosthetically enhanced human body. Knowing is a matter of inter-acting. Knowing entails specific practices through which the world is differentially articulated and accounted for. In some instances, "nonhumans" (even being without brains) emerge as partaking in the world's active engagement in practices of knowing. ... Knowing entails differential responsiveness and accountability as part of a network of performances. Knowing is not a bounded or closed practice but an ongoing performance of the world.[22]

Historical and biological agents are bound together in new assemblages that allow for new conceptualizations of power-knowledge. But this goes further than the assertion that power largely determines the body through various forces such as the archive, as Foucault suggests. Rather, the boundaries of human and nonhuman agency—distributive agency—are challenged by the acceptance of power as demonstrating a broader view of materiality and the dynamic qualities of matter.[23] Material agency includes bodies and social structures but also technologies as in the case of computational nano-/biotechnologies presenting new human machine assemblages and hybrid forms. Barad's claim is that agency is emergent through the "inter-action" of elements and signals different and distinct agencies acting together. In the case of the archive we need to account for social practices and human bodies, but also various nonhuman agents as part of the wider apparatus. Our example is the way a program can be understood only as part of a distributive agency that includes the programmer, computer, network, factory worker, and wider scientific, military, economic, medical, political system within which data is materialized.

Kurenniemi's archive contains an impressive body of documents (audio, video, written diaries, audio diaries) that describe his various thought processes, not least related to one of his main obsessions, sexuality. Visibility seems to be a precondition of this. If sex was unrecorded, Kurenniemi seemed barely interested in the experience, and yet the paradox is that only a few of these documents can now be seen. To be made public, the rights that are attached to the documents have to be cleared, and many of the people pictured in these documents object to their public dissemination. In *Preliminary Work*, these legal documents, the agreements between the different stakeholders, do not represent obstacles to overcome to obtain the most complete visibility of the documents gathered and produced by Kurenniemi, but rather an epistemological challenge to which other kinds of visibility can be explored. In this respect, we are forced to reach beyond the limits of the human eye and the predominance of the visuality in knowledge production.

The strength of the usual legal approach is that it forces us to consider images as sites of attachment for various people. Therefore when legal protocols forbid the publication of the images until the people who are pictured have been contacted and agree to their public dissemination, the effect is not only one of closure but also one of disclosure.

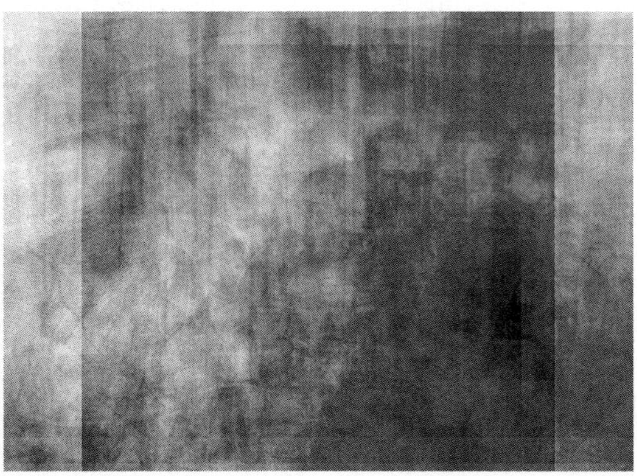

Figure 9.2
Average image: http://kurenniemi.activearchives.org/logbook/?p=215.

It forces us to reconsider the image not as a given object that has been captured and framed, but as a relation and inter-active assemblage of distributed agents that all have "rights." Attributing "rights" to any one agent when agency is shared between diverse objects would seem to be an oversimplification that the legal system blindly attempts to enforce, ultimately without recourse to the complex materialities in operation. If the images cannot be displayed, then other information can be disclosed to us by agents who are not subject to normative legal or human visual protocols. This also opens up a broader ethics of open content and the ideologies related to cultures of sharing and reworking materials (e.g., free software and open licenses such as GNU GPL).[24] Furthermore, the various legal instruments that constrain practices through the regulation of intellectual property extend to life itself (in the case of patents over DNA) with overt biopolitical dimensions. Hence what is demonstrated is a lack of sophistication of the legal system to account for more complex materialities as they remain locked into conservative normativity and a limited worldview related to ownership and property rights.

Leaving aside the "retinal" approach to the image, other agents open up alternative possibilities, for instance in the use of probes and experiments to expose computerized traces of Kurenniemi's life, allowing us to feel temporal intensities, other carnal distances and proximities, other lines of desire. Agency in this sense is attributed to more than the actions of human bodies but also the tools, technical devices, algorithms, the performativity of the network itself, the archive as apparatus, and so on.[25] In such ways, the probes begin to uncover aspects of what is not directly apparent in the material, revealing aspects of what is not yet known.

Active Forensics

When we use the term *active,* it is not by contrast to an archive that would otherwise remain passive, but to indicate inter-active assemblages. No archive remains passive as such. If one looks carefully at the work of an archivist, one can see that he or she is always working on the archive for it to "remain the same," attempting to halt the inevitable decaying processes of all objects, and even taking acts of preservation to be a form of vandalism.[26]

Documents are kept in conditions that need constant care and attention, and even for it to remain a fixed point of reference, it changes constantly. Furthermore, just as no two things in the physical world are precisely the same, data forensics reveals that documents are also different at the scale of discreet bits of information.[27] They exist as complex "media" forms that are subject to "forensic materiality" and processes of mediality. As Giorgio Agamben has elsewhere explained with respect to gesture, mediality is "the process of making a means visible as such."[28] It is also worth noting that the etymology of *media* reveals its roots in anatomy with its "medical" understanding as *cure from diseases*, as well as the more contemporary meaning of *translation from one media to another*, and thus takes on a forensic register. By creating connections between documents and excluding other connections, by contextualizing and decontextualizing them, the archive rewrites its contents, changes the way the documents are seen or read and interpreted. There is a content-authoring inherent in any archival work. If we use the term *active archive* to describe our activities, it is meant more to emphasize processes that are inherent to the archive and to push its actions to the extreme. If an archive is in a permanent state of mediality, always in temporality, always rewriting itself, an active archive is an attempt to describe strategies and tools that amplify and diversify this process, to reveal medialities—indeed we would stress the archive as a software machine: as readable, writable, and executable. And therefore the material is provided with the ability to "speak" for-itself. In a sense, code always speaks, inasmuch as it is a particular form of language that says and does what it says at the same time.[29]

More precisely, we understand active archival practices as something close to the way that Eyal Weizman and Thomas Keenan define *forensics* in *Mengele's Skull*, as more than simply the scientific method of gathering and examining information about the past but as opening up an ethical dimension.

> Forensics is, of course, not simply about science but also about the presentation of scientific findings, about science as an art of persuasion. Derived from the Latin *forensis*, the word's root refers to the "forum," and thus to the practices and skill of making an argument before a professional, political or legal gathering.
>
> In classical rhetoric, one such skill involved having objects address the forum. Because they do not speak for themselves, there is a need for a translation, mediation, or interpretation between the "language of things" and that of people. This involves the trope of prosopopeia—the figure in which a speaker artificially endows inanimate objects with a voice.[30]

The rise of forensics thereby gives an insight into how inanimate objects have been ventriloquized, their testimonies voiced by human witnesses on behalf of the objects. Kurenniemi, who, since suffering a stroke has lost the ability to speak, might be similarly "animated" through the various agents that constitute his archive well before 2048.

Algorithmic Archiving

Even if our goal is not to make an argument about Kurenniemi's material as such, the idea of the archive as a forum where the material has a chance to speak seems productive. Much of *Preliminary Work* has been devoted to developing the conditions for various conversations to unfold between Kurenniemi's materials, tools, and scripts, between different authorial agents, and in this way to extend agency toward a "flat ontology" as a technique to examine the various materialisms at work. By using the phrase "flat ontology," we adopt the terminology of object-oriented ontology to suggest that all things exist equally and that humans are not necessarily at the center of knowledge production. But this is not our position as such, as clearly the various agents, including the archive, are structured in ways in which power-knowledge is unevenly distributed, and the figure of Kurenniemi, too, seems to persist as a centralizing force. Rather, our point is that the act of archiving can be performed by other nonhuman objects that coconstitute the archive as a whole, and that recognition of this exposes subtle dynamics of power. As part of the developmental process, the following blog post describes some early ideas about computational tools as interlocutors in this way, not least as a pragmatic solution to the problem of scale and the limits of human perception.

We can't access the elements of the archive individually. Too many of them. We need intermediaries. People to tour us through. Tools, filters, sensors. That will listen, see, aggregate and separate, connect and disconnect, assemble and disassemble.

With the intermediaries, we will have to learn and speak the same language, accept the gaps, sense the priorities. The tools. They won't see as we see through our eyes, they won't listen as we listen, they will perceive through different dimensions, they will count time with another anxiety.

As our intermediaries, *our tools will be our interlocutors*.[31]

The algorithms become conversational agents that perform forensic operations and then explain phenomena in their own informational terms, uncovering the archive for what it is: essentially, a collection of data. For example, *data gallery* is an attempt to give a first form to this "conversation" beyond the limits of visual representation.

Imagine a picture.

An horizontal picture 2592 pixels wide and 1944 pixels high.

The picture was taken on the 06th of November 2004 at 21h56:37. The document set contains 45732 pictures by Erkki Kurenniemi for the year 2004. Erkki took 223 pictures in 2004 between the hours of 9 and 10pm. Of the 45732 pictures present in the dataset, Erkki took 33712 at night.

In the folder where this file is located, there are 28 other pictures. They have been taken between 21h56:32 and the next day at 19h21:18. The folder *Harrin bileet* can be seen as a sequence of 21 hours 24 minutes 46 seconds of the life of a man of 63 years 4 months at the date the picture was taken.

It took 1/40th second for the camera to take the picture. The blink of an eye.[32]

The interface gives access not to the images gathered and produced by Kurenniemi in their usual form but to different agential points of view on the images. Computer vision algorithms like face detection, contour detection, color measurement, and metadata extraction are combined to give a multifaceted commentary on the image file.

The data gallery interface is a grid that contains the different outputs of the algorithms. The first cell contains a threshold filtered representation of the dominant RGB component of the image, as digital images, such as those recorded by Kurenniemi's digital cameras, are typically represented as streams of pixels of red, green, and blue (RGB) components.[33] Next to this, a cell displays a representation of the contours detected in the image. The detection is done using the library OpenCV using the Hough transform.[34] The next cell contains an image showing rectangles where faces have been detected, or an indication of their absence according to the algorithm.[35] In the second row, the date of capture of the photograph is displayed on a field of the average color of the image. Next to this, a description appears that is produced from additional information recorded by the digital camera and stored in the image as metadata.[36] A final cell displays, when available, a histogram of the image's relative values of red, green, and blue components.

The point is that none of these outputs have the final word on the image. They complement, question, confirm, and sometimes contradict each other. The image is not what is shown on the page but what exists between knowledge produced by the different outputs. One can see the contours of bodylike forms and the location of faces, can see the zones with the dominant red color. The EXIF information reports on the time of the day, and the average color gives a background mood for the scene. The original image doesn't appear on the viewer's retina but it begins to exist in the imagination, and each image can be understood to contain its own archive.

Algorithms do not simply "read" information in images or sound files, they do not only "detect" features in data; they also generate new forms, new shapes, or new sounds. For instance, when working on the collection of audio cassette recordings made by Kurenniemi in the 1970s, as part of the *DATA Radio* project (a second installment of the *Preliminary Work* project focusing on the Kurenniemi's audio cassette diaries),[37] numerous ways to reorganize the material according to its internal structure were explored. That is how Spectrum sort is conceptualized.[38]

Spectrum sort applies a fast-Fourier transform (FFT) to Kurenniemi's digitized cassette recordings, breaking the sound into one-tenth of a second fragments that are each then characterized by their predominant (loudest) frequency band. Moments where the predominant frequency is low are grouped together, as are moments with a predominantly middle or higher frequency. The fragments are reassembled and played as a continuous movement through the material of a single recording. When the algorithm reports back, it sequentially reorganizes its findings. Applied to a particular cassette, C4004, the reassembled recording begins with moments of mostly quiet fragments of ambience just before or between moments of speech. One hears the ambience of a quiet room or the low rumbling of Kurenniemi's automobile and background traffic. Near the end of the rearrangement, bursts of predominantly tonal fragments combine, producing a striking stream of singing/chanting syllables. Besides a certain uncanny sense of artificial sound synthesis, the algorithmic arrangement reveals a use and quality of Kurenniemi's voice that is characteristic of the recordings. The algorithm's and the human's voices combine.

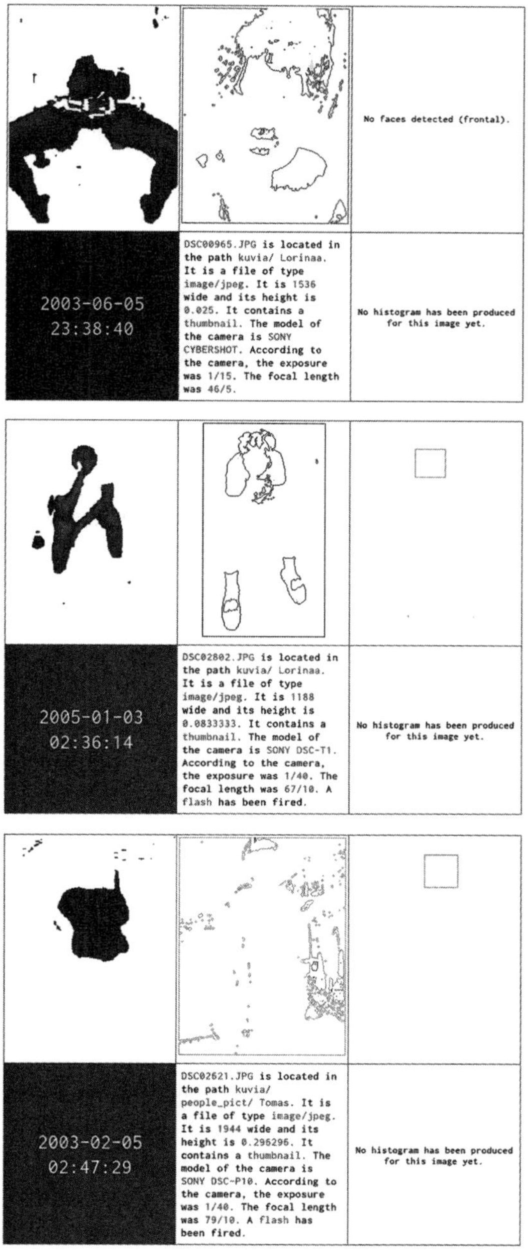

Figure 9.3

Example from the Data gallery: http://www.kurenniemi.constantvzw.org/db/records/images/listim.

Postscript

According to Kurenniemi, algorithms hold the potential to regenerate materials and life itself. Whereas the body is taken to be an ephemeral hardware product at an early evolutionary stage, quantum computers are imagined to be able to store our subjectivities and liberate us, once the databody can be fully executed. Somewhat like Kurenniemi, and building on her notion of "quantum entanglements," Barad also imagines the future possibilities afforded by quantum computation, in which previously unsolvable problems will be able to be solved by computers yet to be discovered. She points to the way that this requires a metaphysical leap of imagination in understanding cells as computational devices able to self-organize and adapt like biological systems.[39]

For Barad, this represents more than simply a description of ever more computational power; it also represents pressing concerns about control, governance, and security. It is perhaps the first steps in quantum cryptography that we currently witness in the revelations about the extent of data surveillance by government agencies such as the NSA (National Security Agency) in the Edward Snowden case and others—sniffing and sneaking around our archives covertly under the pretense of counterintelligence.[40] More to the point, *Preliminary Work* reminds us of the dubious ethics of such practices and an expanded sense of the archive as both a technology of subjection and potential freedom from constraints. As Foucault explains, the archive is not simply a tool of repression but one that is active in the constitution of the subject under the regime of "governmentality."[41] In addressing the archive with algorithms, rather than seeing an algorithm as a "pure" transformation from one representation to another, the algorithm can be understood as something more complex that amplifies and creates other objects with their own properties and qualities, further generating possibilities for the production of knowledge derived from archives in the broadest sense. This also demonstrates the distributive agency of things and new epistemological challenges for data made active as part of the body politic.

Notes

1. Brussels-based art and media collective Constant's engagement with Kurenniemi's archive materials began with a commission as part of *Kurenniemi: In 2048* as part of Documenta 13 (2012) and continued with *DATA radio* as part of *Systemics #1* for Kunsthal Aarhus (2013), and as part of *Erkki Kurenniemi: Toward 2048* for Kiasma, Museum of Contemporary Art, Helsinki (2013–2014). See http://activearchives.org/wiki/Erkki_Kurenniemi:_In_2048.

2. http://kurenniemi.activearchives.org/logbook/.

3. http://activearchives.org/wiki/Manifesto_for_an_Active_Archive.

4. http://activearchives.org/.

5. Reification is the "thingification" of social relations, or the extent to which social relationships are expressed between objects in the market.

6. Thomas Keenan and Eyal Weizman, *Mengele's Skull: The Advent of a Forensic Aesthetics* (Berlin: Sternberg Press, 2012).

7. As Morten Søndergaard also mentions in his essay for this volume, "Kurenniemi's archive is not about what happened in the past, but about what might happen in the future."

8. Michel Foucault, *Archaeology of Knowledge*, trans. A. M. Sheridan Smith (London: Routledge, 2002). First written in 1969 as *L'archéologie du savoir*.

9. See Friedrich Kittler, "There Is No Software," 1995, http://www.ctheory.net/articles.aspx?id=74.

10. In its wiki form, the essay can be read at http://activearchives.org/mw/index.php?title=Archiving_the_Data-body:_human_and_nonhuman_agency_in_the_documents_of_Kurenniemi&action=edit.

11. http://gitorious.org/kurenniemi.

12. Editors' note: We have redacted the female first name to ensure anonymity.

13. http://www.kurenniemi.constantvzw.org/logbook/.

14. Warren Sack, "Memory," in *Software Studies*, ed. Matthew Fuller (Cambridge, MA: MIT Press, 2008), 188.

15. In computer science, "garbage collection" refers to a form of automatic memory management. See http://en.wikipedia.org/wiki/Garbage_collection_%28computer_science%29.

16. Sack, "Memory," 190.

17. Geoff Cox and Alex McLean, *Speaking Code: Coding as Aesthetic and Political Expression* (Cambridge, MA: MIT Press, 2012), 103.

18. As does Allan Sekula's "The Body and the Archive," first published in *October* 39 (winter 1986): 3–64, http://www.jstor.org/stable/778312.

19. Transcript from the audio diary cassette C4076, April 1973. Furthermore, this reminds us of the artist Orlan's posthuman mantra "my body is my software"; http://www.orlan.eu/.

20. Wolfgang Ernst, "E-Kurenniemics: Becoming Archive in Electronic Devices," in this vol.

21. See http://activearchives.org/wiki/Description_of_a_folder.

22. Karen Barad, *Meeting the Universe Halfway* (Durham, NC: Duke University Press, 2007), 149.

23. For instance, as described by Jane Bennett in *Vibrant Matter: A Political Ecology of Things* (Durham, NC: Duke University Press, 2010).

24. http://www.gnu.org/licenses/gpl.html.

25. In addition to agency, the root of *agencer* is to arrange, as an assemblage, even curate. For more on this, see Iain Hardie and Donald Mackenzie, "Assembling an Economic Actor," in *Sociological Review* 55: 1 (2007): 58. Thanks to Magnus Lawrie for this reference.

26. The Museum of Ordure make this explicit in their "preservation policy" statement, to explain the acceleration of "decay" in their online archive:

> Everything that is represented in the Museum of Ordure is subject to the vagaries of an uncontrolled internal process which slowly deforms and disables all information held in the museum. This is comparable to the decaying processes which affect all artifacts in museums, regardless of all attempts at preservation: the retouching, repainting, cleaning, etc., which are incorporated risks to the purity of artifacts when first acquired by museums. Even "successful" renovations are subject to periodic changes resulting from shifts in conservation policies. Eventually (and in accordance with the fallibility of memory) artifacts are institutionally, progressively, determinedly and inadvertently altered by acts of conservation (sometimes unintentional acts of institutional vandalism) until they cease to be recognizable as the objects first acquired. Of course in both cases—in the virtual environment and in the material world—the processes of generation, decay, and entropy are paramount. Museums are by this definition charged with achieving the impossible. (http://www.ordure.org/collection/preservation/)

27. An extensive discussion of "forensic materiality" and its relation to digital storage media can be found in Matthew Kirschenbaum's *Mechanisms: New Media and the Forensic Imagination* (Cambridge, MA: MIT Press, 2012).

28. Giorgio Agamben, *Means without End: Notes on Politics* (Minneapolis: University of Minnesota Press, 2000), 58.

29. For more on this connection to speech, see Cox and McLean, *Speaking Code*.

30. Weizman and Keenan, *Mengele's Skull*, 28.

31. http://kurenniemi.activearchives.org/logbook/?p=308.

32. http://kurenniemi.activearchives.org/logbook/?page_id=521.

33. Dominant color: When the difference between the color components of a pixel results in a value higher than 30 for the dominant color, the pixel is printed. For example, when an image has red as the dominant component, all the pixels of the image where the red value is 30 above the green and the blue components are printed.

34. Because of imperfections in the image, it is often nontrivial to group the extracted edge features to an appropriate set of lines, circles, or ellipses. The purpose of the Hough transform is to address this problem by making it possible to perform groupings of edge points into object candidates by performing an explicit voting procedure over a set of parameterized image objects (Linda G. Shapiro and George Stockman, *Computer Vision* [Upper Saddle River, NJ: Prentice Hall, 2001], 304).

35. Face recognition is done using a Haar classifier, first proposed by Paul Viola and improved by Rainer Lienhart.

36. Exchangeable image file format (EXIF): https://en.wikipedia.org/wiki/Exchangeable_image_file_format.

37. *DATA Radio*, an interactive stream of audio and text: http://kurenniemi.activearchives.org/dataradio/.

38. See Spectrum sort: http://activearchives.org/wiki/Spectrum_sort.

39. Barad, *Meeting the Universe Halfway*, 384.

40. See http://en.wikipedia.org/wiki/National_Security_Agency and http://en.wikipedia.org/wiki/Edward_Snowden.

41. See Michel Foucault, *The Government of Self and Others: Lectures at the Collège de France 1982–1983* (Basingstoke: Palgrave, 2010).

IV Science/Technology

Foreword

Jussi Parikka

This part focuses on science and technology in Kurenniemi's work, a theme that has come up in various ways already in relation to his artistic work and is constantly present in his archive. In addition to including Kurenniemi's own writing, the chapters place Kurenniemi in relation to the scientific and technological developments in media culture. This includes, as in part 1, references to the archive and technical media (Ernst) and also to cybernetics (Søndergaard) as well as to social media culture (Røssaak). Overall, the chapters underline the peculiar nature of Kurenniemi's relation to scientific institutions and discourse. Despite at times being branded as a singular thinker and scientific mind, he was often more of a multitasker interested in a multiplicity of topics across a spectrum of media, art, science, and technology. He is not an academic in the usual sense, and his relation to technology was often branded by the unsuccessful and the incomplete.

Erkki Kurenniemi's work led to innovative products but also to failures. This is meant not as a negative critique but to underline that his ideas and products that were intended as commercial, for instance his music synthesizers, were never very successful in the business sense of the term. His scientific writings on neural networks, algorithms, and theory of the mind spanned many areas but were never published to a wide international scientific peer group of academics. Kurenniemi dropped out early on from academic work but continued to write about related topics. His early 1968 paper "Programmable Space" mobilized physicist David Finkelstein's idea of the conflation of physical theory and computer language. Physical events are computational events: computers are not only tools in understanding advanced physics, but also the "world is a digital computer."[1] His later scientific writings engaged with graph theory, for instance in "Graph Field Theory" (1990), a paper that exists in various versions and drafts,[2] and which is included here as a historical document of its time to place Kurenniemi's scientific writing as part of the interdisciplinary media art historical discussions.

While discussing digital notions of space that is programmable and suitable for manipulation with a computer, Kurenniemi produced a set of ideas that were speculative and yet grounded in engineering. This includes projects that experimented with possibilities of computing; his aesthetic ideas and practice, elaborated in the previous section, started off from his understanding of science and technology. The developing ideas concerning physics and the digital were later boosted with science fiction of the likes of Greg Bear, giving a specific flavor to Kurenniemi's style. The media archeological implications of Kurenniemi's work are present in how astutely he was aware of the role of technologies in shifting social coordinates, preferences, and art practices since the 1960s and 1970s. Some of the possible connections to such projects as Buckminster Fuller's Dymaxion Chronofile (the archival project from 1920s to 1983) are left only implicated; both engineer-artists were thinking of leaving their miscellaneous notes as an archival monument of a specific historical period through their eyes, ears and writing. Such connections have sometimes to be teased out by the reader of Kurenniemi's texts and projects.

Interviews with Kurenniemi (see part 5) place him in the media technological context of past decades and help us to decipher his way of understanding technological change. The emergence of computing and the shift from the analog to the digital was inscribed in the activities of the likes of Kurenniemi, who can be considered a symptom of sociotechnological change.

Kurenniemi is less interested in the past than the present and its potentials to unfold speculative futures. The past is, in Kurenniemi's texts, audio, and other meditations, primarily part of a history of evolution that he approaches with scientific texts. He is keen to discuss Carl Sagan, Richard Feynman, and others' cosmological, physics, and evolutionary theories to see how the human being developed into a technological toolmaking and programmable entity. Kurenniemi speculates whether it is in fact the human that is a Turing machine: we did not only create such machines that are able to simulate any computable sequences granting there is enough of memory available, we ourselves are also such simulation machines, referring in this context to Ronald Fox's "Energy and the Evolution of Life" from 1988.[3] The text included here, "The Origins of Life, Intelligence, and Technology" (1994), is exemplary of Kurenniemi's interest in such big cosmological questions as to our place in the universe and the basic structures of physical reality.

His grand cosmological speculations still often home in on concrete machines, which are his ways of conducting "laboratory experiments." The future unfolds through an inventive spirit infused in engineering and science. The experiments started already in the 1960s at the Department of Musicology in Helsinki with access to the studio and tape edit and montage possibilities: "demo materials on tapes and then combined them live without a prerecorded montage, using my legs to flip different dials and switches and turning the tape reels back and forth with my hands."[4] Artistic practice became a method to investigate new technologies. Kurenniemi began to develop as an artistic scientist who mobilizes his ideas in science-fictional speculations as much as in media artistic experiments of tinkering. This dual aspect has always been present in his work. Kurenniemi's notebooks are often a mix of the two poles of this oscillation: inspiration from such scientists as Edward de Bono's writings concerning the mind together with meditations on flip-flop circuits and computer architecture, all within the space of one paragraph.[5]

It is this artistic vision and engineer's soul that makes Kurenniemi's thinking regarding past technological futures interesting. He was not a lone prophet as such.[6] The 1970s were full of predictions of the future: governmental and expert reports on the changing conditions of factory production, ways of life, ecological implications, technological changes—from Alvin Toffler to Daniel Bell to various reports also in Finland (mentioned in the introduction). But Kurenniemi was able to synthetically pull in experimental ideas from a variety of directions and use the artistic, audiovisual gadgets as one driving motor for writing a history of an imagined future: such as, in 1971, speaking of a personal device that was to be the cinema, typewriter, work place, calendar, clock, camera, and post office of our life.[7] Instead of the usual suspects of Alan Kay and others, we can begin alternative genealogies of media culture from such sidekicks as Kurenniemi. So instead of a mistaken media historical apophenia, seeing connections and premonitions through the eyes of forgotten geniuses, we should be systematically looking at the slow, gradual emergence of ideas, inventions, and technological practices.

Kurenniemi is not a modest thinker, as Erkki Huhtamo reminds in the foreword to this volume. The various chapters throughout the book carve out this immodesty in his archival mania and its relation to the big themes of his generation. Kurenniemi's quasi-media archaeology is complemented by a sci-fi futurism that comes out in "Supermegatechnologies" (1999–2000) included in this part. The text expresses a positivist belief in the increase in computing/storage power as itself determining a cultural change, a theme that also characterizes the (primarily) Anglo-American cyberculture since 1980s and 1990s. This sort of an attitude is also present in the general meaning of the hyperbolic term *supermegatechnology*: the convergence of IBN, info-bio-nano (instead of IBM, one might add). As we know, the new relations of information-, bio-, and nanotechnologies are in a position that cannot be overestimated when it comes to the strategic importance of their impact; they present a situation where technology meets chemistry and organic bodies in the context of new corporations of informational control. Besides imaginary future visions, Kurenniemi tapped into his surroundings with a constant curiosity, like a scientifically tuned investigative gonzo-scientist. It is hence only apt to include a chapter on themes that are very current, and to discuss social media and control societies (Røssaak), as doing so provides a new context to something that is implicit in his work.

Chapters by Morten Søndergaard and Wolfgang Ernst articulate that this futuristic vision, also of Kurenniemi's personal IBN-style 2048 resurrection, is conditioned by the technological archive. The archive performs the future. But the archive becomes also a tool for a critical contextualization of Kurenniemi's role as resonant with wider themes in cybernetics and postcybernetic art of post–World War II period: art processes understood through the ubiquity of information. The world can be regulated through feedback loops and managed as a calculable entity even if probabilistic. Eivind Røssaak argues importantly that both "Kurenniemi and contemporary informational capitalism are perfect children of cybernetics," demonstrating the parallel lines and fruitful readings of cybernetic capitalism and Kurenniemi's imagination. In Søndergaard's chapter, the perspective moves toward the performative elements that enable to understand two things: that Kurenniemi's archival task was one of a future now—a speculative performance of future archival resurrection that forces us to consider his wider life as an artistic event. And also: that this fits in with other notions of (late) cybernetics that characterized the past decades of technological arts internationally. Ernst's chapter emphasizes slightly differently the technological conditions of Kurenniemi's thoughts. Ernst's media archaeological argument produces an invaluable angle to understanding Kurenniemi's technology and synthesizers in relation to memory and time. They become indicators of an epistemological rupture, which Ernst places in conversation with the media theorist Friedrich Kittler's synthesizer, recently reconstructed in Germany as an artistic-engineering performance project. The synthesizers, perhaps main features in the section following this one, are here discussed as symptomatic technological objects and part of a twentieth-century technical media culture gearing toward the coming computer age. We are dealing with symptoms, not with prophesies.

Notes

1. David Finkelstein, quoted in Erkki Kurenniemi, "Programmable Space." University of Helsinki, Department of Nuclear Physics, 1968.

2. Erkki Kurenniemi, "Graph Field Theory," unpublished draft, 1990 (in this vol.).

3. Erkki Kurenniemi, "Elämän, älyn ja teknologian alkuperä," in *SETI—vieraan älyn etsintä*, ed. Heikki Oja (Helsinki: Tähtitieteellinen yhdistys Ursa, 1994), 156–157. See the next chapter for the English version in this vol.: "The Origins of Life, Intelligence, and Technology."

4. "Drifting Golf Balls in Monasteries: Mika Taanila in Conversation with Erkki Kurenniemi," 2001, in this vol.

5. Erkki Kurenniemi's archive, Central Art Archives, Finnish National Gallery (EKA, CAA, FNG). Kurenniemi DRY 1974 1.nb—notebooks from 1974.

6. Jyrki Siukonen, "Dead Computers Tell No Tales—Remarks on the Futures behind Kurenniemi's 2048 Resurrection." In *Erkki Kurenniemi—A Man from the Future*, ed. Maritta Mellais (Helsinki: Finnish National Gallery, Central Art Archives, 2013), http://www.lahteilla.fi/kurenniemi/julkaisu/Erkki_Kurenniemi_A_Man_From_The_Future.pdf.

7. Erkki Kurenniemi, "Message Is Massage" (1971), in this vol. See also Siukonen, "Dead Computers Tell No Tales."

10 The Origins of Life, Intelligence, and Technology

Erkki Kurenniemi
Translation by Teo Välimäki

Estimates of the amount of technological civilizations existing within the Milky Way have been attempted using the Drake equation.

$$N = R_* \cdot f_p \cdot n_e \cdot f_l \cdot f_i \cdot f_c \cdot L$$

The equation includes a lot of disinformation as well as several dubious assumptions. The physicist Freeman J. Dyson has written: "I reject as worthless all attempts to calculate from theoretical principles the frequency of occurrence of intelligent life forms in the universe. Our ignorance of the chemical processes by which life arose on Earth makes such calculation meaningless. Depending on the details of the chemistry, life may be abundant in the universe, or it may be rare, or it may not exist at all outside our own planet."[1] I would add that even if we knew exactly how life began, it would be of little use in estimating the value of *N*.

The Origins of Life in the Universe

If the universe has its origins in the Big Bang, a thunderous blast, there are at least two possibilities. Either (1) life was created from nonliving matter at least once, on Earth or elsewhere, or (2) life has been bootstrapped into existence in a precondition, i.e., the original state of the universe. Since no theory of physics exists about this precondition, I'll ignore the second possibility, since it belongs within the sphere of theology rather than science.

There are other possibilities: (3) Life could have seeped into our universe through a wormhole or a white hole from some parallel universe.[2] The problem with this explanation is the same as in all theories of panspermia: the moment of origin is merely pushed farther into the past. (4) Time travel hasn't been proven impossible with metaphysical certainty. Perhaps our distant descendants have sent the seeds of life backward in a time capsule. Counterargument: Why then would we have needed an evolution spanning billions of years? Counter-counterargument: perhaps the wormholes available are so narrow that only a few organic molecules can be sent that far back in time.

The origin of life on Earth and its possible origin elsewhere are two completely separate problems. Both are difficult, but as has been shown in previous articles, the evolution of life on Earth is largely understood apart from the very early stages. The problem of the very early stages is complex because, for one thing, we don't actually understand what life really is. We have only one precedent: all life on Earth is based on a common chemistry. Generalizing from such an isolated case would be dangerous.

It might be surprising that the definition of life still escapes us. Carl Sagan examines attempts to define life from physiological, metabolical, biochemical, genetic, and thermodynamical bases and comes to the conclusion that none of these provides satisfactory results.[3] If life was defined through thermodynamics as a localized reduction of entropy, refrigerators would have to be defined as organic. It's been suggested that an intelligent living being is "*self-replicating, computer-controlled heat engines ... able to play survival games.*"[4]

Several relatively new research trends might shed new light on the problem, such as chaos theory, automata theory, complexity theory, and artificial life research. On the other hand, it is possible that an appropriate theory hasn't been found yet. Or life on Earth is such an unlikely and random anomaly that it cannot be generalized.

As my first attempt at defining life I suggest that a living being is a *self-organizing system that has the potential for exponential development.* The level of development is measured by the amount of usable information acquired by the system (in bytes).[5] The evolutionary level of the human race could be estimated to be around 10^{18}–10^{23} bytes,[6] which includes information both stored in DNA as well as recorded and actively stored by the species throughout its history.

A growing crystal is a self-organizing system. However, the information included in its regular crystalline structure is limited. The information within the structure flaws can grow indefinitely, though, and information existing on the surface of the crystal is usable. Its surface area grows only to the second power of its diameter, so the crystal is not a living being (except perhaps in a very large supercool environment). Instead, the linear elongation of the DNA of living organisms signifies exponential growth of the alternatives. The word "potential" is included in the definition because otherwise dinosaurs would not have qualified as living beings.

This definition has at least two obvious weaknesses. First, if we consider ourselves as living organisms, the original state of the universe would have had to be living, as it contained the potential for our creation. Second, we don't qualify as living beings unless we have the potential for spreading throughout the Milky Way and eventually to the entire universe. The definition can be efficiently applied only after the fact, and at the moment of a possible apocalypse the last observer would have to face the realization that all phenomena were and are ultimately dead.

These observations show that the fatal flaw in the first definition is the attempt to define the organic quality as an intrinsic and absolute property of the system. Instead, we should define it as a relative attribute.

The second definition: Subsystem *E* of closed dynamic system *M* is living in relation to its environment *M* if, and only if, *E* succeeds in spreading to cover almost all of *M* from almost all primal states.

This version is equally inexact, because the attributes of the system, the subsystem, and their isolation remain undefined. Some of the problems of the earlier definition have been solved, though. The question of the organic qualities of the original state of the universe has reverted back to an open question about the entire structure of the universe. The biochemical coherence of Earth's biosphere is in harmony with the definition, as is ecologic compartmentalization. The definition allows as many life forms as it does closed environments within the universe.

Life on Earth has concentrated in areas where the three states of matter collide: gas (the atmosphere), liquid (the oceans) and solid (the Earth's crust). According to our current understanding, life began on Earth in the liminal area between matter and liquid.[7] Cells consist of liquid that is organized by a complex nested structure of closed membranes. The membranes are a cross of solid matter and liquid. Other special aspects of life on our planet that are not found at the level of geologic systems are linear molecules containing information (DNA and RNA) and extremely complex proteins, which share the distinctive quality of a solid structure with the crystal. Unlike crystals, they manifest a group of specific active loci and adjustment points. Which of these structural elements can be generalized to apply to all life forms? We don't know. However, we do know something about the relationship between the two basic principles of evolution: divergence and convergence. Despite their hereditary heterogeneity, most fish are very similar in structure. The eye has developed independently numerous times (convergence), but so have almost all of its technical implementations (the camera, the compound eye, etc.).

Other aspects of life on Earth that may be universal are mutation and natural selection, sexuality, the origin of species, competition within and between species, predation, parasitic behavior, and symbiosis.[8]

It's also probably pertinent that matter has (at least) two parallel states, in our case semisolid and liquid. "Solid" states are necessary for the storage of information, because they enable an efficient storage code using very few basic components (written text or numbers in a decimal system wouldn't work if the letters would float around freely). "Liquid" states make (the changes in) evolution possible. On the other hand, all these attributes might apply only to pretechnological existence. In a few decades, the human race will possess artificial intelligence and genetic technologies that may free us from being enslaved by our own genes.

The Origin of Intelligence and Technology on Earth

Intelligence and technology are as difficult to define as life is. Both are common in the animal kingdom, and it's impossible to determine a clear starting point for either. When it comes to technology, it's also difficult to choose a time frame. Evolution based on natural selection has produced ingenious technologies during its lengthy reign, which it has then honed into such a level of optimal efficiency that the short history of human technology pales in comparison.

The relationship between intelligence and consciousness is problematic. Advanced chess computers play the game in an intelligent manner, but they don't indicate any signs of sentience apart from following the flow of the game. The question of whether a universal intelligence exists remains open.

Ronald Fox, the American physicist and researcher of the origins of life, has presented an interesting idea that might be pivotal in explaining consciousness.[9] He begins by stating that the world immediately perceived by a living being is nonlinear and therefore often a chaotic system. This results in an individual being able to predict the behavior of his environment only through simulation, much like using computer simulations when no other shortcut is available. In a simulation, a model structure of the scrutinized system is created from elements small enough for their interactions to be imitated.

If evolution had created an environmental simulator such as this in the brain of a living being, it would naturally aid in the survival of the species. Maybe our brains do contain this kind of a simulation program. For it to have any real benefits, it would have to be immensely fast, similar to a weather simulator: a computer model of weather prediction is useful only if it calculates changes faster than the actual weather changes—otherwise the predictions would be outdated before completion.

It's also worth noting that this kind of simulation model would have to include a model of the being itself within its surroundings. Fox's model suggests that we house a scale model of ourselves in our brains, which constantly functions slightly ahead of ourselves! Am I wrong in assuming that Fox's theory explains the problem of consciousness? If I may be so bold, I'll also suggest that this predictive model of the environment resides in the cerebellum. If this proves to be true, it would have consequences, for example, in the ethical treatment of animals, since anything possessing a cerebellum would have to be considered conscious.

What separates man from other animals? Something unique occurred about two million years ago, but there is no consensus on what the pivotal moment was. There have been suggestions of bipedalism, appropriation of tools, development of agriculture, the invention of speech, or mere growth in brain size. Many dinosaurs walked on two legs, yet in over a hundred millennia they never thought of using tools, or at least the thought didn't garner widespread appeal. Dolphins possess an obviously versatile ability to communicate, but we cannot understand it. Even ants practice agriculture. Let's try to search for a better answer using computing theories.

The Turing machine is a simplified mathematical model of the modern computer, created by the English computing pioneer Alan Turing. Turing machines cannot be applied to practical computing, but they can be used to solve principal mathematical problems. It's been shown that there are so-called universal Turing machines, which can be externally programmed to perform any calculation possible for an ordinary Turing machine (modern computers can be considered universal provided they have an unlimited storage capacity, e.g., unlimited disks). I suggest that this attribute is the defining difference between human and animal intelligence. For some reason, our brains evolved to contain connections that made us externally programmable automatons.

According to this theory, the birth of technological civilization has social roots. The language of animals evolved into a programming language. It allowed the leaders of packs to give out functional instructions: not only to order others around, but to explain how the orders should be followed.

Intelligence and Technology in the Universe

So far we've loosely defined life and suggested interpretations about the birth of conscious intelligence and technological civilizations. To be able to reasonably generalize Drake's equation on the amount of technological civilizations in the Milky Way, we'll have to ask some additional questions.

(1) How many (*H*) galactic structural leaps do we need for common materials to evolve into radio communication media?
(2) What's the mean interval between two of these leaps (*T*), and what does the interval depend on? (How much time is needed for a specific leap [innovation] to become viable by displacing other options in its field?)
(3) What's the probability (*K*) for an external or internal disruption (mutation, catastrophe) to cause a significant leap in evolution (either forward or backward)?

These questions are unfortunately not independent of each other. On Earth, progress from the condensation of the planet to the first radio telescope took dozens of separately seemingly very unlikely leaps. Question 2 seems the easiest, because calculating the time constant doesn't appear to be any harder than calculating chemical reaction speeds. Regarding question 3, the second given definition of life remains lacking, unless the quality of the dynamic system is specified. There are at least two problematic factors: first external catastrophes, such as showers of comets created by the Sun's neighboring star, Nemesis,[10] hitting the Earth, as theorized by Richard Müller; or the Seyfert eruptions[11] probably occurring every 10^8 years in the center of the Milky Way.

The other problem is internal catastrophes, such as highly irregular mutations or the possible tendency of living systems toward self-criticality.

The idea of self-criticality is so new that it warrants further scrutiny. Self-criticality or self-organized criticality is the tendency of a complex system to veer toward a critical state on its own accord. A critical state can be defined as a state where even small disruptions can have major effects. For example, the masses of snow gathering on mountain sides are self-criticalizing, which is proved by fairly regular yet surprising avalanches.

If the developmental stage of a biosphere or civilization *E* (in bytes) is represented as a function of time, how does the graph look? In the event of total destruction, the curve drops to zero. It's unlikely that humans could wreak this kind of mayhem on Earth; a possible reason could be a collision with a large celestial object or a nearby star going supernova. On the other hand, we know that the curve hasn't constantly been upward. There have been waves of extinctions on Earth, where the total mass of information has diminished. Still, moderately sized catastrophes seem to expedite progress. Evolution seems to have a tendency to stagnate unless a surprise factor keeps kicking life into it. These disruptions can be depicted on a "catastrophe spectrum," which represents the frequency of the disruptions through a function of their significance. If the elements of catastrophe spectrum *K* were researched enough, if *T* could be estimated and if the natural range of *H* could be proposed as a function of *E*, we might have the prerequisites to solve *E* as a function of time from the resulting random differential equation.

The benefit of this kind of approach is that we don't have to make as many unfounded presumptions about the Earthlike attributes of life as in Drake's equation. In addition, the major uncertainty factor of *L* (the lifespan of a civilization) is automatically taken into account, and in fact becomes a part of the solution.

Internal catastrophes remain problematic, especially if the concept of the self-criticality of complex systems is generally valid. If a system veers toward a critical state, the result can equally likely further progress (enough catastrophes) or hinder it (too many major catastrophes).

This breakdown still proves that the cosmic abundance of technological civilizations is dependent on the velocity of evolution. This velocity is in turn dependent on the speed of basic chemistry and the intensity distribution of external disruptions. Nuclear chemical reactions are about a million times faster than atomic chemical ones, so the existence of intelligent life should be relatively more common on neutron stars than on cool planets,[12] unless stellar quakes or other major catastrophes invalidate this advantage.

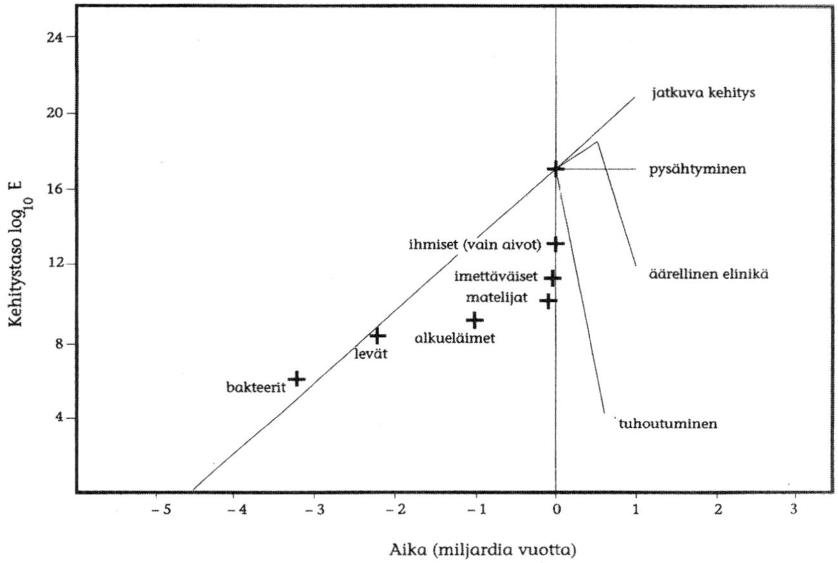

Figure 10.1
The evolutionary stage of different living beings as a logarithm of information content and as a function of time. Adapted from Ball, "Universal Aspects of Biological Evolution."

The Intelligences and Technologies of the Future

Humankind has existed very briefly at the stage of technological evolution where radio contact with other civilizations is possible. The fact that first contact hasn't happened yet can signify several things.

It's commonly thought that contact with other civilizations is possible only through two avenues: space travel or two-way radio communication. If our technological evolution continues at the same pace for the next thousand years as it has done in the past hundred, many other possibilities are likely to appear. I will now explore one of these options.

The radio was once called the "wireless telegraph," because a physical conduit between the transmitter and the receiver was no longer required. But is even the receiver absolutely crucial? The receiver and its antenna are essentially just an amplifier for the signal, which is necessary because radio waves disperse when transmitted. A directional transmitter antenna can limit the dispersion, but not indefinitely over long distances.

As far as we know, interstellar space doesn't contain a medium through which a wavefront would travel undispersed, but it's not impossible that the sent signal could "modulate" the gaseous or dust-based medium in a desired manner. It would probably require an "intelligent wavefront," that is, a very intricately controlled electromagnetic field, which could first ionize and then reorganize matter it encounters into a form that would act as "hardware" for the "software" within the wavefront.

An intelligent wave is a computer program that can transform any medium into a compatible computer and therefore continue traveling.[13] It's likely that even this kind of technology cannot lose energy, and laying a "gaseous optical fiber" or "plasmafiber" won't happen at the full speed of light. The energy needed for launching the signal might prove immense, since it has to propel vast amounts of matter at the tip of the fiber, unless a method of creating energy en route is discovered (an ordinary hydrogen-helium fusion would mean that the plasmafiber would be a one-dimensional hydrogen bomb, i.e., a star).

The previous train of thought attempts to show that it's possible in principle to achieve a well-known theoretical possibility of one-sided interstellar communication and even robotic travel at the speed of light (by programming matter into light and decoding it back into matter at the receiving end). The speed at which the plasmafiber can be laid is still a mystery. This fantasy is still somewhat more realistic than towing a gravitational wormhole (stretching a one-dimensional black hole).

It's become fashionable and almost respectable to present ever more preposterous explanations for "the Great Silence" or the Fermi paradox in the SETI discourse. Earth has been compared to a zoo,[14] a cosmic insane asylum, or a biochemical bacteria culture. My favorite is the inverse insane asylum theory. Perhaps the dominant galactic supercivilizations are still, in some capacity yet unfathomable to us, in their dinosaur stage. We exist solely because we're too insignificant even to snack on.

Notes

Original title: "Elämän, älyn ja teknologian alkuperä." Published in *SETI: Vieraan älyn etsintä* (Helsinki: Tähtitieteellinen yhdistys Ursa, 1994).

1. Freeman J. Dyson, *Disturbing the Universe* (London: Pan Books, 1981), 209.

2. A. Strominger and J. R. Ellis, "Third Quantization [and Discussion," *Philosophical Transactions of the Royal Society of London, series A* (1989): 395–399, reprinted in *Physics and Mathematics of Strings*, ed. M. Atiyah, J. R. Ellis, M. B. Green, and C. H. Llewellyn-Smith (London: The Royal Society, 1989), 77–81.

3. Carl Sagan, "Life," in *Encylopaedia Britannica*, 15th ed. (Chicago: Encyclopedia Britannica, 1987).

4. Jerome Rothstein, "Generalized Life," *Cosmic Search* 1 (2) (1979): 35–45. (Also online at http://www.bigear.org/vol1no2/life.htm.)

5. J. A. Ball, "Universal Aspects of Biological Evolution," in *The Search for Extraterrestrial Life: Recent Developments: Proceedings of the 112th Symposium of the International Astronomical Union*, ed. M. D. Papagiannis (Amsterdam: D. Reidel, 1985), 251.

6. Ibid. See also Freeman J. Dyson, *Infinite in All Directions*, Gifford lectures given at Aberdeen, Scotland, April–November 1985 (London: Penguin, 1988).

7. N. R. Pace, "Origin of Life—Facing Up to the Physical Settings," *Cell* 65 (1991): 531–533.

8. Ball, "Universal Aspects of Biological Evolution."

9. R. F. Fox, *Energy and Evolution of Life* (New York: W. H. Freeman, 1988).

10. Richard Müller, *Nemesis: The Death Star: The Story of a Scientific Revolution* (London: Mandarin, 1990).

11. J. N. Clarke, "Extraterrestrial Intelligence and Galactic Nuclear Activity," *Icarus* 46 (1981): 94–96.

12. F. D. Drake, "Life on a Neutron Star," *Astronomy* 1 (5) (1973): 5; R. L. Forward, *Dragon's Egg* (New York: Ballantine Books, 1980).

13. Tommaso Toffoli and Norman Margolus, *Cellular Automata Machines* (Cambridge, MA: MIT Press, 1987), 136–138.

14. J. A. Ball, "The Zoo Hypothesis," *Icarus* 19 (1973): 347–349.

11 Supermegatechnologies: Some Thoughts on the Future

Erkki Kurenniemi

Information Society

The concept of the information society has become a widely accepted, static ideal. The Great Transcription is in progress. All written, drawn and sculpted structures are being digitised and archived in a scattered server network. The Internet is beginning to cool. Information is beginning spontaneously to self-organize, aggregate, form portals, entities, relations, sediments, crystals.

The phase transition of the Web leads to the Second Encyclopaedia. The first encyclopaedia (Diderot & al.) was a one-dimensional sequence of characters. Only one search field was available: the key word. It is difficult to predict the integrated semantic structure of the free-form Web, because the process of change is part of the "ninth great evolutionary leap"[1] and because it has no historical point of comparison. The new information tool, the free-form network of references, is so flexible that information can find its own, natural organization, if such a thing exists.

The nature of language will change. One-dimensional spoken and written languages are packed codes whose interpretation is based on fragmented, shared linguistic competence. The Web's capacity for transmitting information is sufficient for a "deep language" in which a syntactic and semantic structural description can accompany the message at any depth of significance.[2]

Communications Society

In the communications society, communication has become an end in itself, and information secondary. Non-museal information is no longer worthwhile, and no longer needs to be stored. All information ages too quickly. It is cheaper to create knowledge when necessary *ab ovo*, or to fetch it directly from the original sources. Another reason for the rapid arrival of the communications society is the withholding of knowledge, the debate over copyright. Using an algorithmic generator, a new piece of music or picture can be created in a moment and free of charge. It is a mistake to believe that a structure can be owned. There are trillions of ten-note melodies. If one wishes to be certain, a filter linked to the musical generator can sift out the copyrighted melodies without significant waste. Art is revealed as a cognitive hallucination.

Quantity becomes quality. It is now possible to download on to your own disc, in a single day, more text than you can read in all of the rest of your life. Using bookmarks, the relationship multiplies one thousandfold. Maxwell's and Einstein's equations demand about ten bytes each. Their interpretative code (as Web references) multiplies this by one thousand at most.[3]

An image, too, is worth only as much as a hundred words, effectively coded (I do not mean bit-level compression of raw data, but semantic coding). According to the classic 7+/–2 rule, the conscious mind can process between five and nine different things at one time. Inside the human head, thoughts proceed at slightly less than the pace of conversation.

Control Society

George Orwell wrote his dystopia, *1984*, in 1948. The work forecasts how a totalitarian order could harness the then-new mass medium of television. "Big Brother is watching" entered the language. The "telescreens," which recalled today's plasma screens and were located in every dwelling, were two-directional and could not be switched off.

In 1984, it looked as if the threat of the novel would not be realized. Instead of direct brainwashing, television merely offers harmless sports and other latent violence, soap operas and other latent sex, and political rhetoric. The number of thought police would have been excessive. The control society began to take shape with mobile telephones. Parents control their children, children their parents, and everyone their partners. The traditional clandestine affair is becoming impossible.

When everyone is watching everyone else, there is no need for the thought police. The clients do the dirty work as soon as they have been sold the tools. When GPS positioning, video cameras and electronic signatures are added, soon, to mobile phones, the system will be almost watertight.

The Direction of Societal Evolution

What factors, in the end, define the direction of social change? An easy question, at last. The socio-biological principle is almost a tautology. Evolutionary selection (in an environment whose resources are limited) leaves behind it the fittest evolutionary entities. Fitness, on the other hand, is defined retrospectively as the relative proportion of descendants in the population (the limiting value as generation n tends to ∞). The time constants of genetic and memetic change, which are now very different in value, lead to "aberrations," among them what is known as cultural evolution, in which entities' different conceptions of fitness cause local chaos. Fast flow in a river causes eddies, but nevertheless all the liquid ends up in the sea.

The American humanist psychologist Abraham Maslow, who posited a hierarchy of human needs from the primary (physiological) to higher needs (e.g., self-actualisation) was wrong.[4] Human beings do not have any needs "higher" than the need to reproduce. Killing other people can substitute for sex: in a "linear approximation," fewer of the others is as good as more of us. The hypothetical gene that has men running in a herd to a strange country to kill men and rape women—hmm—flourishes. The tragic additional feature of its success is that the mechanism may hinder genetic stagnation, particularly in mountainous regions where the mobility of genes is limited.

Megatechnologies

Today, megatechnology is the name given to information technology (in its highest developments), biotechnology (whose basic information will be in place in a few decades) and nanotechnology (in the more indefinite future). Space technology may become a megatechnology if certain pressing reasons develop, but when? According to an estimate by the physicist Freeman Dyson, the first real space colony will begin to function around 2085.

Information Technology

The most important measures of the performance capacity of information technology are now expressed in gigafigures (giga = 10^9). Processor frequencies will soon exceed the gigahertz, RAM memories the gigabyte, and discs approach the terabyte (1000 gigabytes). The speed of local networks will soon be in the region of a gigabit per second (one byte = 8 bits).

And nothing is enough, nothing like it. There were 20 years between the mega period and the giga period. The tera (10^{12}) and peta (1015) periods will arrive in between twenty and eighty years. The signals of the peta frequency processor are light-waves. Petahertz technology will make a phase-accurate light-wave front, a perfect hologram, possible. It will be possible to implant a video projector in the vitreous humor of the eye, and data visors will become redundant.

Quantum computation, a potentially revolutionary way to increase the power of computers by many, many orders of magnitude, may become a reality; but it may also merely remain an encryption method.[5] Where the classical bit takes the alternative values 0 and 1, the quantum bit (qubit) can be 0 and 1 simultaneously. A quantum computer could analyze all the possible moves in a chess game at one command.

A networked environment will bring virtual reality everywhere as GPS satellite positioning is complemented with more accurate local positioning networks and portable, personal radio positioning networks. Thus everyone will "radiate" his or her location and position to the network. Your avatar will be as gracious in its movements as you are, or even more so.

Data gloves will not be necessary. The wrist-mouse will be an elastic band which, using ultrasound, will monitor the tension of every tendon that runs through the wrist. Depending on taste and necessity, these bands will be able to be worn on the wrist, ankle, neck etc. The rivets will be able to be cameras.

Alongside virtual reality (which does not yet exist) there will develop the more important augmented reality will develop. Above reality will float synchronised enhanced realities.

Biotechnology

The living cell is a complex machine whose functioning is not yet nearly understood. It is, nevertheless, a finite automaton, and its functioning will be analyzed in a few decades. There are around 80,000 genes in the human genotype, corresponding to the functional units of a computer program, the sub-programs. Their teamwork has been polished over billions of years. Understanding the functioning of the cell thus corresponds to the task of analyzing how a program the size of ten CD-ROMs, written in an unknown programming language and finely tuned, works. At present, the genotypes of a few simple species have been read completely, and human DNA will be unravelled in the five years.

Biochemists are not very willing to speculate in public about practical possibilities. For them, the important thing is to explain and understand the "grand design" of the cell machine. No shortcuts are visible. Such a basic question as the folding of proteins continues to defeat computational modeling.

In time, it will become possible to design living organisms, starting from first principles. It will be possible to design a dwelling or vehicle beginning with molecules, on the Lego principle. Although it is unlikely that dwellings or vehicles will be needed on Earth by that time. It will be a different matter when biospheres are developed in essentially different physical conditions, such as on Mercury or Jupiter.

Before that, genetic manipulation will mature: the reprogramming of living beings by adding or subtracting individual sub-programs. At present, it is possible to move individual genes. The next step is to provide a cell with a wireless connection with whose help genes can be manipulated in real time.

The nucleus of a cell corresponds to the hard disc of a computer. The quantity of information is about the same: a few gigabytes. Regulation of gene expression is a chemical cascade which conveys information from the surface of the cell to its nucleus. As a result, correct genes and their combinations are switched on or off. In this way, cells can specialize according to signals conveyed by the surrounding matter.

In superbiotechnology, new genes are not only added to the genotype: they are also given mobile phones by equipping them (in the last analysis, a few of their coded proteins) with "sensors" through which genes can be turned on or off by rapid chemical or electromagnetic signals. The possibilities offered by the remote control of the functioning of cells are more incomprehensible than those of electricity or the computer were before Charles Babbage's time. The problems that must be solved are not insignificant. The correct, influence-susceptible point in the gene regulation chain must be found. The mechanism should be universal in the sense that the influence can be directed at any gene. It will be necessary to install massive auxiliary machinery in the cell.

Nanotechnology

The term "nanotechnology" is a semi-corrupt word. Many fantasies about universal robots on the molecular level are impossible in terms of energy. The better word "material technology," on the other hand, sounds too tame. The field covers applications involving mechanical movement, with the scale varying a millionfold, from the millimeter to the nanometre (10^{-3} to 10^{-9} m), from micromechanical probes and tools to the atomic level. Individual molecules and atoms are already being probed and moved using atomic force microscopy (AFM).

As an example, let me mention "actively rigid" composite materials, in which buckling or crumpling of buildings is hindered using sensors, feedback and active elements. The conventional rejoinder is, "it won't work; buildings like that will collapse when there's a power cut." The same kinds of argument were used in the early 1970s to suggest that semiconductor memory storage could not replace ferrite ring memory storage because the content of the store would disappear when the electricity was disconnected.

Microelectronics are produced by "lithography," a copying method which is more efficient than assembly, but remains an awkward process for the production of structures. The next challenge of nanotechnology is to replace lithography with self-assembly. Chemically produced nanocomponents will be mixed together and, with the addition of an activator or by heat treatment, will crystallize into structures. The inner structures of living cells are formed by self-assembly.

Self-assembling materials will remain susceptible to aging. In living cells, many critical activities are also self-correcting. For example, DNA is proofed after copying, and most of the copying errors corrected. Self-assembly is a relatively simple process; often, it is sufficient that the building blocks are of the correct shape. Self-correction demands very complex machinery. No technological shortcuts to self-correcting nanotechnology are evident. Nanotechnology will most likely have to copy directly from nature. In the cockpit of a nanorobot there will sit a bacterium listening wirelessly to the Internet.

Supermegatechnologies

The title is a disposable word. I promise that I will not use so banal a term in the future. Technologies are organisms by nature. Sometimes they integrate and universalise. A topical example is the global information network, which is replacing separate technologies by the dozen, from the post-box to printing machines. All technologies which can be used to control a large number of microscopically small things can exploit one another. The same goes for all technologies which are based on the detailed understanding of the functioning of the mind.

IBN (info-bio-nano) technology: information technology, biotechnology and nanotechnology (material technology + chemistry) will merge to form a single supermegatechnology. The linking factor is microscopicality. It would be very natural for the mammoth technologies of the next century to be invisible to the naked eye.

IBN Materials

Plastics, which were once called "man-made materials," or "artificial materials," cf. German "Kunststoff," do not generally reproduce the best characteristics of biological materials, but they are cheap products achieved with rough methods, and even in quality. They can be injection-molded (a primitive casting technique, or copying method) into identical, accurate-dimensioned plastic objects. Will they be replaced in 50 years' time by basic materials of a new type? The development of an IBN composite and its production methods is a great project which has not really yet begun.

It was once naively imagined that an intelligent material could be made by mixing silicon chips with an elastic plastic mass. Nowadays, artificial tissue (bladders, blood vessels) is produced by transplanting cultivated cells into a temporary, biodegradable fiber matrix.

A multi-use IBN material will be a composite. If it does not contain integrated, autonomous, biological species (modified or wild), it will be a simpler IN (info-nano) material. It will have controlled mechanical properties and surface-chemical activities. It will contain an information network, using wires or wireless. It will be capable of mechanical metamorphoses against an opposing force; it expends energy.

For one hundred years, electricity has been the overwhelming way of transporting energy, even though it demands awkward leads and insulation. Many potentials can be used to store energy (with the exception of flywheels, which store kinetic energy directly). In nature, adenosine triphosphate (ATP), a nucleotide found in the mitochondria of all plant and animal cells, is the universal currency of intermediate intracellular storage. ATP gives up its binding energy with little loss in suitable portions, but its synthesis must be isolated from the rest of the content of the cell (mitochondria).

It would be elegant to develop efficient electricity-ATP and ATP-electricity transformers. There are certain means to achieve the latter in electric fish. Even the battery problems of mobile phones would be solved. In the future, there will also be many other uses for artificial organelles (an organelle is an intracellular specialized "machine" like the mitochondrion, the cytoplasmic body which contains enzymes responsible for energy production).

Immortality and Space

All human beings are mortal. This is the price we paid for sex, which was invented a billion years ago. I would not see space technology in itself as a supermegatechnology for 50 to 100 years. Space has proved an environment that is more dangerous to people than had been supposed. Weightlessness has not so far proved to be essential to industrial processes. Because of the costs and risks involved, the exploitation of minerals does not appear attractive. Population pressure must be dealt with by regulation, not expansion. The oceans are a pleasanter environment than a vacuum.

The time bomb will explode soon after the human genome has been read and interpreted. The lengthening of old age could still be controlled by some kind of brain-maintenance tank. If aging can be halted during the years of fertility, the catastrophe will be total. I can see no solution other than a merciless law, the choice between children and immortality. Perhaps this will be made to work, against all the laws of the black market.

There is one loophole, the emigration of either parents or children to space, the moon, Mars, anywhere so as not to add to the burden of this increasingly cramped little sphere. Naturally families will be able to stay together if everyone moves. Their journeys will be paid for eagerly by those who wish to buy additional Earth licenses.

The taboo on genetic manipulation will disappear rapidly. It will be natural to adjust human sub-species to weightlessness, radiation, vacuum. The American science-fiction writer John Varley's idea is of plant-human symbionts which live simply on light, as close to the sun as the skin will bear.[6]

Cognitive Engineering

The Cognitive Revolution Has Already Happened

The challenge of brain research is similar to the understanding of the functioning of the cell. The object system is complex but finite. The development of methods keeps the vanguard in constant motion, but the formation of theory is of equal importance. New concepts are needed for the understanding of thinking.

The most annoying thing is that the computational function of the brain cortex is not yet known. Certain conclusions can nevertheless be drawn as to the nature of this function. About one hundred areas of the cortex are linked to one another through two-directional nerve fiber bundles, forming a parallel processor. The most astonishing thing is that there appears to be a solution to the central "binding problem" of cognitive science. A photograph of consciousness has been obtained. The mechanism may be the synchronisation of nerve impulses on a gamma frequency of 40 Hz.

According to the gamma coherence hypothesis, the language of thought is not a simple frequency modulation of nerve impulses (FM, the principle of the earliest data modems) but, in addition to it, phase modulation (PM, the principle of today's modems). If consciousness can be explained so easily, artificial intelligence is not far off, and collective experiences of consciousness are not inexplicable (cf. techno-acid).[7]

On the other hand, given the existence of phase modulation, no other neural code is necessary. This may mean that thought communication is not possible with any (local) technology, because there is no universal code. Two mutually connected cortical areas may be able to use a completely private coding method formed during childhood experiences. Coding and decoding may be inseparably merged with cortical computation, whatever that may be.

Whatever the details prove to be, the architecture of thought will open up and lead to applications which are described by the general term cognitive engineering. This term is currently used in a limited sense to mean the design of easily learned and easily used equipment and programs. I use the word more broadly to refer to all technological methods employing information yielded by cognitive research.

Fundamental to the information networks is an understanding, developed over a couple of decades, of the stratified nature of communication. In this hierarchy, the application layer ("socket") forms the uppermost layer. In cognition technology, the hierarchy is continued, using the same logic, to thought and consciousness.

The lack of success of artificial intelligence technology requires explanation. It has not proved possible to superimpose the inferential and neural network paradigms. Multi-component models of intelligence (Howard Gardner's multiple intelligence theory, visual intelligence, emotional intelligence, etc.) suggest a much more flexible base mechanism.[8]

The psychologist Arnold Trehub's model, which combines a two-step nervous system and a "winner-takes-all" decision mechanism in the same feedback loop, looks promising because it adds "insight" (learning by single exposition instead of repetitive drill) to the basic level of the machinery. Conclusion: artificial intelligence needs a broader model of cognition as its starting point.[9]

The production of virtual environments has gotten stuck in the limitations of calculation capacity and user interfaces. Immersion in films may be better than a jerky virtual helmet. A critical factor is the efficient manipulation of the window of attention. Cognitive immersion deepens if it is possible to keep the window of attention small. Photorealism in the visual arts invites the attention to direct itself at revealing details. A cognitive approach can combine abstraction and intentional obscurity, a trick of older modern art. The fact that the information contained within the genome is not sufficient to define the detailed connectivity of the brain suggests that virtual technology demands more feedback from the subject than the registration of the position of the head; measuring the direction of the gaze may be an important improvement.

All art is cognitive engineering, and all artists have always been cognitive technicians. The makers of the cave paintings mastered the art of illusory outlines to produce effects of intensity.[10] Aestheticising myths become disposable baggage when evolutionary psychology explains the interdependence of beauty and choice of partner. Fitness rules again. Surprisingly, even geometrical symmetry can fit into the same model. At the basic level, the symmetry of a living being indicates the condition of its genetic inheritance. At the following level, beings develop symmetrical wattles (fleshy protuberances on the neck and head of birds, sexual selection signals) because they are easy to recognize on the grounds of their geometrical invariance.

Music is an interesting case in many senses. Its origins are not understood; it is non-figurative compared to a painting or story; it has its roots in intra-species communication; and both melody and harmony trigger direct emotional reactions. Why are the factors of the number 60, 4:5:6, a cheerful major triad, while 10:12:15 represents a melancholy minor triad?

It is tempting to speculate that the riddle of music may be directly linked to the neural code. The electroencephalogram of the gamma range (40 Hz) is, coincidentally, just an octave distant from the lower frequency limit of hearing, 20 Hz, in harmony with Nyquist's criterion, the principle that in order to represent an analog signal faithfully in digital sampling, one has to take samples at twice the highest frequency of the signal. (For this reason, in compact discs music is sampled at 44.1 kHz, little over twice the 20 kHz frequency of the highest audible pitch.)

Transcutaneous Communication and the Hallucination Mobile Phone

Is it possible, with the visible technology, to build a portable communications device with which the user can move totally to a virtual environment? Room-temperature superconductors would make possible transcranial magnetic stimulation and transcutaneous muscular stimulation, but neither of these is very practical in terms of a hallucination mobile phone. A location-selective way of stimulating the receptors of the peripheral nerves would be necessary. The most direct, if commercially unappealing, method is to use a chemical stimulus whose molecules bind to the receptors in specific cell-membrane channels. These molecules should have a sharply defined vibration spectrum in the infra-red range. Further, there should be so many kinds of molecule that, randomly distributed, they would form a network of sufficient density. If this proves possible, location selectivity will be achieved through the calibration process through which the distribution is charted. This sort of technology would create a chemical body map just under our skins, for lightwave interfacing with the virtual environment.

Other areas of application of cognition technology include religion, politics and many other collective cognitive hallucinations. Alternative therapies and many areas of philosophical speculation similarly offer promising potential.

What Next?

If two supermegatechnologies (cognitive and info-bio-nano) emerge in the next 100 years, what about the following 100 years? The empirical Moore's law (which states that computer performance doubles every 18 months) has been valid now for 50 years. It cannot go on forever; there must be a physical limit to computation, such as the quantum mechanical Bekenstein Bound,[11] which limits the amount of information that can be stored in a given volume with a given amount of energy/mass. The American inventor and visionary Raymond Kurzweil estimates on the basis of Moore's law that by 2040 the power of a home computer will be comparable to that of the human brain, and by 2060 to that of all human brains combined.

Perhaps there is a logical end to the Information Revolution, a point of singularity which in retrospect will be understood to be the goal of it all. Perhaps this singularity is reached when quantum computation and intelligent petahertz light beams make all matter a substrate for computing. (Perhaps the designers of the new IPv6 Internet addressing scheme are preparing for this: IPv6 will provide trillions of net addresses for each square meter on the Earth.)

The material world, the planet and its biosphere, the human body, all the material things, will be conserved in the planet-size museum, but our true descendants will be algorithms and data structures encoded by immaterial bit strings, roaming free of their slimy origins. It will be great fun.

Pity I won't see it.

Notes

Originally published in *things* (winter 1999–2000), London.

1. John Maynard Smith and Eörs Száthmary have listed eight "major evolutionary transitions" during the last few billion years: replicating 1. molecules to cells, 2. unlinked replicators to chromosomes, 3. RNA to DNA + protein, 4. prokaryotes to eukaryotes, 5. asexual clones to sexual populations, 6. unicellular to multicellular, 7. individuals to colonies, 8. primate societies to human societies. The authors observe three features common to several of these transitions: 1. entities that were capable of independent replication before the transition can only replicate as

parts of a larger unit after it; 2. division of labor (specialization); 3. changes in language, information storage and transmission. John Maynard Smith and Eörs Száthmary, *The Major Transitions in Evolution* (Oxford: Freeman, 1995); Eörs Száthmary and John Maynard Smith, "The Major Evolutionary Transitions," *Nature* 374 (1995): 227–232.

2. Hypertext and "semantic tagging" of Web content are two means to remove ambiguity from common language. Ordinary linear language, with its very fuzzy semantics, is an inadequate tool for conveying meanings accurately. Instead of words, sentences and paragraphs as semantic units, hypertext makes it possible to communicate using semantic networks as the basic units of meaning. In a semantic network (a piece of hyperlinked text), words lose their independent meanings. The meaning of an underlined word (a hyperlink) is (defined by) the target of the link. This shift of language brings us nearer to the German philosopher and mathematician Gottfried Wilhelm von Leibniz's (1646–1716) ideal of a Characteristica Universalis, a formal language for all philosophical problems. "Let's calculate" instead of quarrelling.

3. Mathematical formulae often compress enormous amounts of knowledge, for example Newton's law F = ma (force equals mass times acceleration). An intelligent Web browser should be able to expand compressed knowledge in accordance with the preference profile set by the reader.

4. An interesting phenomenon related to Maslow's work is that in spite of a lack of evidence to support his hierarchy, it enjoys wide acceptance (Wahba & Bridgewell, 1976; Soper, Milford & Rosenthal, 1995 [Editors' note: we are unsure which works Kurenniemi is referring to]). Perhaps we just want to feel significant by postulating such innate needs. Similarly the often-repeated urban legend that "people use only 10 per cent of their brain capacity" keeps lingering, despite of lack of any scientific basis. Abraham H. Maslow, *Motivation and Personality* (New York: Harper, 1954).

5. The first quantum components (but not fully fledged computers) have been built. Whether quantum computation will ever become a practical reality, no one can be sure. The idea goes back to the American physicist and specialist in quantum electrodynamics Richard Feynman (1918–1988). Ultimately, the processes of quantum computation could be carried by any solid, liquid or gaseous matter, even in its natural state.

6. Varley has described compellingly how the human race will adapt to living in space. In his plot, humans have been driven from the Earth by extraterrestrials. It may turn out that someday we must leave voluntarily, to spare some remnants of the original Earth biosphere. His symbionts, with genetically engineered humans living inside vegetative and sentient shells, are fully closed material recycling plants, requiring only energy from solar radiation for subsistence. For "symbs," see the short story "Equinoctial" (pages 82–131) in the collection *Picnic on Nearside* (New York: Berkley Books, 1980).

7. It is reported that in acid-house parties a combination of drugs, stroboscopic light effects and loud techno music rhythms with carefully adjusted frequencies may result in experiences of shared collective consciousness. It is attractive to speculate that these effects work their way simply through the synchronisation of electrical brain activity. More generally, the ecstatic ritual dance and loud popular music in mass gatherings await a mechanistic explication.

8. There has been a long-standing debate on whether there is one intelligence or many. The somewhat questionable intelligence quotient (IQ) has been replaced by the "general cognitive ability" or g-factor, which is said to correlate well with success in life. Howard Gardner is a main proponent of the multiple intelligence theory (MI), and he distinguishes seven or more independent components of intelligence: visual/spatial intelligence, musical intelligence, verbal intelligence, logical/mathematical intelligence, interpersonal intelligence, intrapersonal intelligence, bodily/kinesthetic Intelligence.

9. The psychologist Arnold Trehub has developed detailed neuron-level models of brain functions; he claims that his neural networks are able to learn from just one exposition to a new situation whereas most neural networks require several rounds of repetition (like Pavlov's dogs). Much of the early work on artificial intelligence was confident that there is a simple key to intelligence, perhaps just one elegant algorithm. (This was the GPS or "general problem-solver" approach to artificial intelligence of Allen Newell, J. C. Shaw and Herbert A. Simon.) Frustrated by the failure to find this, the conventional wisdom became that intelligence requires a large body of factual knowledge and many specialised and "problem-oriented" algorithms and heuristics ("rules of thumb"). The "beehive paradigm" of artificial intelligence—beehives being the classical examples of "social intelligence," intelligent behaviour arising from a collective of a large number of less intelligent agents—is closely related to the more general study of "complex systems" as formalised, for example, in the notion of a CAS (complex adaptive system). A large city is a CAS: it provides its inhabitants with a continuous supply of very special products and services in the absence of any central control system (for example, a centrally planned socialist economy). The philosopher Daniel C. Dennett's parable of mind is the editing room of a newspaper. Several editors exchange "drafts of reality," but nowhere in the brain there is a master homunculus holding a final synthesis of what is going on. Quite naturally, the ideas of Darwinian selection have been evoked to explain how the brain chooses the things it wishes to believe in (Edelman).

10. According to Donald D. Hoffman, since prehistoric cave painters artists knew (unconsciously) to make use of the effect seen in the "Kanizsa triangle" visual illusion. The ends of hatching lines form an illusory contour, and the area adjacent to the hatched region looks more intense than the background paper or rock, be it white or black.

11. See Tipler, page 407. The Bekenstein inequality follows from the Heisenberg uncertainty principle, and suggests that the maximum amount of information contained in any system of human mass and size is 2.58×10^{45} bits.

Further Reading

Bear, Greg. 1997. *slant.* Tom Doherty.

Burges, Anthony. 1978. *1985*. Hutchinson.

Damasio, Antonio. 1994. *Descartes' Error.* Avon Books.

Egan, Greg. 1998. *Luminous*. Orion Books.

Hoffman, Donald D. 1998. *Visual Intelligence*. Norton.

Kosslyn, Stephen. 1994. *Image and Brain*. MIT Press.

Kurzweil, Raymond. 1999. *The Age of Spiritual Machines*. Penguin.

Naisbitt, John. *Megatrends*, Warner Books, 1982, 1984.

Orwell, George. 1949. *Nineteen Eighty-Four*. New York: Harcourt, Brace.

Tipler, Frank I. 1994. *The Physics of Immortality*. Doubleday.

Trehub, Arnold. 1991. *The Cognitive Brain*. MIT Press.

12 Graph Field Theory

Erkki Kurenniemi

Abstract

In this paper I present a discrete ("digital") model of space, time, and matter. Space is a graph, mostly trivalent (cubic). The field is an edge colored with N colors (e.g., 3 or 61). The edges are considered as cellular automaton processors. They may change the coloring and also the local graph topology. The vacuum is characterized by trivalency and perfect coloring. Fundamental particles are deviations from trivalency and/or perfect coloring. The model is assumed to work at the Planck scale.

Cellular Automaton Model of Finite Physics

The hypothesis that the physical world is finite leads to a discrete fine structure and determinism (because random choice is not a finite process). So we have to abandon the notion of continuum and assume a finite number of space (or spacetime) locations in any volume. We also have to discard relativity and quantum mechanics at the fundamental level. But I see possibilities of recovering these theories at the large number limit. Because the graph model operates at the Planck scale ($\approx 1.6 \cdot 10^{-35}$m), there are about 20 powers of 10 in linear size ($\approx$ 60 powers of 10 in volume) to recover quantum relativity as a "classical" limit.

The ordinary topology (open and closed sets, etc.) does not work in a discrete space. The discrete counterpart for a topological space is the combinatorial notion of a graph, an object built of vertices (points) and edges (lines connecting two points). We use a colored graph to model space (rather than spacetime). We take the edges to be the fundamental objects: their number is absolutely conserved in the "standard model" of graph field theory. We use discrete time (tick length one Planck time, $\approx 10^{-44}$ s). The field is an arbitrary coloring of the edges by three colors (e.g., red, green, blue).

An edge can change its color as a response to a color change in one of its neighbors ("incident edges"). An edge may also decide to change the local graph topology by reconnecting one of its neighbor edges to its other end.

If the physical claims sound too preposterous, graph field theory can be regarded as a pure mathematical study of a particular class: cellular automata.

Color Magnet Soup

To get a hands-on physical picture of the model, first think of a kettleful of ordinary small permanent bar magnets. Their ends tend to stick to each other in such a way that the north end attaches with the south end of another bar magnet. When this has happened, there is not much incentive for a third magnet to attach one of its ends to the vertex. In suitable conditions and with stirring one would expect the magnet soup to contain several closed rings of magnets. In each ring, an even number of magnets forms a stable configuration. The rings interact only residually.

Next imagine another kind of magnet. Instead of having a north pole and a south pole it is uniformly colored. There are red, green, and blue magnets. Their "magnetic" properties are such that whenever a red end and a green end and a blue end meet, they form a stable bond. In all other cases the bond energy is higher and the bond less stable. Take some 10^{220} of these "fundamental strings,"[1] pour them into a Big Pan, and stir well. Then watch.

Shortly after you stop stirring, a great deal of the strings will have found their partners and formed trivalent perfect bonds such that three differently colored magnet ends form stable vertices. But there are bound to be many defects. One reason is that three-coloring the edges of a trivalent graph is NP-complete: the time required to reach a perfect coloring grows exponentially with the size of the graph. The defects or faults I take to represent matter. These defects may be compared to crystal imperfections, for example, dislocations, but with the deep difference that there is no metric order.

The above thought experiment fails in one essential aspect. We visualized the soup as embedded in spacetime. Forget that. The space is the graph. The metric is defined by the instantaneous graph structure. Thus there are no metric limitations for bond formation. In the initial state, the dimensionality of the graph may be quite high. If we assume that pentavalent and higher bonds are unstable and break up, the dimensionality of the graph gets lower and lower. It is natural to conjecture that the graph dimension asymptotically approaches three.

Edge-Conserving Variable Graph

More formally, we have a finite set E of basic objects called *edges* or *fundamental strings*. At each moment of discrete time $T = \{1,2,3,\ldots\}$ the edges form a connected graph U whose vertex set is denoted by V. The number of vertices $|V|$ may change with time, but the number of edges $|E|$ is constant. At each moment each edge possesses a label called its color. The coloring as a function of time is called the *field*:

$$\psi = ExT \rightarrow \{R, G, B\}$$

At this stage I don't care to formalize the fact that the graph topology also varies as a function of time. It is sufficient to think of an edge as a "microprocessor" of the cellular automaton. It receives as inputs the colors of the incident edges at its both ends. The edge can do two things: (1) it can change its own color, and (2) it can reconnect one or more of the incident edge ends to its other end.

I require that each edge is an identical finite automaton without any memory except for that provided by its own color and perhaps some very volatile "dynamic memory." The edges are programmed in such a way that they keep faults in constant motion and handle collisions of faults and their eventual annihilations.

Whenever three different edge colors meet at a vertex, their sum is perfect. Call this White or null 0. This is the neutral element of the color group. The group multiplication table is

0	**R**	**G**	**B**
R	0	B	G
G	B	0	R
B	G	R	0

This is the Abelian Klein four-group of the symmetries of a rectangle.

White is a theoretical notion. The three primary colors R, G, and B are physical: only they can be the colors of the physical edges. The vertex coloring is induced (derived) from the physical edge coloring.

Faults Are Fundamental Particles

We classify vertex types by two "quantum numbers," degree and color. Degree is the number of incident edges. Vertex color is the sum of the colors of the incident edges. According to the degree, we name vertex types as follows:

Degree	vertex name
1.	monon
2.	dion
3.	trion
4.	tetron
5.	penton
	… etc.

Thus we may talk about white dions or colored trions (meaning trions that are red, green, or blue). A white trion is said to be perfect or a vacuum vertex. All other vertex types are called particles.

I shall assume that monons and pentons and higher-degree vertices are so unstable that it is generally not essential to account for them. For example, a penton may disintegrate into one dion and one trion.[2] A monon may transform into a pair of dions. All of this of course depends on how the edge processors are programmed.

We are left with three topological families: dions, trions, and tetrons. If we assume a color-symmetric theory, the fault propagator (edge program) treats all three colors essentially the same way. This means that we have five fundamental particle types:

white dions
white tetrons
colored dions
colored trions
colored tetrons

How exactly these five particles might correspond to the known elementary particles is still unclear. If we write the propagator so that color symmetry is broken, there will be three generations of the five fundamentons. For several years I thought of colored trions as quarks because quite early I discovered their combinatorial confinement property. Dions and tetrons came later and complicated the simplistic picture. For the moment I give just a couple of suggestions.

(1) Colored dions and tetrons have a property that make them neutrino-like: they do not "flip their orbits" (change edge colors as they pass). In practice this means that the motion of colored dions and tetrons (colored "evenons") is strongly affected by the motion of the remaining three particles ("white evenons" and "colored oddons"). On the other hand, the motion of white evenons and colored oddons is only weakly affected by the motion of colored evenons). Only an unprobable head-on collision ("weak interaction") by them will do.
(2) Dions and tetrons are mutual topological antiparticles: when an excess edge (tetron) and an edge defect (dion) meet, the result is trivalent.

Field Lines Are Particle Orbits

Every particle is an entity that moves at the speed of light along its orbit. Particle orbits are instances of field lines.

Define a red field line to be a path consisting of alternating green and blue edges. Similarly for the two other colors:

Red field line = … GBGBGBGBGB …
Green field line = … BRBRBRBRBRB …
Blue field line = … RGRGRGRGRG …

You can create a particle pair in the vacuum by recoloring one edge. If, for example, you change the color of one green edge into blue, you have created a pair of red trions, one in the phase RGG and the other in the phase RBB. As a red trion moves along its orbit, its phase oscillates at the Planck frequency.

A simple experiment with color pen and paper reveals the pleasing fact that red particles move on red field tines, green particles move on green field lines, and blue particles move on blue field lines. We now see the space as a system of three intermingled field line systems (also called *trajectories*, *particle orbits*, *virtual particles*, or *digital vector fields*; graph-theoretically they are two-factor). Because the particles move on their orbits at the speed of light, we, anticipating material to follow, somewhat cryptically say that field lines are on the light cone.

What about white dions and white tetrons? Are there white field lines? No. A more detailed analysis shows that they also move on standard colored field lines. These particles have a *hidden color*, which is the same as the color of the orbit. If these two fundamentons turn out to correspond to quarks, it may well be that the colors of quantum chromodynamics are our hidden colors and that our explicit colors are unobservable at the macroscopic level.

In a finite vacuum, each field line is necessarily a closed cycle. If the particle density is low, all particles may possess closed periodic orbits. Each particle disturbs all incident field lines. Some incident field lines simply end, some bifurcate, some stutter.[3] If the particle density is high, only relatively short orbits can be closed. Orbits that are too long necessarily meet other particles and consequently end or bifurcate. Thus we cannot assign a unique orbit length above a certain statistical limit. Above that limit, the notion of orbit length must be replaced by mean free path (or mean free time).

When two colored trions collide, they annihilate fully or partially according to the color group. If two red trions moving in opposite directions on a common orbit meet, they annihilate completely: $R + R = 0$. If a red trion and a green trion meet, they annihilate partially into one blue trion: $R + G = B$. It is again a simple pen-and-paper experiment which shows that particle annihilations obey the color group. On the other hand, dions annihilate tetrons topologically. These two principles together will make up the general annihilation rule.

The Propagator and the Interactor

The particle propagator is the edge program. First, we consider cases in which the initial state at both ends of an edge is perfect (a vacuum). In this simple case, one particle arrives at time t to one end of the edge. The propagator transfers the particle to its opposite end if it belongs to the orbit of the particle; otherwise it takes no action.

An example. The color of the edge is red. At its "left" end were one green and one blue incident edge in the perfect state. Suddenly the green left incident changes its color to red. This signals that a blue trion has arrived at the left vertex. The edge responds by changing its own color from red to green. Hereby it has propagated the blue trion from its left end to its right end. It also changed the phase of the particle from BRR to BGG. All of this took one Planck time ($\approx 10^{-44}$s).

If, on the other hand, the left green incident changed to blue, our edge should take no action. The incoming trion is now red and should be handled by the left blue incident edge. Our edge has just observed the passing of a red trion as a passive onlooker.[4]

The action illustrated above is called the *vacuum propagator*. The vacuum propagator could also be called the *conservation of particle momentum*. The full edge (cellular automaton) program could be called the *full propagator* or *interactor*.[5] As soon as we can write down the interactor, we can simulate graph field theory on a computer. I foresee that very little in graph field theory can be calculated compared to what can be observed from massive computer simulations.

In the general case, several particles may arrive simultaneously at both ends of the edge. I haven't been able to program the full interactor because I have tumbled into combinatorial complexities, although the problem seems to be finite and well defined. Difficulties arose in situations where no unique rule could be found for the edge response. I am stubbornly sticking to the principle of finiteness, which predicts that edges cannot throw dice.

Let's look now at the full interactor program to see what is involved. One possible formalism for the interactor is a set of if–then rules. If an edge perceives a certain input pattern at its two ends, it will take a corresponding output action. First let us establish a notation for input patterns and output actions.

Consider one fixed edge e. Because e is fixed, it is not necessary to indicate the dependence of the following items on e explicitly. Denote its color at the generic time instance t by c. Thus we write simply c instead of $c(e,t)$. Call one end of e its left end and the other end its right end (think of e as a horizontal line). Denote the number of edges incident with the left end of e by n_l. Call them $l_1, l_2, \ldots l_{n_l}$. Similarly, at the right end of e we have n_r incident neighbors named $r_1, r_2, \ldots r_{n_r}$. If both ends of e are a vacuum, $n_l = 2$ and $n_r = 2$.

Denote the colors of the left incident edges by $cl_1, cl_2, \ldots cl_{n_l}$ and the colors of the right incident edges by $cr_1, cr_2, \ldots cr_{n_r}$. By convention, we may standardize the vacuum initial state denotation as follows:[6]

$$n_l = 2\,,\ n_r = 2$$
$$c = \mathrm{R}$$
$$cl_1 = \mathrm{G}\,,\ cl_2 = \mathrm{B}$$
$$cr_1 = \mathrm{G}\,,\ cr_2 = \mathrm{B}$$

Thus we look at a red edge with green and blue left neighbors and green and blue right neighbors. The arrival of, say, a dion from the left means that n_l goes from 2 to 1. The other left edge may or may not change its color.

There are at least four different output action types available to an edge. We formalize them using an operator notation. Operators in graph theory are not to be confused with quantum operators. Graph operators operate on discrete subgraphs, and they are quite deterministic. We next list four graph operators:

Type Example

(1) Color flip: $c \to cl_1 + cl_2$
(2) Direction flip: $d \to -d$
(3) Edge transfer: $*l_3$
(4) Disconnect: $\mathrm{X}l_1l_2$

(1) The color flip is the most basic edge process. The edge processor changes its own color, that is, its memory contents. The color of an edge is its only long-term memory content.
(2) The edge changes its basic left–right orientation. This basic symmetry break will be introduced later in this paper, so bypass this action for now.
(3) This is the fundamental topological operation. The example above glides the third left incident edge end to the right end. The net effect is that n_l will be decremented by one and n_r will be incremented by one. Introduce the terms *left list* and *right list* for the lists of left and right neighbors and their color attributes. Note the relevance of this for a Lisp simulation.
(4) The ability to disconnect neighbors is a new feature that makes it possible to dispose of pentons and higher-degree collisions. Assume that two tetrons happen to arrive simultaneously at the left end of e. Before n_l was 2; now it is 4. The degree of the left vertex is now 5. There is a penton at the left. We intend to kill the penton by disconnecting two or three edges from the left vertex. But how, exactly? This is the problem.

I face with horror the combinatorial task of drawing up all possible cases, finding unique output actions for all input patterns, and finally writing a long simulation program case by case. There has to be a more elegant way out!

Combinatorial Invariants: Parity Theorem

The parity theorem says that by examining the boundary of a subgraph, one may infer something about the numbers of various particles inside that region. One can tell whether the number of certainly colored particles is even or odd. An odd number of particles means there is at least one particle.

The parity theorem shows how particles are just the singularities of the field. A particle may have combinatorial effects on the field far away. The parity theorem is an analogue of the Gauss–Stokes theorem in a finite context.

Consider first the special case of a trivalent graph (no topological particles like dions or tetrons present). Choose an arbitrary subgraph. Assume that the subgraph contains r red trions, g green trions, b blue trions, and w vacuum vertices. Assume that the subgraph is connected with its environment by r′ red edges, g′ green edges, and b′ blue edges (the boundary of the subgraph). Now the parity theorem says that

$$r' \equiv r + w \pmod{2}$$

$$g' \equiv g + w \pmod{2}$$

$$b' \equiv b + w \pmod{2}$$

In words, the parities of variously colored trion numbers inside the region are determined by the parities of correspondingly colored boundary edges (up to the common Catch-22 term w).

The proof is by a trivial counting of edge ends ("half-edges"). Each green and blue trion contributes an even number of red half-edges; these terms cancel each other out when we combine half-edges into full edges. Each red trion and each white vacuum vertex contribute one red half-edge each. Adding these contributions we obtain the right-hand side of the first congruence. An even number cancels again at half-edge recombination. The remaining parity must equal that of the unmatched boundary half-edges. The first congruence follows, and similarly the other two.

A partial result ($r = g = b = 0$) is known in graph theory as Isaacs' lemma or the Parity lemma. The structure of the color group implies this result. In particle annihilations ($R + R = 0,\ R + G = B$), the parities of particle numbers are the same before and after. Either the number of one species changes by two, or the numbers of all species changes by one. Setting $r' = g' = b' = 0$ we see that in a finite graph the total color is always zero.

Next the more general case. Denote particles by a color letter with a degree superscript:

Fundamenton	Symbol
white dion	w^2
colored dions	r^2, g^2, b^2
white trion (vacuum)	w^3
colored trions	r^3, g^3, b^3
white tetron	w^4
colored tetrons	r^4, g^4, b^4

Use vertical bars to indicate set size. In the subgraph under consideration $|b^4|$ denotes the number of blue tetrons. We count half-edges as above. First count the number of red half-edges N_r ignoring particles whose contributions are known to be even.

$$N_r = |g^2| + |b^2| + |w^3| + |r^3| + |g^4| + |b^4|$$

Now we see how the restricted result above generalizes in the presence of dions and tetrons. On the right-hand side of the first congruence above there will be four additional terms due to non-red dions and tetrons. To obtain a concise form we introduce a symbolic complementary color notation. Call the set complement of a color its anticolor: $-R = \{G, B\} = C$ (cyan), $-G = \{B, R\} = M$ (magenta), $-B = \{R, G\} = Y$ (yellow).

Introduce the symbol c^2 to stand for non-red or cyan dion and so on. (In the present three-color theory there are no physical cyan vertices; the symbol stands collectively for green and blue dions.) The generalized parity theorem now says that

$$r' \equiv c^2 + r^3 + c^4 + w \pmod 2$$

$$g' \equiv m^2 + g^3 + m^4 + w \pmod 2$$

$$b' \equiv y^2 + b^3 + y^4 + w \pmod 2$$

We have omitted the vertical bars and the vacuum trion superscript for readability. The first line of the parity theorem says that the parity of red boundary edges is the same as the parity of the sum of the numbers of red trions, non-red dions, non-red tetrons, and vacuum vertices. The theorem can obviously be generalized to arbitrary degree particles, but the present form suffices for us.

The corresponding topological conservation law needs a bit of clarification. We took as a basic ingredient of the theory that the total number of edges is absolutely conserved. In a finite trivalent graph, $2|E| = 3|V|$. (Proof again by half-edge counting.) Thus the number of edges is a multiple of three. If we add one or two extra edges into a trivalent graph, it is impossible to recover trivalency by edge-conserving topological transformations, local or global. But if we add three edges, it is possible: the above equation is again satisfied.

Adding one edge may result in one dion ("series connection" with some existing edge) or two tetrons (if we add the edge between any two trivalent vertices). In the latter case, the two tetrons will have quite independent lives of their own. The conservation of edges modulo 3 is equivalent to conservation of half-edges modulo 6. Consider a subgraph and assume no particle flow across the boundary. Count the tetrons inside. Add to this number twice the number of dions inside. Call the result the tetron number. Now the topological conservation rule says that the tetron number is conserved modulo 6. Zero remainder is the topological vacuum state. Thus an isolated non-vacuum system stays forever in one of five possible topological states. This is the topological conservation law.

With respect to the topological conservation law, one dion is equivalent to two tetrons, and similarly two dions are equivalent to four tetrons. In defining the conserved tetron number (modulo 6) as a sum $2D + T$ (D = number of dions, T = number of tetrons), we threw away a piece of information about the relative abundancies of dions and tetrons. I have a hunch that we threw away the electric charge. Should the hunch turn out to be valid, it would lead to the experimentally testable prediction that electric charge obeys modular arithmetic: a small number[7] of positive charges brought into sufficiently close interaction will yield a negative charge.

Evolution Principle

Before proceeding, let us review some features of the model developed so far. The physics of particles as vacuum defects seems to resemble classical particle motion in random media. I have very little idea about the degree of regularity of the graph and the coloring, but it seems to be the most natural starting point to assume them to be completely random at small scales. The particles interact directly only via head-on collisions, and this is accompanied with matter annihilation. We have to assume that the matter loss rate is below our current precision of measurement. As we shall see, it is quite possible that enormous amounts of real but weakly interacting matter hide in the physical vacuum and can serve as an invisible source for processes involving annihilations.

In our model there is just one real force with zero range.[8] This does not contradict the observation of long-range forces. Long-range forces can be mediated by particles like photons, vector bosons, and gluons. But there still seems to be a contradiction. If the graph structure is at the Planck scale, the probability of the intermediate particle hitting anything material is terribly low. The contradiction disappears as one realizes the workings of the path-flipping mechanism. In the three-color model, three particles (colored trions and white evenons) move by flipping their orbits: they change the color of each edge they pass and thus leave a trace in the vacuum. This change affects the orbits of all other particles whose orbits share a common edge with the particle track. I propose: pointlike zero-dimensional particles that leave a one-dimensional trail interact just the right amount to stop the dimensionality of space from shrinking below three.[9]

Consider a path-flipping (= track-leaving) particle p with a short orbit of length L. If the effective length[10] of an edge is the Planck length $l_P \approx 10^{-35}$m, L might be anything from 2 to 10^{20} and still the orbit would be pointlike by our current standards. Interpret the direction of the particle motion on its orbit as its spin. As p traces its orbit just once, it changes the colors of all orbit edges. At the next round it flips them back. Thus we can associate with p a frequency $1/2\,L$ in Planck units. Multiply by the Planck constant and get the energy of the particle.

This concept of energy is not related directly to the particle itself; rather, it is a property of its orbit: thus a property of the field (=coloring). In other words, the orbit energy $h/2\,L$ of a particle is a property of the particular vacuum the particle happens to inhabit. We have arrived at a deep conceptual difference between the graph theory and traditional theories. The orbit possesses the energy and most probably the mass as well. The particle as a graph defect is quite a featureless visitor on an orbit full with physical properties. But not fully featureless: all the important quantum numbers required to identify which elementary particle it is are attributes of the fault itself.

When I studied physics in the 1960s none of my teachers were able to explain what the word "virtual" really means in the context of "virtual particles." Here I give an explanation. Virtual particles are vacuum field line systems. Only when a virtual particle is inhabited by a fault, a singularity, a fundamental particle, we have a real particle doing real dynamics. Virtual particles without singularities are fully happy and fully dead. The fundamental particles are simple: say, the number of colors and decide which additional symmetries are broken, and you readily get a spectrum of naked particles. But to compute the properties of virtual particles (field configurations) is a huge task, perhaps fundamentally unsolvable. Although the virtual orbits are empty, they directly and concretely affect physics because faults are bound to their colored orbits.

What are the bosons of this theory? The preliminary answer is that fermions are systems consisting of an odd number of particles (faults) and bosons are either virtual (even-length closed vacuum field lines) or real: white systems of an even number of particles.

Most of the material in the first part of this paper was essentially hardcore combinatorial: assume such-and-such cellular automaton and derive such-and-such properties (like the parity theorem). The rest of the paper will be more speculative.

My present central guess is the principle of cosmic evolution. The idea is borrowed from biology but is actually much simpler. In bare bones it proposes that the configurations of matter we observe today are those that have survived statistically. Assume a moment of Big Bang with a random coloring. Since then most particles have been annihilated. Which ones survive today? Those whose probability of annihilation is low.

The principle of evolution applies both to virtual and real particles: field configurations and defects. Look at the evolution from the perspective of a single particle:

(1) Because the particle still exists, it has clearly escaped annihilation. How come? Because it had a happy childhood?
(2) Maybe because of its conserved quantum numbers it has not met a suitable annihilation partner: they may be rare. Here we may have a clue to the antisymmetry between matter and antimatter.[11]
(3) It may currently possess a well-protected orbit.
(4) There might be ways to swap orbits safely: there might even be safe orbit systems.
(5) It is a young particle born recently from a partial annihilation process.

What is a well-protected orbit? To this question we can give a nice recursive answer. The root of the recursion says that an orbit is safe if it is short. The longer the orbit, the more other field lines it meets, and the greater the probability that an annihilation particle has been met. In the recursion step we realize that in comparing two equally long orbits, shortness of intersecting trajectories adds to the safety. And so on.

The evolution principle thus predicts that the orbits of many existing particles are surrounded recursively by short field line cycles. As we learned above, short real or virtual orbits have a high associated energy, $h/2L$. Now we have found an explanation for the concept of potential. High potential[12] means a region of space where the field orbits are short.

Next we shall apply these ideas concretely. The vacuum might be full of matter well shielded against annihilation. If so, that matter reserve may be the potential for much of the observed hard physics. The observed physics above the Planck scale might be the residual effects of processes that tap into these huge vacuum potentials.

So let's plunge into the microworlds, toward the Planck scale, and have a look at the field line systems surrounding a single particle.

Particle Confinement

Take a snapshot of one solitary particle, for example a red trion, surrounded by vacuum. The particle (a vertex) disturbs all three field lines through that vertex. The red orbit of the red trion suffers a "stutter fault": depending on its phase there are two consecutive green or blue edges just where the trion is. The other two field lines also suffer: one terminates and the other bifurcates. Assume that in the snapshot the green field line terminates and the blue bifurcates. According to the survival principle,[13] these security-threatening field lines should be as short closed trajectories as possible. But it is impossible to connect the four ends of one green line and three blue lines in vacuum. You don't need the power of the parity theorem to see this. So if the green and blue trajectories are short, there must be other faults nearby, for instance another red trion. Thus we have confinement of colored trions. The existence of the other particles is a purely combinatorial result; the shortness of the confinement length follows from the survival principle.

The "meson" formed by a red trion pair is still quite vulnerable: a small disturbance in the field line system may lead to their mutual annihilation. What about the "hadron" formed by three differently colored trions? I don't have any hard data yet, but a simple argument gives plausibility to the conjecture that the hadron might be very much more long-lived than a meson. Before going into that, note how intricate a safe field line system between the orbits of three differently colored trions has to be. From each vertex of each orbit, both directions of the two intersecting field lines must end at another such orbit vertex before wandering too far into the dangerous outer world.

To annihilate the hadron by field disturbance one first has to get two of them to annihilate mutually into the color of the third and then wait for the annihilation of the resulting meson. Thus hadron annihilation is a higher-order process.

I don't think the explanation above to be even nearly sufficient to explain the longevity of the proton. Other mechanisms must be at work there.

The example above was in terms of trions, but exactly the same applies to colored dions and colored tetrons as well. All these particles possess a pair of unmatching field lines from each vertex of their orbit. The color sum of this field line pair is that of the particle itself. White dions and white tetrons are different. Although they also disturb field lines by stuttering, termination and bifurcation, the numbers of emanating field lines of each color are even and thus may close in the vicinity of the orbit; white dions and tetrons exhibit no confinement.

Phase Transition and Modes of Motion

First we sketch a naive graph cosmology. Assume an initial time t=0 random graph (the Big Pan). In a computer simulation, first generate a big random graph (or an almost trivalent random graph), then color its edges randomly. Then turn on the cosmic clock: start the automaton. After the first few ticks (Planck times) most of the particles have been annihilated. What are the main qualitative aspects of the evolution after that?

Does the annihilation continue automatically until there are no more particles left? Of course not. There are no more annihilation events as soon as the following two conditions are met:

(1) The orbits of remaining faults are disjoint; or,
(2) If the orbits of any set of particles meet, their lengths possess common divisors in such a way that the particles never meet.[14]

This is the ultimate phase transition that freezes the universe. There may be lots of real matter in autistic orbits but no interactions, no annihilations. Nothing remarkable will happen after this. Because things seem to happen, I conclude that we are still far away from that phase transition. Instead I conjecture that we are now in an era of deep subcooling.

The subcooling conjecture has several consequences. The vacuum energy is high.[15] The majority of particles trapped in stable configurations furnishes an energy source for the dissipative young particles.

Absolute Space and Absolute Time

The graph field theory is obviously absurd[16] as a physical theory because the graph is an absolute space and because the requirement of finiteness leads us to assume absolute time in the sense that all edge processors are clocked by a common cosmic clock. (In other words, the cellular automaton operates synchronously.)

The second absurdity is that the theory is nonquantum. Every particle has a definite deterministic trajectory. So this is a hidden-variable theory. Hard-core quantum mechanists know that hidden-variable theories have been proved impossible. (Others may know better.)

The natural way out would seem to go from a spatial graph to a spacetime graph. Then we could throw away the idea of an edge as a functional processor and instead formulate the propagator as a set of static connection rules (which kinds of colors and valences [= edge degrees] may be neighbors).

The natural way out of the second absurdity would be to canonically quantize the edge processors, to throw away the determinism.

I am afraid that this route has been already tried too many times. Aesthetically, to me, this approach would mean reaccepting all the mess of relativity and quantum mechanics. These two great theories are, perhaps, incompatible: somehow like the old wave–particle duality. They might well be (despite the initial success of superstrings) two sides of the same coin. You cannot unify them and keep the shape of the coin.

But there is another way out. If the cellular automaton works at the Planck scale, there are about twenty powers of ten separating this scale from the nuclear scale. Could it be possible to rescue relativity and quantum behavior as "classical limits" in cellular automata that are practically infinite? Twenty powers in linear scale imply some 10^{60} edges in the volume of a nucleus. Although the space currently would be extremely dilute (density of particles extremely low), their actual number in a nuclear volume still may be so huge that the continuous loss by annihilation may have gone undetected.

Recovery of Relativity

In relativity, the speed of light is the maximal speed of a physical particle, attainable only by massless particles. In the present model c is the only possible speed.[17]

So, all matter moves at the speed of light. We, as material objects, have to anchor all physical frames of reference to matter. This means that in all physical frames the graph seems to move at the speed of light in the opposite direction. Has anyone heard of an actual Michelson–Morley experiment performed by massless angels using massless instruments all flying at c? This is the first indication that there might be a way out.

The second insight is more technical. The graph is a metric space. The distance between two vertices x and y is the nonnegative integer that is the length of a shortest path joining x and y. It is customary to denote this by $d(x,y)$.

In the presence of a coloring, and when we regard the graph from a frame of moving particles, the basic (topological) metric is not the relevant one. Our idea of locality requires that particles have to be nearby in order to interact, and conversely, they are prone to interact if they are nearby. In the presence of the colored trajectories, two particles may well pass each other by a Planck distance and still not interact if the trajectories do not meet.

To obtain a physically relevant metric in the graph we have to measure distance along field lines (trajectories).[18] Simplistically this would lead to infinities: if vertices x and y are not connected by a field line, their distance would be infinite. The way out might be to define a "helper metric." The helper distance $\Delta(x,y)$ in a colored graph is the minimal number of colored trajectory segments[19] required to connect x and y. If x and y lie on a common field line, $\Delta(x,y)$ is zero. Its value is one if the trajectories of x and y meet. It is two if none of the above applies, but there is a trajectory that meets the trajectories of x and y. And so on.

So this is a well-defined number. If the helper distance of two particles is one, they possibly annihilate mutually, but only on number-theoretic conditions that relate the lengths of the orbits and the phases of the two particles. If the helper distance is greater than one, the annihilation will require the participation of other particles. Thus we see that high values of helper distance are relevant only in multiparticle configurations: at very high temperatures. Suddenly, we see that the observable geometry of the slightly faulty graph depends on the actual dynamics of particles.

The point raised above is relevant in assessing the Planck-level meaning of the helper metric. High values of the helper distance were meaningful only in the very early high-temperature times. Today the helper distance of particles has statistically meaningful values perhaps only in the range of the first few integers. Thus, the helper distance today is a small quantum number.

We have seen now how energy (temperature) and metric are connected in a finite model. The present task is to derive relativity from the absolute graph approach. Now we have two metrics, the graph topological distance and the helper (quantum) distance. Can we combine these two notions of distance into one that would somehow approximate the Minkowski metric?

To recover relativity, we have to show a Minkowski $(-+++)$ metric in some classical limit. We have two graph notions, the topological distance d and the helper distance Δ. The first is a purely spatial distance, evaluated at a hyperplane of simultaneity. The second is a distance evaluated on the light cone.

The next step to approximate a physical metric is to combine the basic topological metric and the helper metric by counting path lengths along as few field lines as possible. First determine $\Delta(x,y)$. Then look at all paths connecting x and y consisting of $\Delta(x,y)$ field line segments, add up their lengths, and finally choose the smallest such number. But this would be quite a fragile measure of physical distance because the procedure does not take into account the time-dependent nature of the helper distance. In fact, each track-change means the implicit assumption of an annihilation process, that is, the fortuitous presence of another particle.

Now we are beginning to see how time enters the notion of distance, and how the originally separate time and space on the graph level are going to mix together. The original cellular automaton discrete time will be the proper time. The cellular level is hopelessly random, and all fully discrete notions of distance will lead to combinatorial explosions. The relativistic mixing of space and time is a result of our desire to describe complex structures in simple and manageable ways.

Recovery of Quantum Mechanics

The recovery of quantum mechanics is built in the graph approach. The simplest way to see this is to look at the path integrals of QED. Consider the emission of an electron. The particle moves at the speed of light on its orbit. There are two possibilities. The particle either meets a particle of its earlier orbit or not. In the first case, it possesses an orbit and thus possesses a definite energy (inverse of the orbit length). Otherwise, it crosses a variable edge and changes its course. The probability of meeting another particle head on is still minuscule. So it switches onto another track. Later, it may either switch to new tracks or return by the motion of other particles moving and joining its track to its original track.

On the next non-Planckian level of description the particle behaves genuinely quantum-mechanically. We see that the recovery of quantum mechanics is no problem at all in our graph model.

Computer Simulation

I have written a Lisp simulation program that contains a basic vacuum propagator for trions. The well-commented program code is available to anyone interested but for the time being this code is not usable for realistic simulations. I have plans to code an initial version of a full propagator sometime in the summer of 1991, and this might be realistic. The data structures have been a problem, but now I think I have a working solution. The basic problem to me was how to represent mostly trivalent graphs in Lisp. The present idea seems to be correct and reasonably efficient.

More Symmetries for Breaking

The model presented so far is maybe not rich enough to account for all observed physics. In this section I shall list some of the symmetries unbroken so far. Because these descriptions are on a basic level, it should be understood that their consequences might be far-reaching.

First a list of symmetries of the basic three-color model, which may turn out to be breakable for profit:

(1) The colors are symmetric, equivalent.
(2) The edges are symmetric: both ends are similar.
(3) The discrete time is symmetric: all moments are equivalent.
(4) There may be more than three colors.
(5) The vertices do not orient the incident edges into a circular list. If the processors were attached to vertices instead of edges, they might want to.
(6) The graph may be a hypergraph.

Here we shall indicate ways to enrich our model by breaking the symmetries (1)–(6). Case (4) will be studied under a separate heading.

...

For (1): the propagator might have to treat the three basic colors somewhat differently. Why not begin with two edge colors? Of course there are 0 and 1 edges (binary automata) and three-colored edges are the results of a type-4 generalization. So color symmetry is almost certainly broken, and hopefully this will explain the three generations of leptons and quarks.

For (2): For this one I am almost well-prepared now. Assume "conical edges." This means that each fundamental string is asymmetric. It is able to distinguish between its two ends. The formal way to incorporate this almost unavoidable symmetry breakup is to shift from graphs to digraphs = directed graphs. An edge is not simply a line connecting two vertices but an arrow. (To the physicist: directed graphs imply currents.)

So far I have only analyzed what happens to the three white particles (white dion, white trion, white tetron) when we break the edge end symmetry. The result is: there will be twelve particles instead of the original three. There will be three directed dions, four directed trions, and five directed tetrons. 3+4+5=12. Maybe "directed" means "charged"; I don't know. Other consideration have suggested to me that these twelve directed particles might be classified as follows (all of this is still quite intuitive):

(1) The three directed dions might be the Higgs and the antidown/antiup first generation quark pair.
(2) The four directed trions might represent the graviton and an electron–neutrino–positron triplet.
(3) The five remaining tetrons might contain the up/down quark pair and three mystery particles I haven't been able to identify.

This is quite promising for a finite and easy theory. Getting serious: compare the ease of getting results in a continuum theory versus getting results in a finite theory.

Notes

Unpublished draft, Heureka, Vantaa, Finland, 1990.

1. The number is an approximation to the current volume of the universe in cubic Planck lengths.

2. Note that this is an irreversible process. The graph model is time asymmetric.

3. These concepts will be explained later.

4. The interesting open question is whether the information obtained in this kind of passive observer situation might be used to process right-end events that are close together in time. The propagator problem is still open.

5. The propagator is thus the subroutine that handles interactions with the vacuum.

6. Assuming color symmetry.

7. I have not worked out whether the number is 3, 4, 5, or 6. Note that there will be another completely different possible way to introduce chromoelectric charge.

8. Or possibly one lp (Planck length).

9. Space dimensionality will be treated later. Penton disintegration seems to lower graph dimensionality. Whenever two tetrons meet, they form a momentous penton.

10. The idea that edges do not have actual lengths but can still be assigned effective lengths will be explained later.

11. But there is another perhaps better way to account for the notion of antiparticle. This will be developed later on.

12. There is still some confusion in signs. A region of short orbits obviously corresponds to a potential well.

13. By this I mean the core of the evolution principle.

14. The phases of the particles on these composites are thus partly determined.

15. Here I mean vacuum matter content. The idea of potential introduced later works well with virtual particles (empty orbits) as well.

16. That is, classical.

17. There is still the slight possibility of zero speed: if the edge cannot solve a combinatorial problem uniquely, a fault might get stuck. But for now I assume that the edges are able to keep all faults in motion.

18. This may have connections with the light-cone gauge.

19. Now I think I should raise the notion of "segment" to a higher conceptual status. A new classification of the field line system is almost visible at the horizon.

13 Interfaces of Future Authenticity: Erkki Kurenniemi's Media Archives (From a Postcybernetic Perspective)

Morten Søndergaard

The media archives of Erkki Kurenniemi are multifaceted and extremely versatile. They mostly document Kurenniemi's life and actions, alone and with others that he shared his life with. The production of the archive is not strictly about art processes, but neither is it autobiographical. Moreover, it has been recorded and produced on many different media platforms during more than five decades from the early 1960s until today, involving sound, images, and footage in various formats, including the more ordinary analog (audiotapes, 8 and 18 mm film) and digital (voc, png, mpg, etc.) and lesser-known or unknown formats (Apple Newton file formats and others). Therefore, Kurenniemi's archives serve other purposes or have another status than the archives you find in museums or libraries.

I will claim that Erkki Kurenniemi's media archives are playing with the role of authenticity while highlighting and questioning the status of "real" life, as it is lived, while it is lived, with sex, drugs, art, anxieties, car drives, food, and a long list of other meticulously recorded, seemingly insignificant events of life.

Thus, I am proposing that the media archive of Erkki Kurenniemi is not to be seen as a result of documentation processes in any traditional sense. Underneath the media archive, or rather ahead of the production of archive material, which is primarily Kurenniemi's own artistic and personal action-space, another and different "submedia" stage is unfolding. Thus, there is a submedia archive running parallel to (and pointing ahead of) the media archive of Kurenniemi, which conceptually is evolving from the postcybernetic metaphors he lived by, charting out a "future space" between storing-machines and memory technologies. This "future space," as we shall see, may be understood as directed toward a construction of interfaces of future authenticity. And, even more radically, since the archive is intended as a submedia reservoir for the construction of the future human version of Erkki Kurenniemi, then the real "sender" is in fact only a real "self" in a future biotechnological configuration. So, how does Erkki authenticate Kurenniemi in the future?

This chapter contextualizes Erkki Kurenniemi's media archive as a practice that borrows heavily from cybernetic and postcybernetic theories and art. I will argue that

Kurenniemi's own and unique position evolves from this context, and with it, the issue of authenticity emerges. The media archive raises the question of whether and how the media archives relate to or reflect the "real" Erkki Kurenniemi at all. Or should they be considered as bits and pieces of processes to be fed back into other archives or machines in some point in the future?

Authenticity Revisited

In *The Work of Art in the Age of Mechanical Reproduction*, Walter Benjamin famously addresses the question of authenticity of art in a situation where its "aura," by which he means its ritualistic function and traditionalistic epistemology, is obsolete.[1] What Benjamin is addressing is not only a change in the ways that art is representing reality, but also how our way of thinking about the world is transforming.

The authenticity of a thing is the essence of all that is transmissible from its beginning, ranging from its substantive duration to its testimony to the history it has experienced.[2]

From this, it would be possible to claim that, in modern culture, the idea of authenticity is based on a notion of a "transmissable" relationship between objects and their representations. Whereas Benjamin took this as a sign of art moving into another field than the one where it had a ritual function and purpose, the issue of authenticity is a far more complicated matter.

According to Benjamin, the criterion of authenticity ceases to be applicable to the production of art.[3] Furthermore, it follows that, in a modern context, a "true" transmission is not possible and therefore is replaced by its "aura"; thus, the "true" relationship is perceived as "lost" (albeit by some, longed for) in the constant transformations and "many faces" of modern culture—decadence, kitsch, modernism, and avant-garde among them.[4] So, a new and modern idea of authenticity emerges that is evolving around a paradoxical dialectics of "aura" and the destruction of presence as witnessed by Benjamin—an exchange between soft memories of things and relations lost, and a culture of perpetual change.

Even the most perfect reproduction of a work of art is lacking in one element: its presence in time and space, its unique existence at the place where it happens to be.[5]

From this, it is possible to claim that art and art criticism in the twentieth century develop the notion of a momentarily "optimized" action reality, a "now" that only belongs to the individual artwork referring to itself as "a reality" in its own right, deciding what is authentic or not.

This "optimized" action reality is commonly termed the "autonomous" space of the artwork. The idea of "autonomy" worked within modernist and avant-gardist artistic practices for a while, as a way to claim authenticity to their work, which did not adhere to "aura" or a cultural industry (at least not intendedly so), but created a new and more free action-space for art.

However, according to the Danish philosopher Jørgen Dehs, for academic theoretical reasons and because politics did not replace the ritual as authenticator of the future as Benjamin predicted,[6] art today is denied autonomy.[7] Thus, artists in a "postcontemporary" landscape are left with the task of finding new ways of reclaiming authenticity; one way to do that according to Dehs is to reclaim an open acting space, albeit in a very different configuration than the one found in the autonomous artwork (and I have no room to go further into this here).

Another way is the one that I am claiming Kurenniemi is paving, claiming authenticity as a probable feature of a future reality—and not only for art. I claim that the media archive is an attempt to reconstruct a real biotechnological "live" consciousness in the future. Thus, Kurenniemi's archiving strategy seems to be borrowing heavily from ideas concerned with human–computer interaction (HCI) and postcybernetics.

A Postcybernetic Condition

In *The Cybernetic Brain* Andrew Pickering surveys the landscape of ideas, specifically from the British cybernetic scene, that according to him is drawing up the cybernetic imagination today. According to Pickering,

> The history of cybernetics is something that provides us with "an imaginative model of open-ended experimentation in stark opposition to the modern urge to achieve domination over nature and each other."[8]

First-order cybernetics, following the work of Turing, Veblen, and Wiener,[9] focused on the study of HCI as dynamic and complex systems designed around computed, dynamic probabilities of behavior based on feedback. Erkki Kurenniemi learned to program at a young age, and the issues of cybernetics and HCI were familiar to him. Kurenniemi's interest is in art's involvement in the key scientific problems in science, including ontology and, as Pickering explains,

> questions of what the world is like, including ourselves, human beings. …
>
> The history of science shows us that we actually live in a non-Cartesian, non-dualist world—an endlessly lively world of flux and becoming, that we are just a part of, linked into it at the level of performance, never reliably in control, ourselves part of the flow. That is the nonmodern ontological vision, which has emerged in science and technology studies.[10]

Pickering's words implicitly also address the very core of Kurenniemi's media archive: it is nonmodern, nonhistoric, and, moreover, it is a way to consider another future for the human being "in the flow" of technology and the computer.

Pickering navigates cybernetics out of the military industrial complex and into a much more succinct and futuristic discourse that relies heavily on artistic practices and methodologies. This outlines what I call a *postcybernetic condition*: it is exactly this condition that Kurenniemi's archive is living by/was produced for. Let me take a brief look at this postcybernetic condition.

In the 1960s, the epistemology of a new science was formulated, which was then termed a *second-order Cybernetics* (a phrase coined by Von Foerster). The idea of the "cybernetics of cybernetics" focuses on the systems that "surround "the dynamic processes and feedback systems—and what is controlling those systems?

This second-order epistemology brings together philosophical relativism and probability. It connects several related propositions; what becomes the focus of inquiry is the dynamic process by which the transfer of information among machines and/or humans alters behavior at the systems level (here paraphrased from Ed Shanken):[11]

1. Phenomena are fundamentally contingent.
2. The behavior of systems can be determined probabilistically.
3. Animals and machines function in quite similar ways with regard to the transfer of information, so a unified theory of this process can be articulated.
4. The behavior of humans and machines can be automated and controlled by regulating the transfer of information.

A quick survey of selected early cybernetic art represents elements of modern technological phenomena. In the 1960s cybernetics is absorbed in popular culture and art begins working with systems on different levels, and becoming systems themselves. Influenced by second-order cybernetics, artists such as Nam June Paik work on a double-focus on art and Buddhism; Stockhausen works on electronic feedback sound systems; Steina and Woody Vasulka investigate feedback systems in video. All of these artists, different as they are and mentioned as selected examples, share a systemic quality and method in their artistic practice; and they all experiment within and with the system.

> The bridge between art and cybernetics had to be constructed by creating metaphorical parallels. … The application of cybernetics to artistic concerns depended on the desire and ability of artists to draw conceptual correspondences that joined the scientific discipline with contemporary aesthetic discourse. … The merging of cybernetics and art must be understood in the context of on-going aesthetic experiments with duration, movement, and process.[12]

Thus, in this movement of ideas, artists play an important role in making diverse fields feed into each other and thereby actively facilitating those ideas being concretized and historicized into cultural configurations; in this way, the discourses of science and technology coming out of second-order cybernetics facilitate and radically transform artistic production.

But what is perhaps less clear in all this, and what leads to a postcybernetic condition, and what I am claiming in this essay, is that parallel to this movement of ideas emerges a discourse about future authenticity. It is not a discourse about solid states and frozen moments, the unique object and true soul, which in the epistemology of second-order cybernetics is deemed unauthentic; it is a discourse of the future of technology and art, where the submedia future-space of the media archive is played out for real. It is this discourse in particular that Kurenniemi's archive is feeding into, messaging the imagination with metaphors and other figurative representations of a future postcybernetic "authentic" consciousness.

The human life is not reducible to files or bits, yet in Kurenniemi's archive it seems we are reduced to authenticate ourselves by a variety of technological "stylistics" or "formats." Even so, he records and records, incessantly, throughout his active career as scientist and artist, on all accessible platforms and in mainstream and obscure analog or digital formats. The implicit human–computer or human–machine interfaces, which Kurenniemi develops and unfolds incessantly throughout his work with the archive, are not focused on creating interfaces to his biography, which seems irrelevant to him; they are focused on how relations may occur and develop between digital and human phenomenological systems; and how they, in turn, react to one another based on implicit patterns. His goal is, I would claim, an interface to a future human-technological existence.

The issue about future human-technological existence is important to stress because it points toward a completely different way of looking at the world as a nonintended, probabilistic set of possible interactions. It is a symptom of a search for future authenticity.

Symptoms of Future Authenticity

Kurenniemi's archive contains about ninety hours of audio tape recordings that appear to be private. Intercourses, masturbation, strange private conversations, meals, or car drives. On the surface these appear not to be intended for the public. But they become public through the archive. Is it possible they are symptoms of a search for a future biotechnological construction of "Kurenniemi," which is more "real" than the present human (collecting) version?

The recordings in the media-archive are made on a great variety of machines that encompass decades of technology in hardware and software, making them hard to access today—in some cases even inaccessible. The media archive is as much about archiving recordings as it is about storing machines. And the latter, the configuration of the storing of the machines, is also configuring our future access to the archive. The machines, in some intricate ways, control the output and organization of the archive. Because we need to interpret and "rewrite" both hardware and software in a restoration process to get access, the relation between "original" and "future copy" is being reconfigured in that very same process.

What, we may ask, is the purpose of this reconfiguration of the media in the media archive?

In a recent online publication on Kurenniemi, Jussi Parikka argues that Kurenniemi has a tendency to oscillate wildly between technological thinking and doing, and that this constant "movement" in his practice points toward a general methodology "in which an interesting tension between the way in which Kurenniemi constructs his discursive position and his expertise in technological practices" emerges.[13] Thus, Kurenniemi, according to Parikka, is a *symptomatologist* (to use the term of Deleuze)—one whose artistic practice is followed closely and intricately by a violent outbreak of "archive fever." Following Derrida's notion, archives are to be seen as "modern desires," which we cannot completely control. By operating subconsciously, archives feed into our ways of understanding and perceiving our cultural realities as "real." Kurenniemi's media archives are bringing testimony to this pattern and to the ways art and technology interfere in our culture, albeit through some unconventional methods and ideas.

Susanna Paasonen develops this point further, arguing that "the overall horizontality of the archive can be explained through the principles and aims of the documentation, which is premised on the inseparability of perception and consciousness."[14] She further quotes Kurenniemi, who "wanted to record his consciousness for future access (and sharing) by recording the flow of his everyday life: memories, logs and documentations became data to be stored." The critical point concerning authenticity, then, becomes the matter of future access to, and processing of, the *stored* data, not memory (or "history"). According to Kurenniemi (here, again, quoted from Paasonen), in order to access the database, "the algorithmic structure of one's persona would need to be mapped and relocated, or installed on some computational substrate such as a computer or brain tissue grown from stem cells."[15]

However, I am arguing that a different level of the archive, and of authenticity, emerges from the ideas and practices of Kurenniemi. The "archive fever" is not only about storing data and reconstructing a memory space in the future that matches the old one. It is a symptom of a submedia level in the media archive that, I would argue, combines computational calculations and artistic executions with biological processes.

From this emerges a tension between the status of (the technologies of) memory and stored material, which seems to run intertwine with a postcybernetic archiving strategy. What, we may well ask, constitutes the concept and use of authenticity in this postcybernetic condition?

Media Archives of Probable Futures

There is a scene from Mika Taanila's film *The Future Is Not What It Used to Be* (2002) where we see Kurenniemi sitting in front of a computer explaining his archiving project. It appears that his intention is to collect as much as possible for future purposes—because, as stated in the film, in 2048, a quantum computer will be able to do what no man can: put it all together and create a "new" Erkki Kurenniemi based on the archive. Thus, the first thing to note is that Kurenniemi's archive is not about what happened in the past, but about what might happen in the future. Second, the archive is involved in a dialogue with a machine, a computer, which does not yet exist. The point is that it *probably* will exist at some point—Kurenniemi assumes that the probability for it to appear in the future (in 2048) is high. So, third, it is this field of probability that he is playing with the media archive, like a game with a future life that is not a recreation of his biological body and not determined by a certain logic goal.

However, this totally different kind of "memory" of "life" and "human existence" he is aiming at is not without constraints; creating the outline of an interface to future "posthuman" authenticity is a game that involves the cross-over of different techno-philosophical discussions—not least those that are inspired by cybernetic philosophical ideas of automata and the possibility of generating "life" from even a few elements of "stored" information.

As the very issues of "memory," "life," "authenticity," "game," "storage," and "archive" are each huge enough to fill several chapters, I will attempt to position those issues as concretely as possible inside the vernacular of Kurenniemi himself—which is to say, as derived from the symptoms of his media archive.

In 1944, John von Neumann and Oskar Morgenstern used the term "memory" for the first time in relation to the computer,[16] borrowing a term from biology to describe something, which is essentially not biological. As recent research indicates, the metaphorical connection between computers and humans was misleading our understanding of computers and the human brain for decades.[17] However, before von Neumann conflated the computer with the biological metaphor of memory, computing devices were envisioned as "storage" devices.[18] This denotes a salient difference in the way the human brain and the computer is understood; whereas memory is a psychological entity often known to be *fallible* (even potentially destructive) that points backward in both time and logic, "storage" connotes something more "reliable," which is pointing toward a future and future uses.

A store—like an archive—is both what is stored and its location. Stores look toward a future: we put something in storage in order to use it again; we buy things in stores in order to use them. This erasure of the difference between memory and storage thus erases the difference between the past and the future, positing digital media an ever-increasing archive in which no piece of data is lost. Memory-as-storage hardens information—turning it from a measure of possibility into a "thing"—while also erasing the difference between instruction and data and the vicissitudes of execution.[19]

The central premise of Kurenniemi's media archive seems to be formulated around a kind of "conflict" or debate between "memory" and "stored" data as sources of "real life" and a "consciousness." As Chun is also pointing out, there is an "enduring ephemerality" at play in this core issue of human–computer interaction.[20]

Both terms show the complexity of matters at play in the media archive, and further tests in the favor of a future perspective. How do we ever get a clear grip on what is "real" or what is "life"? That kind of "godly" insight and power is not available to anyone; we only have the traces of patterns and symptoms of what we leave behind to build our case on. The question then becomes, how to interpret those as "real"? Are they real because they are true to some inner processes, which make them authentic to the individual subject, as some schools in modern psychology would claim (i.e., Erich Fromm)? Or, are they real because they converge with outer, cultural processes, where authenticity emerges in the transsubjective construction of the real (e.g., Hal Foster)?

Kurenniemi takes a different position than the above, it seems. His position develops apparently in a (playful) dialogue with cybernetics and human–computer philosophy. That is to say—the "construction" of the real moves from a question of authenticity based on psychologically and culturally motivated inner or outer processes to one based on the relation between determinism and chance. To reach authenticity, the media archive of Kurenniemi drifts into a gray zone between mediated and documented biological memory (of the past) and machine storage (for future uses).

Storing could be seen as an activity that makes a promise: that you will be "gaining" something, which would otherwise not be there. Also, by inference, the very activity of storing and looking toward the future is implying that the "now" is somehow (experienced as) faulty and lacking, and based on a flawed definition of "authenticity." This leads to the interesting observation that, seen from the perspective of a nondeterministic ontology, authenticity will only be possible in a future perspective. It is something that may occur if the right configuration of mediated memory archive and machine storage is achieved on a technological future stage.

Interfaces of Future Authenticity

I have proposed that we should see Kurenniemi's active collecting and storing large quantities of diverse material in as diverse media formats as a way to make a future authenticity probable. With Pickering, that scope was widened, and we may now understand that effort as coming from a postcybernetic epistemology where the issue is not to maintain an outdated (modern and dualist) worldview and to find new ways for humans to interface with the future that are nondeterministic. There is no doubt that Kurenniemi is influenced by second-order cybernetics; but he interprets this influence in his own, unique way.

Up until now, I have looked into the transient area of authenticity in the field of media archive and machine storage; I have placed and discussed the peculiar practices in the auto-archiving of Erkki Kurenniemi in a postcybernetic perspective. All this raises the question: Did he use his archive to achieve a future authenticity of art or, for instance, human life? Who decides who will control the production of the archive into a feedback system?

I propose that the activity of Erkki Kurenniemi, and his media-archiving project, is part of a methodology that allows him to define and operate in a future acting space, a future-space, emerging in between chance and deterministic systems on the one hand, and between memory and storage, on the other; a position that works against the very notion of contemporary artifacts as a singular positions only existing in one particular time and space.

This is quite evident from the interview with him in the film, where he explains—and the film further illustrates—how all the fragmented parts of media recordings (video, sound), diaries, and data storage devices are bits and pieces to be collected and reassembled in the future—to produce the *real* Erkki. The use of the word "real" in Kurenniemi's self-narration is striking. But in the context of authenticity as an open acting space, it begins to come together in a—literally quite logically motivated—conversation with his future self; or at least with a "different self" than the Kurenniemi from 1968, 1971, or any other year. The date is not important—because it is not an archive about something, explaining the life of a man. It is a model kit for a future man, perhaps, but not of an existing one. The real, authentic Kurenniemi is a speculative, futuristic "self" that does not yet exist—and only probably will. Yet, this futuristic "self" is somehow more "real" than the present one.

This becomes even more evident when we go further into the methods necessary to be able to discuss or operate with the concept of authenticity as more than "just" a theoretical concept. It is, as Dehs also argues, necessary to use "quite ordinary, familiar experience types" to be able to get near the situation of an open acting space and analyze it from that perspective. This is because "their pattern greatly implies that it is not myself (the experiencing) that sets the stage."[21] However, it is experiences such as to face "something that is consistent with itself," or "something that in an unpredictable manner is as it should be," or "something that it was not intended that anyone should see (or hear)."[22]

Kurenniemi is playing with the role of "self" in authenticity, bringing into question the relationship between the archive, the storage, and our perception of what is "real." One cannot simply call these familiar things from the everyday life of Kurenniemi an experience of authenticity, "but probably experiences of various symptoms of the fact that it is not me who sets the stage."[23]

The content and configuration of the media archive should be seen as symptoms; symptoms of the fact that it is not Kurenniemi who sets the stage and decides what is authentic. However, they are also symptoms of his intending that it was not himself who should set the stage. It is marking out a nondeterministic, futuristic ontology.

It may be stated, then, that the archives of Kurenniemi especially highlight the yet unwritten parts of human–computer interaction. They constitute a kind of postcybernetic view on his time and himself, which could be considered as an implicit contribution to posthuman themes as well as "body art."

In effect, what is apparent is that Kurenniemi reconfigures art production, any production, in fact, as well as the archive in general in such a way that it transforms into a representational system of bio-inputs and bio-outputs of "human life" as we do not (yet) know it. It is this future ontological status of human life that he attempts to compute into a reactive interface—a new media that involves the experience of the artist and the experiences of the possible future viewers in an active communication on the edge between past and future values of authenticity. Thus, in Kurenniemi's media archive the future playable field between storage and technology is expanding—into nonsensory, unwritten, ubiquitous fields and systems of (post)human life; and emerging from it are probable human–computer interfaces of future authenticity.

Notes

1. Walter Benjamin, *The Work of Art in the Age of Mechanical Reproduction* (New York: Penguin Great Ideas), 3.

2. Ibid.

3. Ibid., 5.

4. Matei Calinescu, *Five Faces of Modernity: Modernism, Avant-Garde, Decadence, Kitsch, Postmodernism* (Durham: Duke University Press), 12.

5. Benjamin, *The Work of Art in the Age of Mechanical Reproduction*, 7.

6. Ibid.

7. Jørgen Dehs, *Det autentiske: Fortællinger om nutidens kunstbegreb* (Copenhagen: Edition Vandkunsten), 14. (My translation.)

8. Andrew Pickering, *The Cybernetic Brain: Sketches of Another Future* (Chicago: University of Chicago Press, 2010), 23.

9. Norbert Wiener, *Cybernetics, or Control and Communication in the Animal and the Machine* (Cambridge, MA: MIT Press, 1948), 17–18.

10. Andrew Pickering, *The Mangle of Practice: Time, Agency, and Science* (Chicago: University of Chicago Press, 1995), 67.

11. Edward A. Shanken, "From Cybernetics to Telematics: The Art, Pedagogy, and Theory of Roy Ascott," in *Telematic Embrace: Visionary Theories of Art, Technology, and Consciousness*, ed. Edward A. Shanken (Berkeley: University of California Press, 2003), 19.

12. Ibid., 21.

13. Jussi Parikka, "DIY Futurology: Kurenniemi's Signal Based Cosmology," in *Erkki Kurenniemi—A Man from the Future*, ed. Maritta Mellais (Helsinki: Finnish National Gallery, Central Art Archives, 2013), 17, http://www.lahteilla.fi/kurenniemi/fi/a-man-from-the-future.

14. Susanna Paasonen, "Slimy Traces: Memory, Technology and the Archive," in *Erkki Kurenniemi—A Man from the Future*, 8.

15. Erkki Kurenniemi, quoted in *Erkki Kurenniemi—A Man from the Future*, 8.

16. Oskar Morgenstern and John von Neumann, *Theory of Games and Economic Behavior* (Princeton: Princeton University Press, 1944), quoted in Wendy Hui Kyong Chun, *Programmed Visions: Software and Memory* (Cambridge, MA: MIT Press, 2011), 36:

> There is a good deal of evidence that memory is static, inerasable, resulting from an irreversible change. (This is of course the very opposite of a "reverberating," dynamic, erasable memory.) Isn't there some physical evidence for this? If this is correct, then no memory, once acquired, can be truly forgotten. Once a memory-storage place is occupied, it is occupied forever, the memory capacity that it represents is lost; it will never be possible to store anything else there. What appears as forgetting is then not true forgetting, but merely the removal of that particular memory-storage region from a condition of rapid and easy availability to one of lower availability. It is not like the destruction of a system of files, but rather like the removal of a filing cabinet into the cellar. Indeed, this process in many cases seems to be reversible. Various situations may bring the "filing cabinet" up from the "cellar" and make it rapidly and easily available again.

17. Chun, *Programmed Visions*, 47.

18. Cf. Charles Babbage, "The Store," and J. P. Eckert, "Disclosure of Magnetic Calculating Machine," quoted in Chun, *Programmed Visions*, 8.

19. Chun, *Programmed Visions*, 47.

20. Ibid.

21. Dehs, *Det autentiske*, 22. (My translation.)

22. Ibid., 23.

23. Ibid.

Bibliography

Walter, Benjamin. *The Work of Art in the Age of Mechanical Reproduction*. New York: Penguin Great Ideas, 1987.

Calinescu, Matei. *Five Faces of Modernity: Modernism, Avant-Garde, Decadence, Kitsch, Postmodernism*. Durham: Duke University Press, 1987.

Chun, Wendy Hui Kyong. *Programmed Visions: Software and Memory*. Cambridge, MA: MIT Press, 2011.

Dehs, Jørgen. *Det autentiske: Fortællinger om nutidens kunstbegreb*. Copenhagen: Edition Vandkunsten, 2012.

Morgenstern, Oskar, and John von Neumann. *Theory of Games and Economic Behavior*. Princeton University Press, 1944.

Paasonen, Susanna. "Slimy Traces: Memory, Technology and the Archive." In *Erkki Kurenniemi——A Man from the Future*, ed. Maritta Mellais. Helsinki: Finnish National Gallery, Central Art Archives, 2013, Available at http://www.lahteilla.fi/kurenniemi/fi/a-man-from-the-future, accessed January 28, 2014.

Parikka, Jussi. "DIY Futurology: Kurenniemi's Signal Based Cosmology." In Erkki *Kurenniemi—A Man from the Future*, ed. Maritta Mellais. Helsinki: Finnish National Gallery, Central Art Archives, 2013. Available at http://www.lahteilla.fi/kurenniemi/fi/a-man-from-the-future, accessed January 28, 2014.

Pickering, Andrew. *The Mangle of Practice: Time, Agency, and Science*. Chicago: University of Chicago Press, 1995.

Pickering, Andrew. *The Cybernetic Brain: Sketches of Another Future*. Chicago: University of Chicago Press, 2010.

Shanken, Edward A. "From Cybernetics to Telematics: The Art, Pedagogy, and Theory of Roy Ascott." In *Telematic Embrace: Visionary Theories of Art, Technology, and Consciousness*, ed. Edward A. Shanken, 1–97. Berkeley: University of California Press, 2003.

Wiener, Norbert. *Cybernetics, or Control and Communication in the Animal and the Machine*. Cambridge, MA: MIT Press, 1948.

14 E-Kurenniemics: Becoming Archive in Electronic Devices

Wolfgang Ernst

Kurenniemi-as-Archive: Resistance against Hallucinations of "Life"

Is Erkki Kurenniemi a Ulysses of the electronic age, still waiting for his Homer to become immortal in technocultural memory?[1] Kurenniemi is "an unsung" pioneer of electronic art, the endorsement of the DVD *The Dawn of DIMI* articulates: "Viewed from a historical perspective, Kurenniemi's music foretold digital directions in rhythm, noise and jumpcut editing, only back then no-one was listening."[2] There is a nonhistorically delayed listening and understanding of such knowledge, a kind of technoarchival phase shifting. Can one archive a life in real time when the protagonist is still alive? Kurenniemi (like Johann Wolfgang von Goethe's prestructuring of his own archive to become an enduring "classic") is known for having created his own idiosyncratic electronic life archive to be posthumously released to the public. But in order to resist the (auto)biographical impulse, the relational vector between the author and his works has to shift from the composer to the composed, from engineer to the electronic artifact, from alphabetic writing to the computer program.

A special case of archivization arises with the recording of visual movements and acoustic voices, which are affectively associated with "life" itself. Such ephemeral articulations of the instant moment have become almost immortal by media of signal recording like photography, phonograph, film, and magnetic tape. The relationship between the classical archive as an institution of frozen symbolic memory and technological storage is just a metaphorical one. Signal recording actually allows for addressing the dead as something alive. The humanist desire to speak with the dead is the phantasma of a rhetorical feedback channel between past and present. While the classical archive that consists of alphabetic records remains silent, recorded audiovisual signals evoke a different interaction from records from the past: "What remains of people is what media can store and communicate."[3] A well-known trope still governs our cultural discourse on archives, that is: hallucinating speech and life in what is actually mute and mechanically dead. In signal-recording devices, the ancient rhetorical figure of *prosopopeia* returns. Exorcizing such archival phantasms goes with the media-archeological aesthetics and ethics of research.

Instead of historicist hermeneutics, the media archeologist suppresses the passion to hallucinate "life" when he listens to recorded sound, voices, or noise. To resist the biographical impulse, media archeology rather analyzes the subliminal reality below and beyond the sensual and cognitive thresholds of sight and sound—a level that is not directly accessible to human senses because of its subliminal physiological nature and its electronic and calculating speed. A spectrographic image of a tiny section of previously recorded bit of speech provides a direct look into a different kind of archive. For logocentric humans it is difficult to resist auditory hallucinations; therefore, a close reading of its underlying technological apparatus is mandatory. With a techno-physical close reading, the materiality of the recording medium itself turns out to be poetical, dissolving any semantically meaningful archival unit into discrete blocks of micro-memory.

Articulations of living beings are recorded by technological media either as analog signals (physically) or by symbolically sampled data. The very act of analog-to-digital conversion is already an essentially archivographical act, prior to all subsequent classifications. "Live"—which is the discursive term for time-continuous processes—thus transforms into time-discrete data series, mapping live into a micro-archival order. Will the electronics developer, music theorist, and computer expert Erkki Kurenniemi thus become digitally *im-mediate*?

From 1972 to 1974, Kurenniemi recorded diary entries on tape cassettes, just as Samuel Beckett's one-act drama *Krapp's Last Tape* from 1958 has the protagonist keep a phonetic diary on magnetophone. The last remark in Beckett's script declares: "Tape runs on in silence"—an endless loop, against which engineers developed the auto-stop. From the early 1980s onward, Kurenniemi also kept a constant video and photograph log of his surroundings and personal events with the aim of producing material to be digitally compiled some time after his death when information technology would be able to negentropically reconstruct life from data.

By definition, only what is actively set apart from the present in administrative terms can be called an archive, just like historicizing means narrative temporalizing, demanding a proper beginning and a proper end. In terms of systems theory, archiving the present in real time is impossible—unless the micro-archive is defined in its time-discrete sample-and-hold procedures of digital presence. Kurenniemi documented his life, but in contrast to Goethe he did not conceive this to "govern" his future memory in anticipatory ways as archival future-in-the-past. The data files are rather loosely coupled and thus become a real "medium" of memory.[4] How to cope with such an abundant mass of audiovisual and textual data in terms of an open archive with multivariant access, multiple interfaces, no filtering by metadata, no unifying index, not reducing the raw data to taxonomy, with no tagging for grouping and retrieval?

Kurenniemi attends to the quantum computer to make sense of it all. But a different kind of memory emerges when not primarily the personal writings and recordings of an individual is considered as his archive but the devices he constructed. Some of Kurenniemi's digital music instruments (DIMIs) today are kept in the collection of the Finnish National Gallery; others are deposited in the Electronic Music Studio of the University of Helsinki Musicology Department, where they are intended to be made accessible for replay. Historians traditionally use documents rather than technological artifacts in their archival reconstructions of past lives.[5] Let us not miss this opportunity to note a different kind of "source" in the case of Kurenniemi's synthesizers. To keep them as an active archive, in contrast with textual and audiovisual data that can always be read, heard and seen, these pieces of electronic hardware themselves must be kept running in order to prevent their simple historicizing and musealization. When an electronic image from a cathode ray tube successfully transmits an event, the fabrication year of the device does not really matter. Only the switched-off TV is a historical object. When a technical device from the past is functionally operative, it is beyond history; its historicity is of a different media-archeological kind.

Operative Archaeo(porno)graphy and Knowledge

Instead of a verbal reduplication of Kurenniemi's artistic work, which is genuinely intellectual already, let us rather follow the bifurcative "Y" model. Both academic media theory and media-artistic research art have roots in technological preconditions (the media apparatus and the *dispositif*) from which one branch arises as art-aesthetic research and another branch as its academic analysis. Both branches articulate themselves in parallel ways. Supposing this model of communicating branches, let us develop theoretical models instead of discursive comments on Kurenniemi's technological works that actually resonate with them[6]—even at the risk of creating an echo of enunciations that were never intended. Media analysis points out the epistemologically implicit value in technological objects created by idiosyncratic techno-thinkers (and tinkerers) like Kurenniemi that deserves to be expressed.

It has been said that "the evolution of Kurenniemi's instrument designs may be regarded as a logical and gradual transition from analog to 'quasi-digital' to digital electronics."[7] But if we somewhat deconstruct what looks like a smooth transition in terms of the linear history of technology,[8] Kurenniemi himself turns out to be a moving and dynamic focus, a kind of flying spot scanner, a *medium* of both current technopoetical discourses of his epoch and a context-independent, autopoetical Romantic subject. The individuality of Kurenniemi's works needs an explanation that cannot be reduced to a function of its contemporary electronics but is also a function of biographical contingencies, anecdotes, obsessions, and dramas that build up to the forces driving research energies. Such forces are individual idiosyncrasies to be preserved in Kurenniemi's life archive. Parallel to his abundant pornographic recordings runs the transitive relation Kurenniemi has with naked electronics and mathematics. Kurenniemi does not perform in a media theater; he *is* media theater himself, acting the (ob)scene of technologies of the self. Not simply subjected to Latourian nonhuman agencies, he always remains a protagonist, driven by impulses that are in the best sense idiosyncratically anarchival. To describe this, proper archaeo(porno)graphy is required.

Philological hermeneutics looks at an author's collected works, while materialist discourse analysis investigates the Kantian a priori, the conditions that made such articulations possible at all, the rules that make their appearances and articulations. This *dispositif* is technomathematical when it comes to digital media culture. Kurenniemi defined this Foucauldian archive himself, by changing the rules—that is: creating his own electronic circuits and equivalently his own mathematical theory of musical harmonics. Technologically the digital archive represents a dynamic format which keeps its records in updated migration. Algorithmic operations at the speed of electricity transform archival memory implicitly into multiple temporalities that are traditionally known from musical and sonic remembrance. This is part of a textuality that cannot be expressed by narrative historiography but only by the diagrammatic archive itself.[9] Even if in a Heideggerean sense the essence of technology does not reveal itself directly, what can be revealed is the coupling between technological elements.[10] Kurenniemi defines himself as media-theatrical actor of techno-aesthetic knowledge, as expressed by the Active Archives project *Erkki Kurenniemi (In 2048)*: Intermediaries function as detectors of relationships.[11]

One paradigm of electronic music as composed in studios (such as Pierre Schaeffer's *musique concrète*) has been tape recording and manual tape editing; some composers experimented with fully electronic sound generation, haptic interfaces, and motion control. In that sense, Kurenniemi has not been unique; but with his obsessive inquiry into the relation between sound and mathematics he still belongs to the earliest protagonists of digital sound-control technologies. The construction of such instruments results not only from his fascination with electronic music but also from the implicit musicality in technology itself—both materially (*techné*) and mathematically (*logos*). Kurenniemi defines music strictly mathematically—thus closer to a structural archeology of knowledge than to aesthetics.[12] He turns out to be a true Pythagorean: for him, all the tonal structures in music result from number-theoretical coincidences. This allows him to visualize complex harmonic structures by diagrams that become operative when embodied as electronic circuits. In its almost time-invariant knowledge sphere, Kurenniemi's musico-mathematical theory correlates with the logical circuits devised for his early electroacoustic instruments. Underneath the simple intersection between physiological hearing and electroacoustics, a deeper *sonic* dimension is implicit in Kurenniemi's electronic devices.

K-U-R-E-N-N-I-E-M-I, or How Does One Become Archive?

Kurenniemi's obsessive self-recording, starting with his audio cassette tapes and video diaries, has been meant to result in an algorithmically orchestrated, posthumous digital reincarnation of his life on July 10, 2048 (his 107th birthday). But the real archive of a techno-visionary is to concentrate on his electronic devices, which are not body- but mind-related embodiments of his thoughts—diagrammatic in both operative (hard wiring) and archival forms. The challenge to archive Erkki Kurenniemi's electronic synthesizers is not just a museological one but is correlated with a shift in emphasis regarding what is considered an archive. Kurenniemi's synthesizers *are* archives themselves already, being the material embodiment of symbolical machines: blueprints and diagrams, just like his design for new musical scores.

While the symbolical order refers to the archive, the actually operative electrotechnological media inscribe into cultural memory by processing signals instead of symbols. Therefore, it is mandatory to record (and sample) what Kurenniemi's synthesizers actually sound like, preserving the articulation of the medium as sonic memory worthy of tradition. Any "historic" recording of music produced on such devices, in fact, is at the same time a technological witness.[13] It is not sufficient to archive the circuit diagrams of electronic devices; we must preserve their actual signal operations as well. The term *operative diagrammatics* refers to the specific media-archeological way of literally understanding technology: materiality and algorithms in action. The diagram in its media-technical sense refers to information patterns, circuits, and relations that give an idea of how complex machines actually work.[14] That is exactly what Kurenniemi's film *Electronics in the World of Tomorrow* (1964) shows:[15] a labyrinth of electric cables combined with close-ups of prefabricated and integrated circuits, ranging from analog to digital. This film reveals Kurenniemi's closeness to the technological device. The film itself is interlaced with light patterns generated by the very same machine that is the visual object of the film: the media-archeological insight of the DIMI synthesizer as a self-referential, truly media-dramatic argument.

The concept of operative diagrammatics relies heavily on regarding technical apparatuses through the approach of an engineer as specialist in the diagrammatics of circuits.[16] Thus the diagram becomes what could be described as a "literal crossing-point between epistemologically wired humanities analysis of technical media and the engineering enabled understanding of and tinkering with operationally."[17] It is through the diagram that time-criticality is being executed on the various micro-levels of technology; humans operate through the diagrams of the machine. The operative diagram is not simply something inserted into the machine, but is what truly generates it through its operation—the machine in motion, understood from the micro-temporal point of view.

The Incorporation of Media Memory

The encounter between human and machine is the central drama of modernist life. Even after death this drama continues, with reverse temporal operators. Human memory becomes a function of storage mechanisms itself[18]—the driving idea of Kurenniemi's archival obsession, which relies on a future artificial intelligence algorithm to be resurrected as a virtual existence. Such a kind of reverse engineering is the plot in Lynn Hershman Leeson's film *Conceiving Ada* (1997).

But Kurenniemi's memory and afterlife is already implemented in the wiring, diagrams, and programs of his electroacoustic instruments. Taking into account Kurenniemi's own obsession with automation (e.g., his development of robots), we need to focus not only on the archival but also his machinic memory: his electronic artifacts. So let us not simply turn Kurenniemi's live into a multimedia database and transform Kurenniemi into a virtual databody; let us remember his hardware embodiments as well. Each time a sonic articulation is recreated by a DIMI synthesizer, Kurenniemi is alive, owing to the special power of electronic ways of "re-presencing."[19] The same is true for the retroactivation of IMIO (1971) with its central feature: an optical interface, the original purpose of which was to read sheet music graphically (comparable to Iannis Xenakis's UPIC system). DIMI-O can also be used in tandem with a dancer whose movements are simultaneously transformed into music. Such instruments are truly media-archeological not only in the temporal sense of being the "earliest electronic musical instruments,"[20] but in their archaism, reducing the technological form to the essential hardware function as prototypical aesthetics. Kurenniemi's electronic devices demand engineering skills from the musician to operate them. As experimental prototypes they do not hide the electrical circuits but rather present these themselves as the media-transitive user interface. Here, the technoarchive is opened for access. Kurenniemi's electronic instruments reflect the technical functionality at the hardware level, which is the media-archeological layer. Against the temptation of interfacial metaphors, Kurenniemi severely resisted applying conventional control interfaces like a keyboard in favor of the pure doctrine of electronic synthesizer access. Only Kurenniemi's DIMI-O, which uses a video camera for opto-phonetic signal input, allows for a keyboard interface. The device includes a memory unit with a 32-step sequencer, with the memory locations presented on a television screen from which the player can read the contents of the memory while a cursor on the screen tells which location is being played at any given instant. Even if Kurenniemi put some thought into the user interface design and usability, his media-archeological aesthetics expresses the *dis*continuity that takes place in conventional versus electronic music instruments, instead of hiding this rupture.[21]

Kurenniemi designed instruments based on biofeedback reflecting the cybernetic paradigm, such as the DIMI-S (known as the Sexophone, 1972), "where sound generation is based on the electric conductivity of the skin, and DIMI-T (a.k.a. Electroencephalophone, 1973) where the sound control is based on a signal generated by the electric activity of the brain."[22] These fabrications are electrophysiologically and electronically based on signal processing, not on the symbolic order of the archive.

Brothers in Mind: Hunting for Kittler's "Harmonizer"

The traditional task of the archive is to preserve textual records that are worthy of knowledge, filtering out the redundant material. But the archives of the future are confronted with symbolic and material records of a different kind: alphanumeric codes and electronics. One case similar to Kurenniemi's challenge is that of Friedrich Kittler's "papers" in the National Archive of Literature (Deutsches Literaturarchiv) in Marbach, consisting not just of unpublished manuscripts but also his self-constructed electronic synthesizer (and voice harmonizer) from the 1970s as well as his code written in Assembly on the hard disk of his personal computer. These must be kept running—which cannot be performed by a book, which documents but cannot compile source code.[23] Among Kittler's written source codes "recorded" on his personal computer hard drive is one for generating a simple sinus tone; like a musical score, this can be "interpreted" by a computer at any time as long as the computing environment is being preserved as well.

One option for Kurenniemi's DIMIs is to create a library of tonal grains and samples from the original instruments, by digitally sampling their real sonic articulations instead of simply emulating them. We are reminded of Thomas Wilfred's visual music known as *lumina* and his color organ Clavilux. Wilfred, like Kurenniemi, has been musician, composer, engineer, and inventor; he explicitly objected to being remembered only through recording performances on film, making the survival of his works rather dependent on the material existence of his machines.[24] Whereas we reactivate written archival records by literally reading them and subsequently transforming them into historiography, a technological artifact needs to be operative to be understood as a medium of cultural tradition.

Like Kurenniemi, Kittler experienced the epistemological rupture between analogue and digital electronics in a literally transitional device: the analog-to-digital converter. A voice transposer that does not simply want to produce the Mickey Mouse effect by speeding up tape recordings of a voice must contain a microprocessor. While the phonograph is incapable of achieving real-time frequency shifts, only electronic harmonizers are able to reverse—with considerable electronic effort—the inevitable speed changes, at least to deceivable human ears. All of a sudden, "women can be men and men can be women again."[25] Kittler constructed such a Pitch Shifter, but the object (a wooden case stuffed with microelectronics, attached to a crude computer) is well remembered by Kittler's first assistant Bernhard Siegert but has not been found since.

Within the digital archive, the difference between written papers and audio and video recordings is liquefied into streams of digitized numbers that translate any medium into one system of computing. But there are system-theoretical limits to the musealization of hard drives: the computer can archive all other media but not itself.[26] Kittler's modular synthesizer and his computer hard drive, when simply put onto an archival storage shelf, cannot be analyzed in the same way as a traditional archival record. They rather demand a kind of vivisection—which means not just deconstructing the electronics in an inert state, but to set it under currency, under voltage, in a running platform again: processual philology, an exegesis of Kittler's thoughts by circuit grammatology or rather operative diagrammatics. Similar to Kurenniemi's early film *Electronics in the World of Tomorrow*, Jan-Peter Sonntag audiovisually documented the conceptual "anatomy" of the late Friedrich Kittler's synthesizer modules from the early 1980s. Following Kittler's circuits and codes, operating with and against his intentions, Sonntag's art project *apparatus operandi* enacts a media-archeological argument with hardware close to the process, oriented at its performative essence.[27] The project rejects the simple museum object as obsolete and is more interested in procedural behavior—a critique of institutions like the archive that, in present memory culture, tend to render records inert. To outdate such passive archives, its signals and data must be mobilized in operative media memory. Written as it is in alphabetic symbols, a textual legacy can be reread as long as the present reader shares the code of the past author. But what happens when the archival records are not text but sound machines? Can this legacy resonate? When it comes to electroacoustically generated signals, the reanimation of a dormant modular synthesizer is crucial—either by constructing a replica for curatorial reasons (respecting the unchangeable "original" where not even a rotten condenser might be exchanged) or by reinforming the original hardware virtually, that is, algorithmically within programming platforms like SuperCollider. There are two bodies of archival memory: its material authority (to be kept intact) and its virtual "liveness."

Reactivating Kittler's modular synthesizer allows for the experience of the sonic uniqueness of the technical artifact—its temp*aural* individual articulation. We are dealing here with a nonhistorical mode of represencing: dynamic equitemporality, as known from the concept of performing ancient musical scores on original instruments. This is not "contextual" in the classical sense of historical research, which reduces the context to the ensemble of available textual records in the period's archive, but operative materiality of a continuous past.[28] Kurenniemi's as well as Kittler's technical embodiments are much more alive than any archival record.

Notes

1. This essay takes up arguments of my talk "Liquefying the Archive/Sonification of Memory" given at the University of Oslo on occasion of the workshop Life Archives, organized by Eivind

Røssaak and Ina Blom within the research project Archives in Motion, June 6–7, 2013; it is further connected to the research project Archiving Presence between Humboldt University Berlin and Hebrew University Jerusalem (Amit Pinchevski).

2. *The Dawn of DIMI*, DVD, ed. Mika Taanila (Helsinki: Kinotar Oy and Museum of Contemporary Art Kiasma, 2003). The text refers to a quote from *Wire*: A record review by Stephen Robinson in *Wire* 227 (January 2003), p. 67 of Kurenniemi's "Äänityksiä/Recordings 1963–1974."

3. Friedrich Kittler, *Gramophone—Film—Typewriter* (Stanford, CA: Stanford University Press, 1999), xl.

4. For a definition of "medium" as loosely coupled elements different from tightly coupled "things" (a definition that was updated by Niklas Luhmann's systems theory), see Fritz Heider, "Ding und Medium," *Symposion* 1, no. 2 (1927): 109–157.

5. See Christian Sichau, "Die Replikationsmethode: Zur Rekonstruktion historischer Experimente," in *Im Labor der Physikgeschichte: Zur Untersuchung historischer Experimentalpraxis*, ed. Peter Heering, Falk Rieß, and Christian Sichau (Oldenburg: Bibliotheks- und Informationssystem der Universität Oldenburg, 2000), 9–70, at 9.

6. "The function of a communicator is to achieve a state of resonance with the person receiving visual and auditory stimuli": Tony Schwartz, *The Responsive Chord* (New York: Anchor, 1974), 24.

7. Mikko Ojanen, Jari Suominen, Titti Kallio, and Kai Lassfolk, "Design Principles and User Interfaces of Erkki Kurenniemi's Electronic Musical Instruments of the 1960's and 1970's," in *Proceedings of the 2007 Conference on New Interfaces for Musical Expression* (New York, 2007), 88–93, at 93.

8. The human ear is already the most immediate analog-to-digital device, converting analog sound signals into discrete nerve impulses in coded time patterns (phase locking).

9. See Wolfgang Ernst, *Digital Memory and the Archive*, ed. Jussi Parikka (Minneapolis: University of Minnesota Press, 2013). For a diagrammatic structuring of the Kurenniemi archive, see "Graph for Kurenniemi," https://gitorious.org/kurenniemi/kurenniemi/graph/ae6530b807e73f81f4cce05f3b49b277f3f78bb9?page=1 (accessed December 5, 2013).

10. See Martin Heidegger, "Die Frage nach der Technik," in Heidegger, *Reden und Aufsätze*, 2nd ed. (Pfullingen: Neske, 1959), 13–44; see also Gilbert Simondon, *Du Mode d'Existence des Objets Techniques* (Paris: Aubier, 1958).

11. *Erkki Kurenniemi (In 2048)*. Constant Association for Art and Media, based in Brussels, started their Active Archives project in 2006. Constant members Michael Murtaugh and Nicolas Malevé, in cooperation with the Central Art Archive of the Finnish National Gallery in Helsinki, currently create a dynamic online archive with and about Kurenniemi's works based on innovative algorithm-based options of searching, retrieving, and sorting data.

12. See Erkki Kurenniemi, "Musical Harmonies Are Divisor Sets," in *Proceedings of the Nordic Acoustical Meeting 88*, ed. Matti Karjalainen, Tapio Lahti, and Jukka Linjama (Tampere: Acoustical Society of Finland, 1988), 371–374.

13. Similarly, the recordings of Baroque music played on the Silbermann organ at Freiberg Cathedral in the era of the German Democratic Republic state record label Eterna in the 1980s preserves how the organ as a technical monument sounded in those days.

14. Jussi Parikka, "Operative Media Archaeology: Wolfgang Ernst's Materialist Media Diagrammatics," *Theory, Culture, and Society* 28, no. 5 (September 2011): 52–74, at 62.

15. See http://www.ubuweb.com/film/kur_electronics.html (accessed June 2013).

16. As argued in Morton Riis's PhD thesis, "Machine Music. A Media Archaeological Excavation," submitted to the Royal Academy of Music and to the Department of Aesthetics and Communication at Aarhus University (Aarhus University, Denmark 2012).

17. Parikka, "Operative Media Archaeology," 65.

18. See Matthew Kirschenbaum, *Mechanisms: New Media and the Forensic Imagination* (Cambridge, MA: MIT Press, 2008).

19. See Vivian Sobchack, "Afterword: Media Archaeology and Re-presencing the Past," in *Media Archaeology: Approaches, Applications, and Implications*, ed. Erkki Huhtamo and Jussi Parikka (Berkeley: University of California Press, 2011), 323–333.

20. Ojanen et al., "Design Principles and User Interfaces," 92.

21. According to Marshall McLuhan's diagnosis, the aesthetic content of a new medium is always the previous medium.

22. Wikipedia, "Erkki Kurenniemi," http:/en.wikipedia.org/wiki/Erkki_Kurenniemi (accessed November 2012).

23. See Friedrich Kittler, *Aphrodite* (*Musik und Mathematik*, vol. 1: *Hellas*, book 1) (Munich: Fink, 2006), which displays a recursive function and its final computational result on page 300 and following.

24. Most of the extant works are now in the Epstein Collection: See Wikipedia, "Thomas Wilfred," http:/en.wikipedia.org/wiki/Thomas_Wilfred (accessed April 2013).

25. Kittler, *Gramophone—Film—Typewriter*, 35.

26. See Friedrich Kittler, "Museums on the Digital Frontier," in *The End(s) of the Museum*, ed. Thomas Keenan (Barcelona: Fundació Antoni Tápies, 1996), 67–80, at 78.

27. *Apparatus operandi* was presented at the Transmediale media arts festival, February 2, 2013, in Berlin.

28. See Heering, Rieß, and Sichau, eds., *Im Labor der Physikgeschichte*, esp. 9–23 (on textual versus artifactual evidence), and 142 (on the idiosyncrasies of the experimental setting [*eigenzeit*]).

15 Capturing Life: Biopolitics, Social Media, and Romantic Irony

Eivind Røssaak

As memory becomes big business, closer attention to alternative modes and ways of organizing memory becomes urgent. Erkki Kurenniemi's unusual archival enterprise is of considerable strategic interest in a world of consumer profiling and social media. More than ever, making each individual's memory operable has become a key ingredient in the current postindustrial transformation of capitalism. Whereas Kurenniemi creates expressions and memorabilia from his life *before* they can be inscribed into a future quantum computer, contemporary social media and consumer profiling works in the opposite way: their affordances determine what forms of memorabilia are acceptable—in advance, algorithmically speaking. However, at every step along the way, Kurenniemi and contemporary informational capitalism are perfect children of cybernetics. At the same time, the exuberant unruliness of Kurenniemi's project operates along a subversive trajectory that seems to both fulfill the cybernetic dream of sending a human being "over a telegraph line"[2] and render it untenable.

Targeting the Brain

Kurenniemi's avant-gardism has become the default mode of citizenship in control societies. Software capitalism has doomed every member of the Western world to an eternal life as looped information in the digital media system. It assembles every electronic trace of life in the network. Tiziana Terranova calls it "soft control."[3] We seldom notice it, but in its presence it is ubiquitous. Modern management techniques are grounded on a new application of subjectivity that turns labor into immaterial labor: the worker's soul shall become part of the factory! We shall all become subjects! The worker's personality and subjectivity must be made susceptible to programmability.[4] "First and foremost, we have here a discourse that is authoritarian: one *has to* express oneself, one *has to* speak, communicate, cooperate and so forth."[5]

This management mandate is even more totalitarian than the earlier rigid division between mental and manual labor, because as production is organized around ways of tapping into life itself, subjectivity becomes the raw material. Life as multiplicity is captured, integrated, and differentiated as a commodity. The new microphysics of power registers and usurps even the tiniest forces of minds, brains, and memories. Our most intimate affects have become no longer ideology (a.k.a. Marxism), but raw data. The multitudes are no longer transformed into classes, but into samples; individuals become "dividuals," as Gilles Deleuze called them.[6] Dividuals are actualized by institutions that stabilize them and translate (or mediate) them into forms of usable life. Vital multiplicities are subdivided into dualities like sex, gender, and class. "The thousand tiny becomings of sexuality" are captured and coded by this statistical dispositive.[7] What is confined by this dispositive is the outside, that is, the virtual, becomings. The power of invention is neutralized.

Power has moved from disciplining the masses in closed space to controlling or modulating them in open space (networks). Biopower is a modality of action aimed at this new multiplicity or life, though "action at a distance," that is, through the media system.[8] How can the media system intervene into the affective contagion moving rapidly in the elastic milieu of brains? What kind of forces are at play in this flexible milieu? Biopolitics is aimed at what Tarde and Nietzsche called "the infinitesimal beings which want, feel and think."[9] The media system addresses the glue that actualizes these passions, namely memory itself, the power of actualization of the virtual. Without memory, without this force (a duration that conserves, a power of actualizations and repetitions), there would be no sensation of life.[10] Thus, memory is a force that needs to be captured in order to control and exploit the assemblage of difference and repetition. One has to intervene in the flows of passions. Memories and impressions create undulations that must be regulated.

The Media/Memory Convergence

Memories never resided in the biological system alone. They have progressively been "liberated" by first primitive tools, and now, by more advanced cultural mnemo-techniques such as monuments, artworks, and books. With the technological revolutions around the turn of the twentieth century, external analog memory media like film and the gramophone was added to the list. In the age of digital networks, memories are embedded in informational networks that systematically connect and reorder them according to the logics of layers of algorithms. Contemporary biopolitics is grounded on this digital media/memory convergence. Social media are one of its most advanced products so far.

Social media are not simply a new access to so-called collective memory such as the fact that many now use social media for commemorative purposes or as an alternative news source continuously updated by friends or groups. Social media are memory media in a much more brutal and savage sense—they are the main form of inscription and connection to the life of the multitudes (as formalized language, i.e., digital code).

This is memory as programmable memory, no longer simply memory of the past, but a memory for future use, as we will see below.

Social media can be roughly defined as "a group of Internet-based applications that build on the ideological and technological foundations of Web 2.0, and that allow [for] the creation and exchange of user-generated content."[11] With the development of Web 2.0 around year 2000, we experienced an explosive proliferation of blogs, wikis, folksonomies, video sharing sites, hosted services, Web applications, mashups, and social networking sites: Blogger was created in 1999, Wikipedia in 2001, LinkedIn and Myspace in 2003, Facebook and Flickr in 2004, YouTube in 2005, and Twitter in 2006. In December 2011, 1.2 billion users worldwide—82 percent of the world's Internet population over age fifteen—logged on to a social media site.[12] This is truly the age of archive fever. The implementation of interactivity and user-generated content on the Web challenged the domination of basically analog, top-down archives (State archives) by the digital online bottom-up archives (archives of the multitudes).[13]

A social media site is an impressive application. Its user-friendly interface hides several layers of algorithms that actually instrumentalize and make easily available hundreds of years of experience in archiving huge amounts of information. With social media at hand 24/7, Internet users can suddenly become notorious and potential mega archivists. The multitudes now transform previously private archival practices—the storage of letters, the diary, the family album, the curiosity cabinet, the chest granddad kept in the attic—into public display.[14] However, the way social media partake in making the Web a social arena for memories in the form of life records and the registration of affects (through the like and dislike buttons) comes with a hitch. Making the Web social actually means making "sociality technical."[15] Social media is not simply a platform where people interact and share files. The fact that the Web has become social also means that human connections (any connection, whether it is uttered through sound, images, or language) have become formal inscriptions, or simply: data.

> Sociality coded by technology renders people's activities formal, manageable, and manipulable, enabling platforms to engineer the sociality in people's everyday routines. ... Besides generating content, peer production yields a valuable by-product that users often do not intentionally deliver: behavioral and profiling data.[16]

Users' social capital generates the owners' economic capital. As Steve Goodman and Luciana Parisi put it in their article "Machines of Memory," the archives of the past (i.e., our activity on social media) become the archives of the future as the sum total of our online activity is analyzed by algorithms and used to prescribe our future choices.[17] They call it preemptive capitalism. Even if the new tools of connectedness may promise both a democratic and utopian future, they are also among the primary arenas for manipulation and alienation in our time. The promise of the archives of the multitudes is thwarted.

The new technical sociality of media is a situation we still have not figured out how to resist. Are there any alternatives? We need to assess how memories reorganize themselves in this new landscape between the archive, libraries, social media, and the arts. Indeed, we find alternative ways of organizing memory through participation, sharing, and resistance among many groups on social media, among artists and NGOs (not to be covered here); but one of the most peculiar cases is that of the strategies of Finnish artist-engineer Erkki Kurenniemi.

Grasping Affects

Kurenniemi's idea of archiving his life to create a systematic model for a template of all human life seems to have dawned on him around the time of his first diary cassette recordings in the early 1970s. However, at the time he decided that everything he had ever done or participated in back to the early 1960s should be part of this archive. His archival project implicitly reframed his entire life in one stroke. From now on, everything he did ended up in between two strata: on the one hand, as part of *a special project*, recording life, the *real*, whether it was making music or a synthesizer, having sex, or collecting memorabilia, and on the other, as part of *a more general or ideal project*, a future artificial life, which, I would argue, also involves testing life equations (mathematics) and computer programs as well as other potential ideas around the attempt to implement his systematic model for the future. The relationship between these two levels of life (the real and the ideal) is enigmatic, and the boundary between them seems to be uncontrollable or eroded by slippage. They are like strata mirroring and influencing each other; both have the ability to affect and be affected by the other.

For Kurenniemi it was not important that the thousands of pictures he took, or the hundreds of films he shot, were nice, or artistically interesting. They should document the flow of living. Implicitly this shifts the attention away from the representational to the nonrepresentational, that is, to the invisible or affective processes that ultimately govern his will to reproduce, store, and transmit this content. His life has ended up being a hyphenated or divisible life. He has lived his life while at the same time recording it. The archivists of the Constant group write that the idea of a divisible present is a utopia, a fantasy.

> The nature of what we live changes when we record it. There is no seamless archiving. One never looks enough at the photographer's body. We do not walk the same way when we take pictures every two minutes, we never drive a car the same way when we film and sexual arousal is heavily conditioned by the presence of the camera. The photographer's body is always dancing, always adjusting to a frame, it also acquires a form of transparency, a diminished presence. It is there but to a lesser degree.[18]

This odd and continuous doubling of life also yields a dislocated life. Indeed, the ubiquitous presence of recording devices creates an awareness of a possible future repetition—a return. There is, in other words, a permanent informatization of life taking place at every aspect of his lived life. This permanent archivalization of life through information gathering makes Kurenniemi's life paradigmatic for a certain informational turn in contemporary informational capitalism. He lives his life as an autopoetic system regulating itself through recursive feedback loops, where one level of the strata guides and operates on the other. In this organization of memory and life, we find a panoply of fundamental mathematical, informational, and physical ideas. We encounter wholes and parts, distinctions, pointers and indications, local-global duality, circulations, feedback, recursion, invariance, self-similarity, reentry of forms, paradox, strange loops, topology, knots and weaves, fractal forms, curvature, and imaginary numbers.[19] Like contemporary informational media, these processes do not use the code of a workaday language, but, as Friedrich Kittler writes, they "make use of physical processes which are faster than human perception and are only susceptible to formulation in the code of modern mathematics."[20] Today, information is not simply the content of a message but the environment within which contemporary culture unfolds. Most notably, the body itself belongs to this *informational* milieu, a milieu composed of dynamic and shifting relations between massless flows, and it is this massless flow of a life—affects, I would argue—Kurenniemi wants to tap into with his recordings.[21]

Affects are to a large extent a "biological phenomenon, involving embodiment in its many incarnations, but a phenomenon that is not easily captured."[22] It involves hormonal flux, body language, shared rhythms, and other forms of entrainment to produce an encounter between the body and the particular event. Affect is semiconscious, "a certain sense or form of touch," a "sensation that is registered but not necessarily considered in the thin band of consciousness we now call cognition."[23] Further on again, affect is understood as a set of flows moving through the body of humans and other beings, not least because bodies are not primarily centered repositories of knowledge—originators—but rather are receivers and transmitters, ceaselessly moving on messages of various kinds. This theory of affects challenges the idea that bodies are fixed components of humanity.

It might be more accurate to liken humans to schools of fish briefly stabilized by particular spaces, temporary solidifications that pulse with particular affects, most especially as devices such as books, screens, and Internet act as a kind of neural pathway, transmitting faces and stances as well as discourse and providing myriad opportunities to forge new reflexes.[24]

Kurenniemi's files can be considered as temporary solidifications. They are the supplementary neural pathways of his life. His incentive to live his life as a recorder, receiver, and interpreter of feelings, affects, and attentive energies combines biology and informational science in one huge enterprise, ultimately highlighting the nonhuman aspects of signal processing. However, there are material remains as well as mathematical efforts in this enterprise that threaten to render these affects either ungraspable or superfluous. This is where his endgame becomes truly dramatic and baffling.

Indecent Excess

When I first came across Erkki Kurenniemi and his collection he immediately reminded me of *Askeladden*, the Ash Lad from the Norwegian fairytales. He is smart, unpredictable—and a trickster. One tale in particular comes to mind, "The Princess Who Could Not Be Silenced":

> There was once a King, and he had a daughter who was so cross and crooked in her words that no one could silence her, and so he gave it out that he who could do it should marry the princess and have half the kingdom, too.

Men flocked to the King's palace to try, but none succeeded. To avoid too much flurry and worry the king soon declared that those who failed would have their ears marked with the big red-hot iron with which he marks his sheep. The three brothers arrived on foot. The youngest of them, the Ash Lad, had it with picking up things along the way: a crow, a willow twig, a wedge, a broken saucer, two goat's horns, and even a worn-out shoe sole. When his brothers, dumbfounded, asked, what's the point? He just retorted: "Oh, I haven't much to carry, I might as well carry this."

The two oldest brothers visited the princess first. They couldn't get a word out after they saw the irons glowing in the hearth in the princess' room. Then it was the Ash Lad's turn.

> "Good day," said he.
> "Good day to you," said she, and she twisted and turned again.
> "It's nice and warm in here," said the Ash Lad.
> "It's hotter in the hearth," said she, and she was no sweeter, now the third one had come.
> "That's good, I may bake my crow there, then?" asked he.
> "I'm afraid she'll burst," said the princess.
> "There's no danger; I'll wind this willow twig around," said the lad.
> "It's too loose," said she.
> "I'll stick this wedge in," said the lad, and took out the wedge.
> "The fat will drop off," said the princess.
> "I'll hold this under," said the lad, and pulled out the broken bit of the saucer.
> "You are crooked in your words, that you are," said the princess.
> "No, I'm not crooked, but this is crooked," said the lad, and he showed her the goat's horn.
> "Well, I never saw the equal to that!" cried the princess.
> "Oh, here is the equal to it," said he, and pulled out the other.
> "Now, you think you'll wear out my soul, don't you?" said she.
> "No, I won't wear out your soul, for I have a sole that's worn out already," said the lad, and pulled out the shoe sole.
> Then the princess hadn't a word to say.
> "Now, you're mine," said the Ash Lad.
> And so she was.[25]

The Ash Lad is a pragmatic "archivist." The archive becomes a tool kit. This fairy tale about the Ash Lad is like an allegory of the trouble with Kurenniemi. There is an anecdote in Kurenniemi's diary from 1996 (June 24, 9:44 p.m.). Two women visit him, and he shows them his archive. He writes:

> Will she (either of them) share the love of pornography? Or at least, art? I shall present myself to both of them as a geniality self-flagellation machine. Just one bottle tonight, ok?
>
> I shall invite them on to my journey of change, showing the way ahead. Immortality.
>
> I will let them to sniff and sneak through my archives.[26]

While the Ash Lad uses the archive to succeed in life, Kurenniemi uses it here for more immediate pleasures, it seems, while at the same time documenting it. However, Kurenniemi's pragmatism is also romantic in a very conceptual way, as we will see.

When I entered Kurenniemi's archive in Helsinki I felt somewhat like the Ash Lad's brothers: Why was he collecting so many odd things? Luckily, Kurenniemi has so far been less successful than the Ash Lad in "converging" with the princess and half the kingdom—a.k.a. eternal life in a golf ball in a postbiological utopia.[27] But he managed to silence me when I saw his "real" (also in the Lacanian sense) collection.

Visiting the actual Kurenniemi archives in Helsinki was a jaw dropper. In the storage room of the National Gallery, surrounded by locked doors, hallways, and staircases, there are seventeen meters of archival shelves of boxes filled with everything from shopping receipts, handwritten diaries, science and porn magazines, drawings, and marginalia from unknown activities. These things, their materiality, traces of deterioration and use, marginalia of all sorts, along with for example the collages of pubic hair in the so-called Hairy Diary, seem infested with an ingrained resistance to digitization. They are loaded with the indecency of excess so celebrated by theoreticians like Georges Bataille. They are like a physical sight gag ridiculing our current level of information technology; they make the sharing of memories on social media seem like a stifling prison-house activity. These material vaults travel far beyond the most primitive *and* modern obsession with fetishizing remains, and at the same time, Kurenniemi seems willing, yet again, to cancel them out with one stroke of his pen.

How to Unwrite Things?

Kurenniemi's own computer was deposited at Kiasma and the National Gallery archive in 2006. It contains, among other things, several autobiographical fragments and letters from the 1990s. We do not know if all of the letters were actually sent. Sometimes he writes: "I will not send this letter." One of the longest letters is to Bronwyn, written July 20, 1993. I managed to identify Bronwyn as Bronwyn Bevan, Director of the San Francisco Exploratorium's Institute for Research and Learning. However, she never received the letter. Kurenniemi spent some time there during his stays in San Francisco during the 1980s and 1990s. At the time Kurenniemi was working on solving the "unsolved problems of mathematics," reading and actually trying to write a sci-fi novel.[28] In a curious passage in the unsent letter, he discusses how the evolution of life can be stored in one perfect equation.

> [I] Have to find a way to unwrite things. Mathematics would be the ideal language for compact ideas. Think of Newton's formula $F = ma$. Four letters which govern almost everything we see. Well, not quite. For that you also need Maxwell's equations. I think Maxwell needed something like a full page to write his equations because the compact vector notation was not known in his time. In vector notation you need four short lines only. Then Einstein rewrote the Maxwell equations in four or five letters. Today, they need not be written at all because they follow automatically from the gauge condition. This I mean by unwriting poetry.[29]

It is as if he believed that he could render life, even his life, into, or even replace it by, one perfect equation (discussed in several of his articles). Alas, we could say that to "unwrite things" refers not only to mathematics or the laws governing the relation between things, but also, in this particular connection, to rendering the things in his own collection superfluous. This gesture is both apocalyptic and utopian. In the same gesture he can sublate life as we know it, and simultaneously, lay the grounds for the algorithms programming the next generation of supercomputers, like the quantum computer that would ultimately reanimate his life. Again, we find a peculiar recursive struggle relation between *a special project* (the documentation of daily activities, and the inertia and resistance of matter and marginalia) and *a more general project* (its framing, *raison d'être*, or future reanimation). On the one hand, it seems like the excessive remains from daily life (ironically?) would never be able to be incorporated in and fulfill the primary target of attaining an eternal life in media; on the other hand, he also works with more superior and peremptory solutions inspired by mathematics and programming language that could ultimately cancel out the need for a collection. His different strategies seem at times to be each other's interruption or destruction.

The structural relation between these two strata may resemble what the Jena Romantics[30] and Friedrich Schlegel saw as a relationship between the real and the ideal. Schlegel did not believe in the dialectical sublation of the real and ideal into a new entity. Rather than sublation, he insisted on a new type of poetry uniting art, philosophy, and science—he called it romantic poetry or the novel—that would be fragmentary, incomplete, ironic and digressive (he compares it to marginalia), and still be more "complete" and alive. This program or attitude toward life and living has been called Romantic irony.[31] Schlegel specified it paradoxically thus: "Die ironie ist eine permanente Parekbase" ([This type of] irony is a permanent parabasis).[32] Parabasis (which literally means "to step aside") as a permanent process involves the paradoxical gesture of positing something and canceling it at the same time. It playfully discloses the gap between the real and the ideal. It signifies the highest form of self-reflexivity. As a literary strategy it usually develops through a series of digressions or fragments. For Kurenniemi it takes the form of ever new loads of marginalia, projects, plans, and ideas (ever new solutions).

Some critics have meant that Schlegel's "permanent parabasis" must itself be ironic; one cannot permanently suspend something without losing sight of the suspended element.[33] However, this is exactly what Schlegel wanted, and this is exactly what tends to happen in the cleft between the real and the ideal in Kurenniemi's project. Each real element in his collection both strives toward and cancels out its ideal realization in the same gesture—and likewise, the ideal equations potentially render the "real" things superfluous. The suspension is total, and at the moment of its final realization, we will not be able to take it into consideration at all; most likely, we will not be around (as humans) to witness the end of Kurenniemi's game.

Acknowledgments

Thanks to Jussi Parikka, Perttu Rastas, Ina Blom, and the Archive in Motion project for useful comments on my work on Erkki Kurenniemi.

Notes

1. Erkki Kurenniemi, in "Drifting Golf Balls in Monasteries: A Conversation with Erkki Kurenniemi," in this vol.

2. Norbert Wiener, *God and Golem, Inc.: A Comment on Certain Points where Cybernetics Impinges on Religion* (Cambridge, MA: MIT Press, 1964), 36.

3. Tiziana Terranova, *Network Culture: Politics for the Information Age* (London: Pluto Press, 2004), 98–99.

4. In his graph-theoretical paper "Programmable Space," Kurenniemi indicates that human beings could be the original Turing machine as they are programmable. See E. Kurenniemi, "Programmable Space," University of Helsinki, Department of Nuclear Physics, undated manuscript (possibly from 1990). (File at Kiasma.)

5. Maurizio Lazzarato, "Immaterial Labor," in *Radical Thought in Italy: A Potential Politics*, ed. Paolo Virno and Michael Hardy (Minneapolis: University of Minnesota Press, 2006), 134.

6. Gilles Deleuze, "Postscript on the Societies of Control," *October* 16 (winter 1992): 5.

7. Maurizio Lazzarato, "The Concepts of Life and Living in the Societies of Control," in *Deleuze and the Social*, ed. M. Fuglsang and B. M. Sorensen (Edinburgh: Edinburgh University Press, 2006), 174.

8. Ibid., 179–180. Lazzarato borrows the term "action at a distance" from Gabriel Tarde.

9. Nietzsche and Tarde are quoted in Lazzarato, "The Concepts of Life and Living," 183.

10. Tarde's notion of memory is explained in Lazzarato, "The Concepts of Life and Living," 184.

11. A. M. Kaplan and M. Haenlein, "Users of the World, Unite! The Challenges and Opportunities of Social Media," *Business Horizons* 53, no. 1 (2010): 60.

12. José van Dijck, *The Culture of Connectivity: A Critical History of Social Media,* (Oxford: Oxford University Press, 2013), 4. Furthermore: "In the U.S. alone, total minutes spent on social networking sites has increased 83 percent year-over-year. In fact, total minutes spent on Facebook increased nearly 700 percent year-over-year, growing from 1.7 billion minutes in April 2008 to 13.9 billion in April 2009, making it the No. 1 social networking site for the month" (see Nielsen.com: http://www.nielsen.com/us/en/newswire/2009/time-spent-on-facebook-up-700-but-myspace-still-tops-for-video.html [accessed April 3, 2014]).

13. However, it may be argued that social media's server-client model differs in no significant way from hierarchical, top-down power.

14. As when I as a kid went to Aunt Lovise's down the street to eat goodies and play; the highlight was when she walked over to the corner of the living room and opened the little mahogany cabinet filled to the brim with old toys; I was in heaven.

15. Dijck, *The Culture of Connectivity*, 12.

16. Ibid., 12, 16. One of the biggest consumer profiling agencies, Nielsen, writes about the fine-grained profiling they can make on their website: "Our segmentation solutions answer these questions in as much detail as you desire. We can give you aggregated and customized demographic data to understand geography based on region, state, zip, neighborhood, block-by-block and any level in between" (see Nielsen.com, cited in n. 12).

17. Steven Goodman and Luciana Parisi, "Machines of Memory," in *Memory: Histories, Theories, Debates*, ed. S. Radstone and B. Schwartz (New York: Fordham University Press, 2010), 344.

18. http://kurenniemi.activearchives.org/logbook/?page_id=526 (accessed April 17, 2013). See also Geoff Cox, Nicolas Malevé, and Michael Murtaugh, "Archiving the Databody: Human and Nonhuman Agency in the Documents of Erkki Kurenniemi," in this vol.

19. Louis H. Kauffman, "Self-Reference and Recursive Forms," *Journal of Social and Biological Structures* 10 (1987): 53–72, at 53. For more on cybernetics and Kurenniemi, see also Morten Sondergaard, "Interfaces of Future Authenticity: Erkki Kurenniemi's Media Archives (From a Postcybernetic Perspective)," in this vol.

20. Friedrich Kittler, "The History of Communication Media," *C-theory*, July 30, 1996, http://www.ctheory.net/articles.aspx?id=45 (accessed April 7, 2013).

21. The term "milieu" comes from French, *mi-lieu,* the halfway place, the zone in between, or the surrounding circumstances (for more on this concept, see John Tresh, *The Romantic Machine: Utopian Science and Technology After Napoleon* [Chicago: University of Chicago Press, 2012], xiii). Jussi Parikka talks about Kurenniemi's experiments in this milieu as a "symptomatology," a performative symptom of the information era. See Jussi Parikka, "DIY Futurology: Kurenniemi's Signal Based Cosmology" in *Erkki Kurenniemi—A Man from the Future* (2013), http://www.lahteilla.fi/kurenniemi/a-man-from-the-future.

22. Nigel Thrift, *Non-Representational Theory: Space, Politics, Affect* (London: Routledge, 2008), 236.

23. Ibid.

24. Ibid.

25. "The Princess Who Could Not Be Silenced" is quoted from P. C. Asbjørnsen and Jørgen Moe, *Norwegian Folktales* (New York: Pantheon Books, 1982), http://www.surlalunefairytales.com/books/norway/thornethomsen/princessnotsilenced.html (last accessed January 24, 2014). In this translation the Ash Lad is named Boots. I keep the more literal translation of his name, "Ash Lad" (in Norwegian: *Askeladden*).

26. Constant uses this anecdote from Kurenniemi's diary as an epigraph to their logbook on Kurenniemi: http://kurenniemi.activearchives.org/logbook/?page_id=511/.

27. This is what Kurenniemi more or less ironically talks about in the interview with Mika Taanila (see "Drifting Golf Balls in Monasteries: A Conversation with Erkki Kurenniemi," in this vol.).

28. Jyrki Siukonen has argued, with a reference to Kurenniemi's diary writings from 1989, that "originally Kurenniemi's project was nothing more than a series of failed attempts to write a [sci-fi] novel." See Jyrki Siukonen, "Dead Computers Tell No Tales—Remark on the Future behind Kurenniemi's 2048 Resurrection," in *Erkki Kurenniemi—A Man from the Future* (2013), http://www.lahteilla.fi/kurenniemi/a-man-from-the-future.

29. Erkki Kurenniemi, letter to Bronwyn, July 20, 1993. (File at Kiasma.)

30. A group of influential German writers and critics centered in Jena from 1798 to 1804.

31. For a more elaborate explanation of this program, see Philippe Lacoue-Labarthe and Jean-Luc Nancy, *The Literary Absolute: The Theory of Literature in German Romanticism* (New York: SUNY Press, 1988).

32. Friedrich Schlegel, "Zur Philosophie [1797]," fragment 668, in *Philosophische Lehrjahre I (1796–1806)*, ed. Ernst Behler, in *Kritische Ausgabe*, vol. 18 (Paderborn-Vienna-Munich: Verlag Ferdinand Schöningh, 1963), 85.

33. Paul de Man, "The Concept of Irony," in Paul de Man, *Aesthetic Ideology* (Minneapolis: University of Minnesota Press, 1996), 179.

Bibliography

Asbjørnsen, P. C., and Jørgen Moe. "The Princess Who Could Not Be Silenced." In *Norwegian Folktales*. New York: Pantheon Books, 1982.

Deleuze, Gilles. "Postscript on the Societies of Control." *October* 16 (winter 1992): 3–7.

de Man, Paul. "The Concept of Irony." In Paul de Man, Aesthetic Ideology, ed. A. Warminski, 163–184. Minneapolis: University of Minnesota Press, 1996.

Goodman, Steven, and Luciana Parisi. "Machines of Memory." In Memory: Histories, Theories, Debates, ed. S. Radstone and B. Schwartz. New York: Fordham University Press, 2010.

Kaplan, A. M., and M. Haenlein. "Users of the World Unite! The Challenges and Opportunities of Social Media." *Business Horizons* 53 (1) (2010): 59–68.

Kauffman, Louis H. "Self-Reference and Recursive Forms." *Journal of Social and Biological Structures* 10 (1987): 53–72.

Kittler, Friedrich. "The History of Communication Media." *C-theory*, July 30, 1996, http://www.ctheory.net/articles.aspx?id=45 (accessed April 7, 2013).

Kurenniemi, Erkki. DRY, diary, 9:36 am, March 1, 2001 (computer file).

Kurenniemi, Erkki. Letter to Bronwyn, July 20, 1993 (file at Kiasma).

Kurenniemi, Erkki. "Programmable Space." University of Helsinki. Department of Nuclear Physics. N.d.; possibly from 1990 (file at Kiasma).

Lacoue-Labarthe, Philippe, and Jean-Luc Nancy. *The Literary Absolute: The Theory of Literature in German Romanticism*. New York: SUNY Press, 1988.

Lazzarato, Maurizio. "Immaterial Labor." In Radical Thought in Italy: A Potential Politics, ed. Paolo Virno and Michael Hardt, 133–150. Minneapolis: University of Minnesota Press, 2006.

Lazzarato, Maurizio. "The Concepts of Life and Living in the Societies of Control." In *Deleuze and the Social*, ed. M. Fuglsang and B. M. Sorensen, 171–190. Edinburgh: Edinburgh University Press, 2006.

Parikka, Jussi. "DIY Futurology: Kurenniemi's Signal Based Cosmology." In *Erkki Kurenniemi—A Man from the Future*, 2013, http://www.lahteilla.fi/kurenniemi/a-man-from-the-future (accessed April 7, 2013).

Schlegel, Friedrich. "Zur Philosophie [1797]," Fragment 668. In *Philosophische Lehrjahre I (1796–1806)*, ed. Ernst Behler, in *Kritische Ausgabe*, vol. 18, no. 2. Paderborn-Vienna-Munich: Verlag Ferdinand Schöningh, 1963.

Siukonen, Jyrki. "Dead Computers Tell No Tales—Remark on the Future behind Kurenniemi's 2048 Resurrection." In *Erkki Kurenniemi—A Man from the Future*, 2013, http://www.lahteilla.fi/kurenniemi/a-man-from-the-future (accessed April 7, 2013).

Terranova, Tiziana. *Network Culture: Politics for the Information Age*. London: Pluto Press, 2004.

Thrift, Nigel. *Non-Representational Theory: Space, Politics, Affect*. London: Routledge, 2008.

Tresh, John. *The Romantic Machine: Utopian Science and Technology after Napoleon*. Chicago: University of Chicago Press, 2012.

van Dijck, José. *The Culture of Connectivity: A Critical History of Social Media*. Oxford: Oxford University Press, 2013.

Wiener, Norbert. *God and Golem, Inc.: A Comment on Certain Points Where Cybernetics Impinges on Religion*. Cambridge, MA: MIT Press, 1964.

V Music

Foreword

Petri Kuljuntausta

Erkki Kurenniemi came to the music world by way of science. When he began his studies at Helsinki University in the early 1960s, he was soon focusing more on his own projects and on assisting others with technology than on his academic studies. As a young student he programmed experimental computer graphics using the analog computer of the Nuclear Physics Department and started to build the Electronic Music Studio at the Department of Musicology. He was at home in both departments, but first and foremost he was a nonconformist, whose interest was in digital music, computer arts, and the construction of automated composition systems.

Kurenniemi's work at the crossroads of science and art developed in conjunction with the progress of the digital computers of the 1960s and 1970s. As a result, the lack of traditional musicianship and a distance from the classical music tradition was not a barrier for him. The emerging field of electronic music and its production technology freed the composer from the traditional "composer-score-interpreter" hierarchy, the typical chain of music production.

In 1962 he set up a pioneering small-scale music studio at Helsinki University. The studio was not intended to be a manual electronic studio where each sound is produced, edited, and mixed separately. Instead, Kurenniemi aimed to automate the entire composition process. This led him to design the Integrated Synthesizer (1964), which could combine all the sound processing and controlling units needed in studio work. With this new instrument Kurenniemi composed a piece entitled *Hyppy* (The Jump, 1964).[1] This rhythmic piece predicted the birth of techno music, which started to find its own form decades later.

Kurenniemi's studio included conventional studio equipment, such as tape recorders and the VCS 3 synthesizer (1970–), but the core of the studio was unique. It included devices built by Kurenniemi such as the Integrated Synthesizer and the DIMIX (1972–), which, along with other DIMI synthesizers, in particular the DIMI-A (1970), gave the studio its distinctive sound. At the time when Kurenniemi built his studio, electronic music instruments and studio composition were developing in many places; we can consider, for example, the work of Donald Buchla (Buchla Electronic Musical

Instruments), Ralph Lundsten in Stockholm (Andromeda Studio), Max Mathews (Music synthesis languages and the Groove real-time system), Robert Moog (Moog synthesizers), or Peter Zinovieff in London (EMS Studio and synthesizers).

Kurenniemi was familiar with the electronic studios outside of Finland and he had close connections to classically trained composers, but he didn't compose music based on formal ideas that could be accredited in the circles of international academic music. Like Canadian scientist and composer Hugh Le Caine, he saw himself as a builder of electronic instruments, not as a composer. But, in Kurenniemi's case, this statement is only partly true. He created works intuitively while testing his self-made studio instruments. When listening to Kurenniemi's dozens of unreleased test recordings, we find that his official works are not only sound tests but prestructured performances. There is no traditional score behind these pieces or improvisations, but it is clear that he followed a preliminary plan when he made the recordings. On his view, as a logical continuation of the automated composition process, the score of electronic music work is actually its circuit diagram.

When Kurenniemi's first surviving electronic composition, "On-Off" (1963), was premiered, one of the reviewers stated that in the work, "technical skills of the composer meet intuitive formal thinking."[2] The work was the first example of Kurenniemi's idea of automated sound processing; "On-Off" was composed in real time for tape, without any cuts. "On-Off" was quite a strange piece for its time, and it is still hard to believe just how powerful the sound world was that Kurenniemi succeeded in producing in his studio. The tape work sounds like the marriage of a power plant's generator and Jimi Hendrix's distorted feedback sound of a Stratocaster guitar. Hendrix's loud feedback effects shocked audiences some four years later in the summer of 1967, so it is no surprise how effective the powerful "On-Off" was in its time on Scandinavian soil. The title of the work reflects the idea that in the distant future there will be only one control button in a computer music studio: an on-off switch.

In 1967, Kurenniemi predicted that in the future computer compositions would lose their identity and composers will be seen as an industrial designers or fashion creators. Kurenniemi produced musical material easily with his music automatons, and as he worked with the digital technology and studied its possibilities, he understood what the technology could offer for future composers. Generative music is a good example on this. The composer creates the rules for the composition and leads the direction of the music. The software follows the rules and makes decisions based on the given compositional algorithms. The object-nature of a musical work is meaningless, as generative music is different in every performance. There is no beginning or end. Kurenniemi's soulmate was Italian composer Pietro Grossi, an unsung visionary, who came to the same conclusion in 1969 and questioned the concept of musical authorship.[3]

The radical ideas of the 1960s led to surprising directions among the composers of younger generations. An example of the music style that Kurenniemi developed could be called "functional music," which has been composed, for example, for a commercial television series. The computer generates music that is stylistically correct background music and suits the drama and the plot of the show. The opening theme of the show might be composed manually, but after that the composer just selects the musical style and tempo, presses the on-off button, and saves the new soundtrack on a hard disk. We have software that has been programmed to follow composition routines, but it is hard to find composers who admit that they use these tools and let the machine do the work.

Kurenniemi's vision of computer music has also materialized in DJ culture and techno music. With the latest DJ software you can let the computer to do the work and make music continuously. You don't even have to organize the drum samples according to their beats per minute (b.p.m.), as the software does it on the fly. When you place the sample into the session, the software analyzes it immediately and corrects the tempo to fit the tempo of the present track. And in a fraction of second you can play it out in your live session. In DJ culture, it is not always a question of your own pieces, but of how well you mix these bits together and fulfill the expectations of the dancing audience. The DJ musician challenges the concept of musical authorship.

In 1968, Kurenniemi gave a speech at electronic music conference in Florence, Italy, and introduced the idea of terminal music. He envisioned a mainframe computer-based composition system with terminals distributed around the university. In the network, 50 users could work at the same time, and 200 terminals could serve 5,000 people. His Electric Quartet (Sähkökvartetti), built in 1967–1968, was a small-scale version of the idea. The Electric Quartet is a collective electronic instrument that required four players to control it. The new instrument was evidence of Kurenniemi's innovative thinking. The instrument required advanced electronics and interface technology, and it took Kurenniemi a year to complete the machine.

Immediately after the Electric Quartet, during the fall of 1968, Kurenniemi constructed the Andromatic, a polyphonic synthesizer-sequencer. With the Andromatic's sequencer he composed "Dance of the Anthropoids" (1968), which was released on a record that same year. The American record producer Kim Fowley heard the piece, regarded Kurenniemi as a genius,[4] and included a short excerpt on Wigwam's album *Tombstone Valentine*. On the international releases (in the United States and Italy) of this album, he made Kurenniemi's work the opening track of the album. Of all Fowley's many album productions around that time, this album was the one Fowley gave as the source of inspiration for the iconic rock singer David Bowie.[5] Kurenniemi's Andromatic wasn't only a musical instrument. It was used in art exhibitions in Stockholm and New York, where it made electronic music for the exhibition and also controlled the lights of a sculpture.

In the United States, Don Buchla spent the 1960s on the West Coast developing modular Buchla 100 analog synthesizer. On the East Coast, composer Joel Chadabe commissioned Bob Moog to create an automated analog synthesizer system with eight sequencers. The system, called CEMS (Coordinated Electronic Music Studio), was ready in December 1969. Almost a decade later, in the late 1970s, Chadabe finally moved from an analog-programmable electronic music system to digital equipment, and, according to him, he never looked back to the analog days.

While Bob Moog and Don Buchla built analog synthesizers, Kurenniemi focused on real-time sequencing digital synthesizers. The integrated circuits had been publicly available since the mid-1960s, and he used the new circuits in the construction of the Andromatic and its follower DICO (1969), from "digitally controlled oscillator." Both are programmable synthesizers with a digital memory. After these instruments, Kurenniemi constructed a series of DIMI synthesizers. The starting point "was to study to what extent digital technology could be applied in the production of electronic music."[6] Since 1970, Kurenniemi and his Digelius Electronics company produced DIMI-A, DIMI-O, DIMI-T, DIMI-S, and, as the last one in the series, the microprocessor-based DIMI-6000, in 1975. All these unique synthesizers from the early 1970s showed innovative thinking and introduced a new kind of interface for controlling the instrument. The optical video-synthesizer DIMI-O especially was ahead of its time.

Kurenniemi was a respected inventor among his colleagues in Finland and Sweden, but his instruments didn't spark wide interest among musicians at the time when he invented them. Only the prototype was built; in the best cases, two instruments were made. The instruments were too experimental, too narrow-sounding, and too wild for ordinary musical productions. Instruments without keyboards were not intended for a commercial market, and thus only a few experimental composers used them in Finland and Sweden. The same happened to Triadex Muse (1972), a sequencer-based digital synthesizer that was designed by Edward Fredkin and Marvin Minsky at MIT. Around 280 instruments were built, but there was no market for them.[7]

Kurenniemi writes about his digital instruments from an instrument designer's point of view in his 1971 article "On Electronic Music Instruments." In "Interaction of Music and Technology," Kai Lassfolk, Jari Suominen, and Mikko Ojanen return to the subject and give a profound hands-on analysis of the DICO and DIMI-A instruments.

In 2003, during his Oslo visit, Kurenniemi formulated his music-theoretical research question on tonal space with the following words:[8] "Why [do] all known musical cultures use intervals and scales which are based only on three first prime numbers: 2, 3 and 5?" He asks the same question in his essay "Chords, Scales, and Divisor Lattices" (2003), where he writes about musical harmony and tonal space. As the text shows, Kurenniemi has seven possible explanations for the question why the tonal space is only three-dimensional. He had developed mathematical theory of tonal music for over thirty years, and he wrote several articles on the subject. Another example of his music-theoretical writing is the very short text "Tonal Theory" (2003), which also summarized his thoughts on the subject. Around that time he wrote a book manuscript entitled "Mathematical Foundations of Music"; these two articles are probably from this manuscript, now lost.

In "On Sound and Artificial Neural Networks," artist Florian Hecker and philosopher Robin Mackay discuss neural networks and Kurenniemi's relation to the work of Iannis Xenakis and David Tudor. This text demonstrates some of the inspiration Kurenniemi has produced in contemporary sound arts, as well as illustrating his connections with some of the more internationally established names of past decades.

Notes

1. Eino Ruutsalo: *Hyppy* (The Jump), experimental movie, 35 mm. film, 4:20). The soundtrack was composed in 1964; the film was edited in 1965. A digital copy of the film is kept at the Finnish National Gallery Collections. "Music for the Film *Hyppy*" (6:38) was released on the vinyl album *Rules* by Erkki Kurenniemi (Full Contact Records KRYPT-022, 2012).

2. Kaj Chydenius, "Suomalainen musiikki elää uutta nousukautta," *Kirkko ja Musiikki* (August 1963).

3. David Toop, "Soundcheck," *Wire* 223 (July 2003): 64.

4. Petri Kuljuntausta, interview with Kim Fowley, October 30, 2013, unpublished.

5. Ibid.

6. Kurenniemi, "Elektronisen musiikin instrumenteista [On electronic musical instruments]," *Musiikki* 1, no. 1 (1971). In Finnish.

7. After the first experiments with my own Triadex Muse, I found it is easy to control the instrument. It creates interesting random sequences, but what might have frustrated users is that you can't make your own melodic lines with it. For more on Triadex Muse, see Wikipedia, http://en.wikipedia.org/wiki/Triadex_Muse.

8. *Erkki Kurenniemi on Tonal Music Theory* (video). Erkki Kurenniemi in 2003 in Oslo discussing, with Arild Boman, the theory of tonal music. Original video in the personal collection of Arild Boman in Oslo. A copy of the video is available at the Finnish National Gallery.

16 Tonal Theory

Erkki Kurenniemi

I have been developing a mathematical theory of tonal music for over 30 years, in the footsteps of Leonhard Euler, who in 1729 published a major study on mathematical music theory, where he defined *gradus suavitatis*, a measure of musical consonance of intervals, based on number theory (greatest common divisor).

The study of mathematics of music has a long history since Pythagoras found that nice-sounding musical intervals correspond to simple numerical ratios, like 2:1 (octave), 3:2 (fifth), 4:3 (fourth) and so on.

The interest in numerical relations faded slowly about 100 years ago when it was gradually realized that there is no objective way to assign a rational number (of form m/n, where m and n are integers) to a physical interval. The equally tempered (ET) scale had become the universal standard in music instrument tuning. All intervals in the ET are irrational numbers (except for the octave and its multiples) and there is no unique way to compute a rational approximation for an irrational number.

Iannis Xenakis and Johan Sundberg pointed out to me of this difficulty in 1987–88. At that time I didn't have a good answer. Later I realized that although simple ratios do not exist "out there," in the physical world, they may well exist in the brain. Simple circuits of neurons can act as frequency dividers or phase locked loop (PLL) frequency multipliers.

My research is two-pronged. First, represent rational numbers as lattice points in the "tonal space" or prime number exponent vectors. This space is infinite-dimensional but luckily all of music theory can be done in the first three dimensions corresponding to the prime numbers 2, 3, and 5. (Why it is so, is a major open problem.) In the tonal space view musical chords and scales become geometrical objects, easily visualized with 3D computer graphics.

Second, I try to model the auditory system as a neural network (actually, a dendritic compartment network) where neuronal spikes are phase-locked to the sound waveform or musical rhythm. The emerging hypothesis is that melody notes correspond, in terms of evolution theory, to high peaks of auditory sound signals, and the musical harmony serves as a "glue" to bind together sound components emitted by the same physical sound source.

In the three-dimensional tonal space of the first three prime numbers, one performs an orthogonal linear transformation (or change of coordinates) from the original *prime exponent coordinates* to "*pitch coordinates*" where the first coordinate axis is in the direction of ascending pitch. The remaining two dimensions are called "*enharmonic dimensions*." They characterize the tonal dimensions of musical tones that have about the same pitch but differing tonal functions.

Geometrical objects in the tonal space, like divisor lattices and cylinders in the pitch coordinates, offer a natural way to classify musical chords and scales. Three "magic numbers" appear. The number 60 gives an explanation of the importance of major and minor triads (4:5:6 and 10:12:15). (Actually, tonal space theory proposes that there are better sounding settings of the triads than given by the traditional music theory, 2:3:5 and 6:10:15 living in the divisor lattice of $30 = 2 \times 3 \times 5$.)

The next magic number (known to Euler) is 8,640. Its divisor set gives the diatonic scale ("white keys") with interesting deviations from the standard theory. The last found magic number, 345,600, gives the just chromatic scale.

I think Euler made an error in his definition of *gradus suavitatis* and I am trying to correct that error.

The deep mystery of music hinges on its ability to arouse immediate emotions. Could it be that music reflects, in a culturally formalized fashion, the way brains encode emotions and perhaps cognitions in general? After all, everything that we perceive is simply a temporal distribution of neural action potentials.

Unfinished text published with *The Dawn of DIMI* (DVD, 2003), 42-page picture/text slideshow.

17 Chords, Scales, and Divisor Lattices

Erkki Kurenniemi
February 19–October 9, 2003

File CSDL 1.nb

Abstract

In this paper I study musical harmony from the rational point of view, stressing the significance of integer ratios. The main concept is the divisor set of an integer. A divisor set is the set of all divisors of a number and it has the algebraic structure of a lattice with the greatest common divisor and least common multiple as the lattice operations. A divisor set can also be interpreted as a geometric lattice, a set of points with integer coordinates in the space of prime exponent sequences ("tonal space"), with the shape of a parallelepiped ("shoebox," "brick"). Divisor lattices can also be seen as set intersections of harmonic ("overtone") series with subharmonic ("undertone") sequences.

I shall concentrate in particular on the divisor lattices of three numbers, 60, 8,640, and 345,600 with 12, 56, and 120 divisors, respectively. In the tonal space, they are bricks with side lengths 2×1×1, 6×3×1, and 9×3×2, respectively. Projecting these point sets on the "pitch line" gives the following musically recognizable sets: major and minor triad chords and the tonic seventh chord, the diatonic scale, and the just chromatic scale, respectively. The equation of the pitch line in parametric form is $p \log p_i$ where p_i is the ith prime and p is the pitch parameter.

The Tonal Space

We define the **tonal space** as the set of rational numbers $\mathbb{Q}$ represented as the exponent vectors (sequences) of primes. The fundamental theorem of arithmetic allows us to write every rational $f \in \mathbb{Q}$ uniquely as

$$f = \prod_{i=1}^{\infty} p_i^{ei}$$

where p_i is the ith prime and e_i is the multiplicity of p_i, in the prime factorization of f. The logarithmic pitch corresponding to the frequency f is then

$$\log f = \sum_{i=1}^{\infty} e_i \log p_i$$

and this form suggests that we interpret the pitch as an inner product $e \cdot p$ between a *tonal vector* e and a constant *pitch vector* p whose i^{th} component is the logarithm of the ith prime. Slightly more abstractly, the tonal space can be defined as the infinite-dimensional real vector space $\mathbb{R}^{\infty}$ with a distinguished non-zero vector p. One can then choose a basis in which this vector has the logarithms of primes as components. Finally, the integer lattice points $\mathbb{Z}^{\infty}$ in this basis are called the *tonal points*.

The advantage of this representation is that whereas divisibility relations are lost when taking ordinary logarithms, primewise logarithms preserve them. In particular, for two tonal points a and b, a divides b, $a \mid b \Leftrightarrow a_i \leq b_i$ for all i.

The tonal space geometrizes the sets of rational intervals, chords, and scales, and aids in visualizing complex harmonic structures. The infinitude of primes does not bother us because traditional Western music theory makes use of the first three prime dimensions 2, 3, and 5 only. It will be interesting anyhow to extend the study to the next higher primes 7, 11, and 13 and study whether we can hear the "eerie" harmonic qualities of the higher dimensions. If not, it is an interesting question why just three primes suffice. Does the reason lie in the neural structures of our auditory system, or is it just a historical accident?

The natural "prime basis" of the tonal space consists of the orthogonal unit vectors $\mathbf{e}_2$, $\mathbf{e}_3$, and $\mathbf{e}_5$ corresponding to the prime numbers 2, 3, 5, … . It is useful to introduce another set of basis vectors $\mathbf{e}_p$, $\mathbf{e}_u$, $\mathbf{e}_v$ called the "pitch basis," where the first unit vector is parallel to the pitch vector and the remaining ones orthonormal with it. In the 3D case there is arbitrariness in choosing the direction of the second basis vector, but once that choice is made, the third basis vector is determined by orthogonality.

The graphic in figure 17.1 shows the prime basis, the pitch vector p and one pitch plane orthogonal to p.

The first new base vector $\mathbf{e}_p$ is obtained by dividing the pitch vector by its length: $\mathbf{e}_p = p/|p|$

The second unit vector $\mathbf{e}_u$ is chosen from the plane orthogonal to p, in the direction of the projection of $\mathbf{e}_2$. The third pitch basis vector $\mathbf{e}_v$ is obtained as the vector product of the first two.

We construct the tonal rotation matrix P with the pitch basis vectors as rows.

$$P = \begin{pmatrix} \mathbf{e}_p \\ \mathbf{e}_u \\ \mathbf{e}_v \end{pmatrix} = \begin{pmatrix} 0.335136 & 0.531178 & 0.778161 \\ 0.94217 & -0.188943 & -0.276797 \\ 0 & -0.825924 & 0.563781 \end{pmatrix}$$

The symbolic expressions for the matrix entries are not terribly complicated but here we give only numerical values. Any tonal space vector given in prime coordinates can be represented in the pitch basis by multiplying it with this matrix. The first component

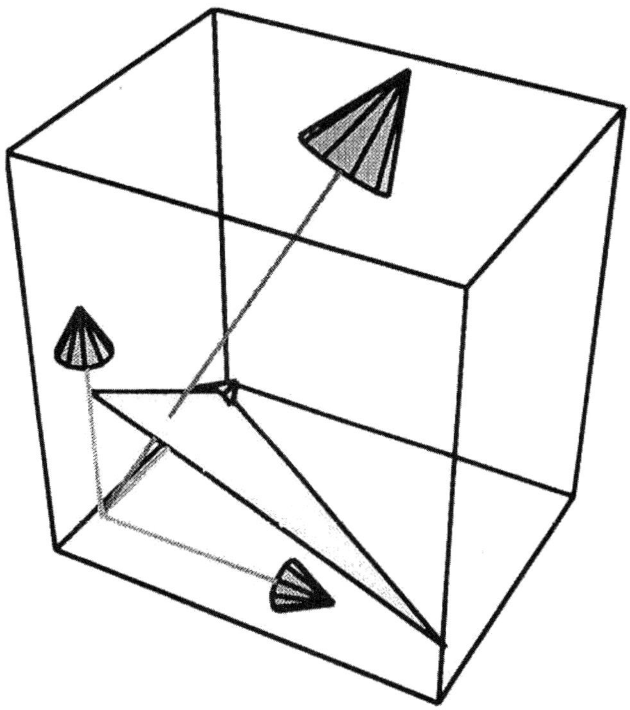

Figure 17.1
Prime basis, pitch vector, and pitch plane. The 2-axis is to lower right, the 5-axis to upper left.

will be the pitch, the second u-component might be called the "octavicity," and the third v-component gives the mixture between "dominantness" and "mediantness." The lone zero in the rotation matrix reflects our particular choice of the u-axis. One can easily calculate the pure dominantness and pure mediantness by first projecting $\mathbf{e}_3$ and $\mathbf{e}_5$ to the enharmonic pitch plane, normalize them, and use inner products with these projections.

Divisor Lattices

The divisor set `Divisors[n]` of a natural number will play a central part in what follows. In mathematics, the term *lattice* has two distinct meanings, an algebraic and a geometric one. In our case of tonal space geometric representation of the algebraic divisor lattice of a natural number, these two notions are unified. In the following, we give graphical images of divisor lattices in the "pitch-up" orientation, transformed by

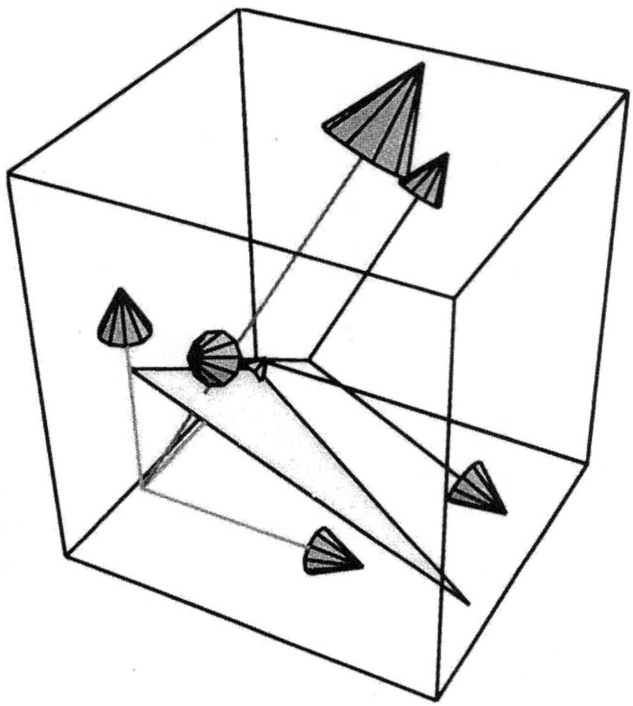

Figure 17.2
Prime basis, pitch vector, pitch plane, and pitch basis. Actually, the origins of the two coordinate systems coincide. Here, for clarity, the pitch basis is drawn shifted to an arbitrary location.

p to pitch coordinates. We connect the points representing two numbers m and n by a line if one of them **divides immediately** the other, i.e., one is a prime times the other.

Figure 17.3 gives an example of a divisor lattice, that of the number 60. This structure will be analyzed in more detail in the next section.

All positive integers make up an infinite divisor lattice $\mathbb{N}^{\infty}$. It could be called the "overtone lattice." The diagram shown in figure 17.4 is the bottom corner of this infinite (and infinite-dimensional) lattice showing the first four dimensions up to 60 (but omitting numbers that are not 7-smooth, i.e., contain a prime divisor larger than 7).

Figure 17.5 gives all divisor lattices up to 100 in musical notation, of 5-smooth numbers (that is, numbers not containing prime divisor higher than 5.)

The Triad Lattice

Major and minor triads in their root positions both appear in the divisor lattice of $60 = 2^2 \cdot 3^1 \cdot 5^1$. This number appears as the number base of ancient Babylonians and

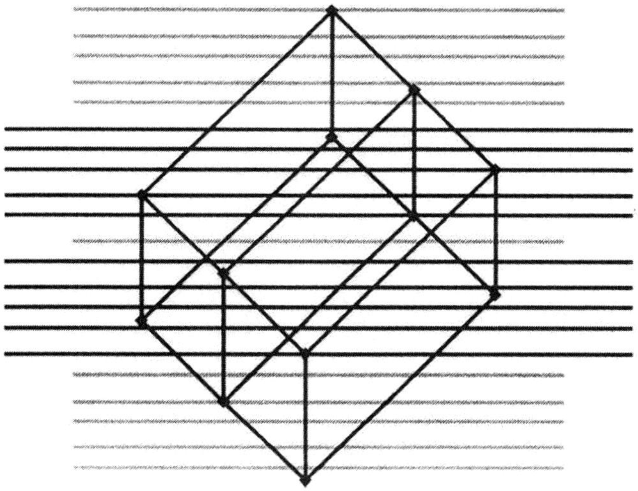

Figure 17.3
The triad divisor lattice of 60 with fundamental C_1 and leading tone B_6. Vertical dimension is strictly logarithmic, thence the diatonic staff and ledger lines are unequally spaced.

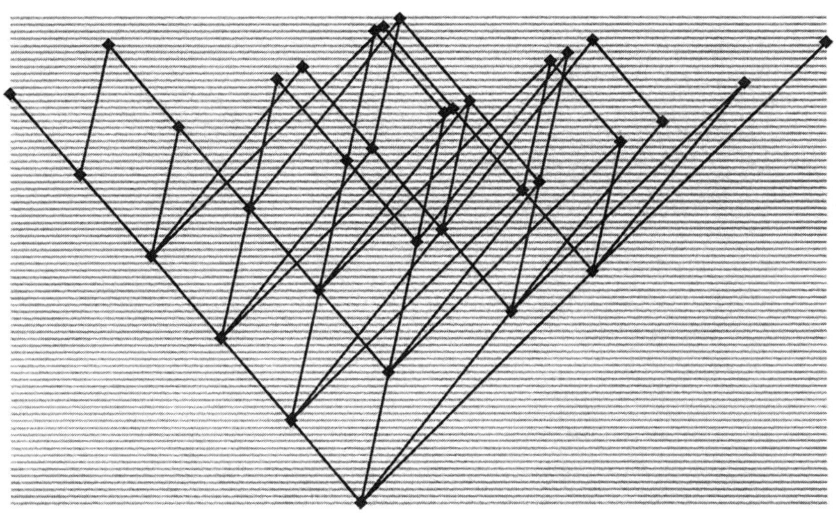

Figure 17.4
"Overtone series" is an infinite divisor lattice. Here is a lattice diagram of it, actually just one corner of it, up to the number 60. The number 49 is situated at top right.

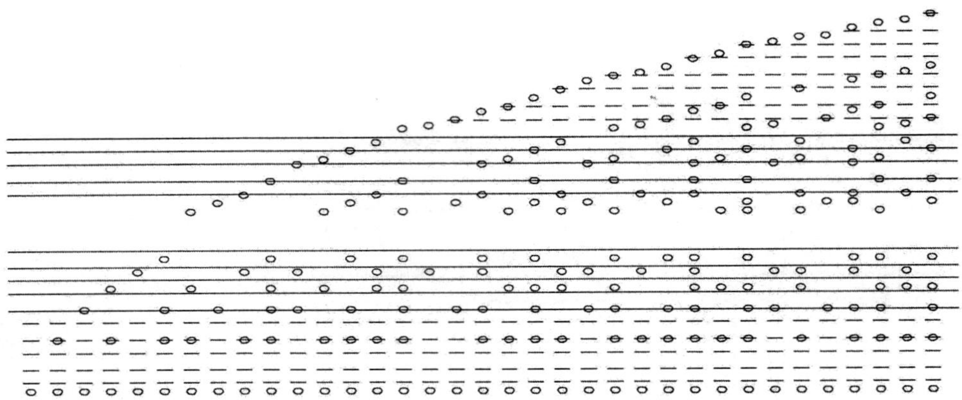

Figure 17.5
Divisor lattices of all 5-smooth numbers up to 100, starting from C_1.

still today in time and angle measurements, probably because it has a large number of divisors. This is due to a good mixture of small prime factors. Observe that powers of 2 have the smallest prime divisors, but all of these being the same, they do not yield particularly many *different* divisors.

The major triad 4.5:6 and the minor triad 10:12:15 appear both symmetrically in the middle of the divisor set of 60: {1,2,3,4,5,6,10,12,15,20,30,60}.

In music notation this combined major/minor chord (with C_1 as the fundamental) appears as shown in figure 17.6.

One can see here also the C seventh chord, in a particular setting, with the third and the fifth doubled and the C-major leading tone well separated from the root. In this case it is natural to consider the C_1 as the fundamental (the greatest common divisor, gcd, of chord frequencies), and equally natural although less obvious, the B_6 (the least common multiple, lcm) as the actual leading tone. Because of two prime factors 2 in 60, the fundamental is an octave doubled twice and, correspondingly, the leading tone (and similarly every other tone).

The difference in sensory qualities of major and minor can be explained nicely if one makes the following (unproven) hypothesis about the mechanism of hearing. *When we hear a tone with reasonably many dominant harmonic spectral components, we compute their common fundamental (gcd) and their common leading tone (lcm) and then "prime" the hearing apparatus to pitches in the divisor set, as if a set of tuning forks in the brain were tuned to all frequencies that are multiples of the fundamental and submultiples of the leading tone.*

This computation is well known in the context of the "missing fundamental" problem (see the research by the Dutch pioneers, Plomp and Levelt). Whether the same phenomenon is known in the context of a "missing leading tone," I am not aware.

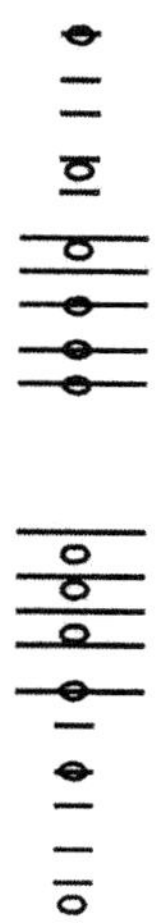

Figure 17.6
The triad lattice of divisors of 60. A note on notation: This system of notation does not use any accidentals. The vertical scale is pure logarithmic, not diatonic (except for a gap just under middle C). Observe that the spaces between staff lines are unequal. In the narrow space there is room for two pitch levels, in the large for three. By using right-sized note heads, these cases are easy to distinguish by seeing if a space note touches the line above it, below it, or neither.

Under this hypothesis, it is easy to explain why major is "bright and light" and minor "dark and subdued." In fact the hypothesis predicts that the C major chord and the e minor chord, as positioned above, are equally bright. But if one compares the C major chord and a c minor chord from the same root, the minor chord activates a set of internal resonators to an identical divisor chord which is tuned lower by a twelfth.

The two standard inversions of the triads do not fit in the divisor set of 60; they both require the larger harmony 120 and are in this sense less consonant. Surprisingly, the divisor theory predicts that there are more consonant settings of the triads than the root positions. Consider the divisor set of 30, $\{1,2,3,5,6,10,15,30\}$, as shown in figure 17.7.

Here one finds the major triad C_2, G_2, E_3 as a major sixth above a fifth, and the minor triad as a fifth above a major sixth. I call these chords the "mellow triads." In the mellow major triad, the third is transposed up by an octave; in the mellow minor triad the third is transposed down by an octave. The reader is asked to aurally test whether my claim of mellowness is true or not. Comparing the standard and mellow triads on a piano is not very conclusive because of the rich spectrum of a piano tone. Use sines instead.

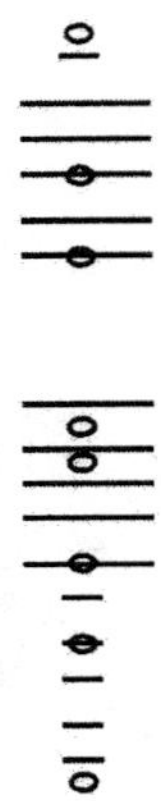

Figure 17.7
The "mellow triads" in the divisor lattice of 30.

The number 30 is actually the smallest three-dimensional number because it contains all three smallest primes 2, 3, and 5, each just once. In the tonal space its divisor lattice is represented by the unit cube. The larger elongated parallelepiped of 60 is simply the "octave doubling" of the "mellow harmony" 30.

The nearest tonal extensions of the lattice 60 are obtained by enlarging the 2×1×1 box of 60 by one unit in each tonal dimension, in turn. Notated versions of the lattices 60, 120 = 2 × 60, 180 = 3 × 60, and 300 = 5 × 60 are as shown in figure 17.8.

Notice that in 120 the seventh chord (C-E-G-B) appears in the middle symmetrically. Being itself symmetrical (major third, minor third, major third), this is the only place in a divisor set where it can exist without being mirrored somewhere else. We conclude that this chord itself is major-minor neutral. We generalize and claim that the attributes "major" and "minor" actually refer to the lower half and the upper half of a divisor lattice, respectively. From the other two extensions 180 and 300 the reader may want to read a few interesting things about what the lattice theory predicts next in terms of "gracious-sounding" chords in a given tonality.

It may also be instructive to look at the minimal extensions of the mellow harmony 30. We get the lattices, 60, 90, and 150, shown in figure 17.9.

The Diatonic Lattice

We turn to the study of the divisor set of one particular number, $8{,}640 = 2^6 \cdot 3^3 \cdot 5^1$. It has 56 divisors. On a linear scale the divisors grow fast as a function of their index, as shown in figure 17.10.

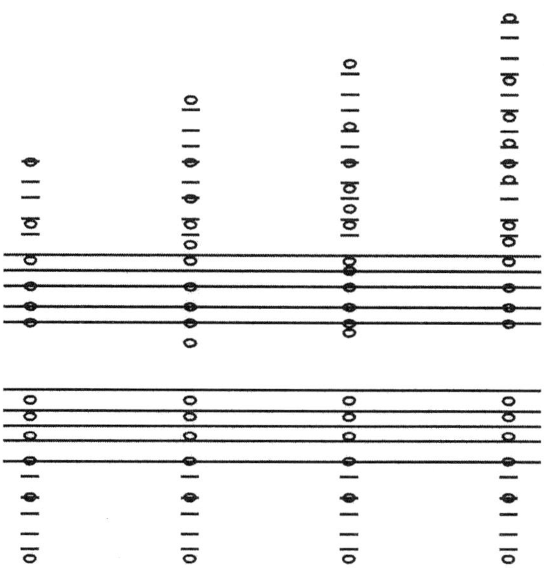

Figure 17.8
Divisor lattices of 120, 180, and 300 give "enrichened" triad harmonies.

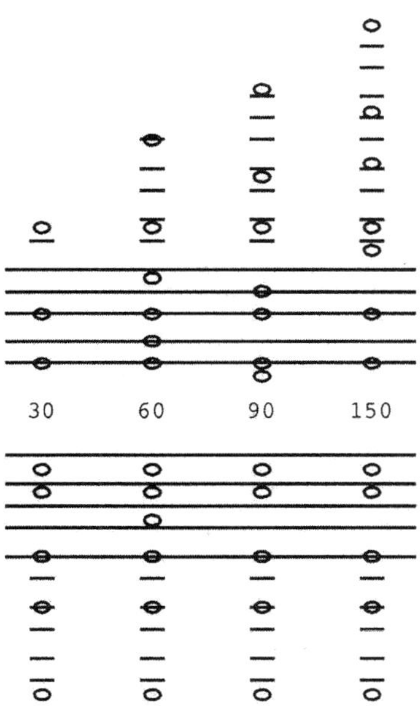

Figure 17.9
Enrichened lattices of the mellow triad lattice 30.

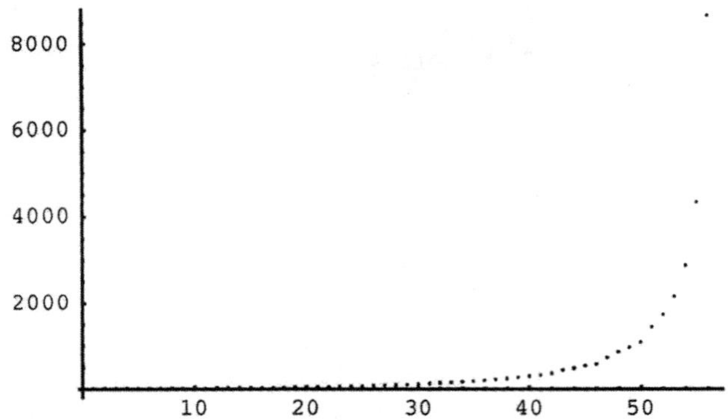

Figure 17.10
Divisors of the diatonic number 8,640.

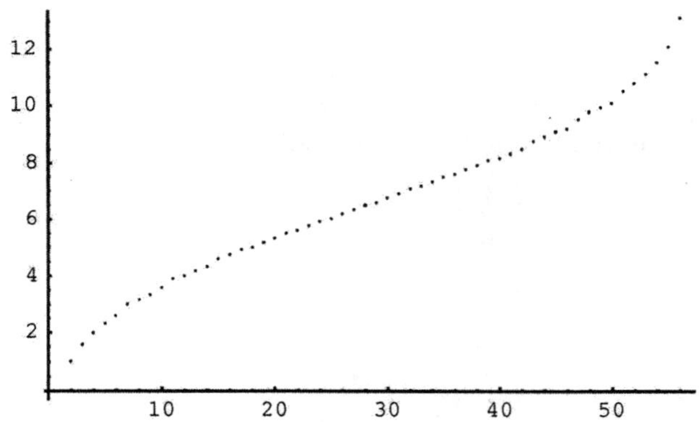

Figure 17.11
Logarithmic plot of the lattice 8,640.

A logarithmic plot reveals a sigmoid shape typical of divisor sets. The central almost linear segment gives something that resembles the diatonic musical scale (fig. 17.11). The vertical scale is given in octaves. As frequencies, the divisor lattice of 8,640 exceeds the human auditory range by three octaves.

A logarithmic plot of the intervals between consecutive divisors shows the central linear segment as the flat bottom of a bathtub-shaped curve. The vertical scale units are now equally tempered semitones (fig. 17.12).

The central tooth pattern reflects the alteration of the diatonic scale intervals 16/15, 10/9, and 9/8.

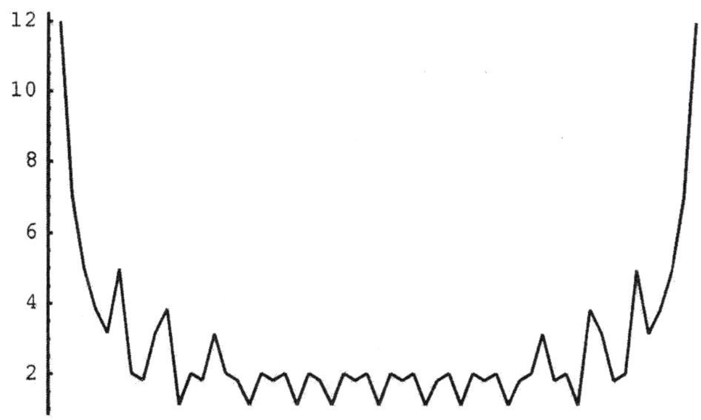

Figure 17.12
Intervals between consecutive divisors of 8,640, in semitones. The central set of "teeth" reproduces the tone/semitone structure of the diatonic scale.

The ordinary just major scale is obtained as the 15th through 22nd divisors: `{24,27,30,32,36,40,45,48}`.

The standard normalized form is obtained by dividing these numbers by the first number 24.

$$\left\{1,\frac{9}{8},\frac{5}{4},\frac{4}{3},\frac{3}{2},\frac{5}{3},\frac{15}{8},2\right\}$$

In analogy with the lattice 60, we expect to obtain the just minor diatonic scale as the inversion mirror of the major scale. We get

`{180,192,216,240,270,288,320,360}`

$$\left\{1,\frac{16}{15},\frac{6}{5},\frac{4}{3},\frac{3}{2},\frac{8}{5},\frac{16}{9},2\right\}$$

The mild surprise here is that we obtain the natural minor scale with its *second degree flattened.*

The graphic shown in figure 17.3 (major scale left, minor scale right) shows how the principal three degrees (I, IV, V) are common to both scales. All other pitches are shifted downward from major to minor, by roughly equal amounts, and as a result, fill in the octave into 12 very roughly equal steps, except for a central gap for the tritonus.

It is instructive to look at the intervals structures of the tonal neighbors of 8,640, obtained by dividing and multiplying it by the primes 2, 3, and 5. These tonal relatives are 1,728, 2,880, 4,320, 17,280, 25,920, and 43,200. We shall not do that analysis here.

The full lattice of 8,640 is reproduced in figure 17.14.

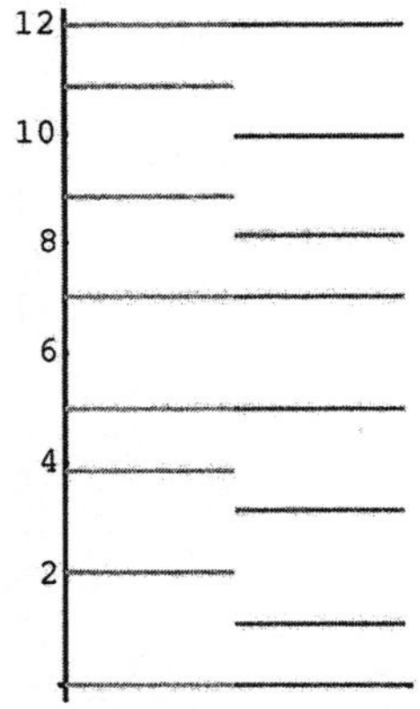

Figure 17.13
The major and minor scales from 8,640 compared. They are inverted mirrors of each other.

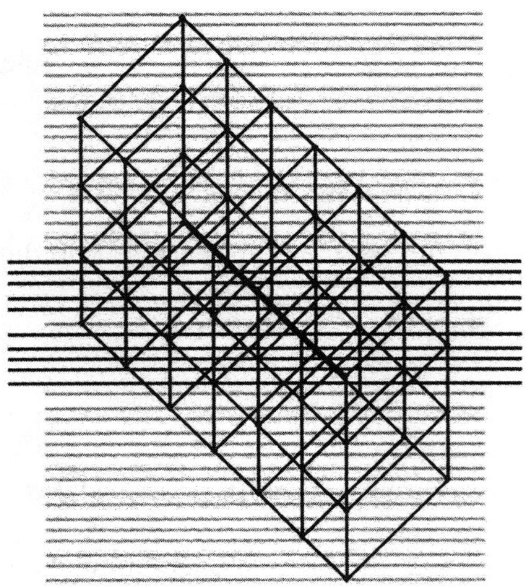

Figure 17.14
Divisor lattice of 8,640 gives the diatonic scale. The three floors' of this "building" correspond to the tonal functions IV, I, and V (counted from bottom up).

The Just Chromatic Scale

What is the just chromatic scale? What are the exact rational pitches of the black keys? Textbooks are generally silent about this and published values of scale degrees are not consistent. The standard answer to this dilemma is that the actual chromatic pitches depend on the tonal environment, but this is not a satisfying answer. Why is it not possible to specify the tonal environment?

I did a systematic computer search for a number among whose divisors there would be a long (>12) sequence of intervals of one semitone (when the intervals were rounded to the nearest ET semitone), and found the smallest such number to be $345{,}600 = 2^9 \cdot 3^3 \cdot 5^2$. I call this the "Donald Duck number" because it looks ridiculous but is not hard to remember.

The ambitus of this divisor set is ~18.4 octaves. If the set is centered around the middle C, the fundamental is at ~0.4 Hz and the leading tone at ~140 kHz.

Taking the lowest sequence of its divisors with semitone intervals, the 52nd to 64th of its 120 divisors, we obtain the following *just chromatic scale*:

{360,384,400,432,450,480,512,540,576,600,640,675,720}

The first number is 360, again familiar from angular measurements. The normalized scale degrees are:

$$\left\{1,\frac{16}{15},\frac{10}{9},\frac{6}{5},\frac{5}{4},\frac{4}{3},\frac{64}{45},\frac{3}{2},\frac{8}{5},\frac{5}{3},\frac{5}{3},\frac{16}{9},\frac{15}{8},2\right\}$$

All intervals are familiar, except for the strange looking 64/45 for the tritonus, a nice rational approximation to $\sqrt{2}$. The scale intervals are

$$\left\{\frac{16}{15},\frac{25}{24},\frac{27}{25},\frac{25}{24},\frac{16}{15},\frac{16}{15},\frac{135}{128},\frac{16}{15},\frac{25}{24},\frac{16}{15},\frac{135}{128},\frac{16}{15}\right\}$$

We have four different semitones here, with sizes measured in ET semitones:

{0.707,0.922,1.117,1.33}

One of them, 16/15, is the *diatonic semitone*. The smallest one, 25/24, is the *chromatic semitone*. The remaining two are 27/25, the *Pythagorean great limma*, and 135/128, the *major chroma*.

We took the divisors 52nd to 64th from the **bottom half** of the divisor set. For this reason the scale obtained is the *major just chromatic scale*. Had we taken the scale from the mirror symmetrical upper part of the divisor set, we would have obtained the *minor just chromatic scale*:

{480,512,540,576,600,640,675,720,768,800,864,900,960}

Figure 17.15
Major (left) and minor (right) just chromatic scales compared. The gray lines give the equally tempered (ET) semitone values. This graphic should be remade for better clarity; there may also be an error in the choice of the minor scale.

The graphic in figure 17.15 gives an indication of how much the just major scale and the just minor scale differ. The gray lines mark ET semitones.

We had chosen the 52nd divisor as the lowest pitch of the major just chromatic scale because in the full divisor set it marks the first divisor in the sequence of 18 divisors with semitone distances. (There are 17 consecutive ones in the following list.) But if you carefully look at the pattern, you may agree that the "tight JCS" begins considerably earlier, there are just a few gaps in it, marked by 2's in the list. The just scale gradually dissolves into a diatonic scale at its both ends.

{12, 7, 5, 4, 3, 5, 2, 2, 3, 4, 1, 2, 2, 3, 1, 1, 2, 1, 2, 2, 2, 1, 1, 1, 2, 1, 2, 1, 1, 2, 1, 1, 1, 2, 1, 1, 1, 1, 1, 2, 1, 1, 1, 1, 1, 1, 1, 1, 1, 2, 1, 1, 1, 1, 1, 1, 1, 1, 1, 1, 1, 1, 1, 1, 1, 1, 2, 1, 1, 1, 1, 1, 1, 1, 1, 1, 1, 2, 1, 1, 1, 1, 1, 2, 1, 1, 1, 2, 1, 1, 2, 1, 2, 1, 1, 1, 2, 2, 2, 1, 2, 1, 1, 3, 2, 2, 1, 4, 3, 2, 2, 5, 3, 4, 5, 7, 12}

The idea that one begins with the lowest continuous stretch of semitones is artificial. Just increasing the size of the lattice will give a different result. The two curves shown in figure 17.16 indicate this. The first is the "bathtub curve" for 345,600, the second for 4 times it (1,382,400).

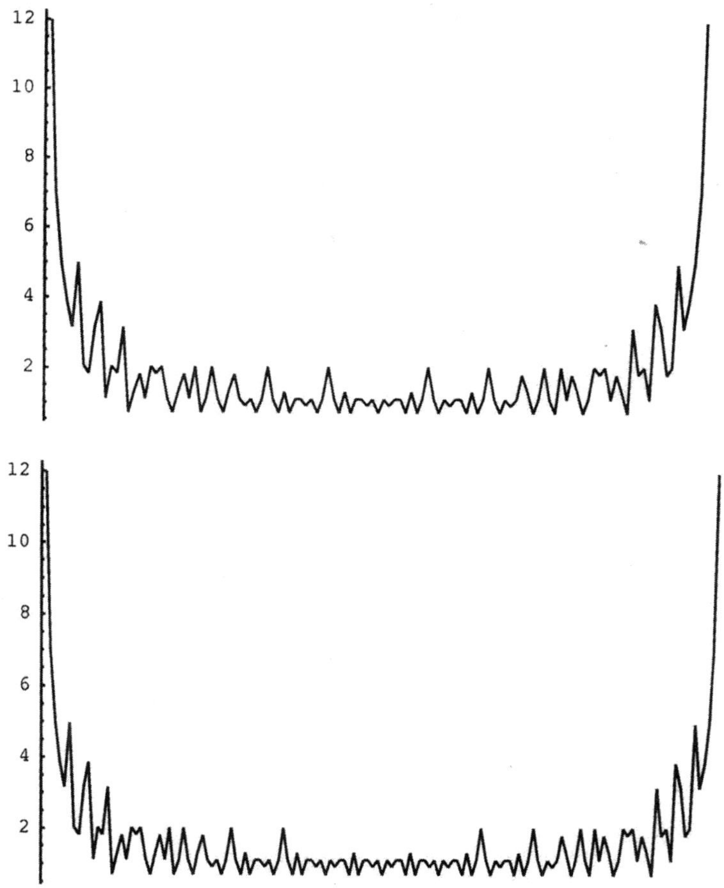

Figure 17.16
Bathtub curves for 345,600 and 1,382,400.

By octave enlarging the set, the upward teeth going to 2 semitones get more separated and we obtain a longer central stretch of chromatic scale. The interval structures of the flanking diatonic stretches look interesting and may deserve separate study.

We have, in several cases, identified majorness with the lower half of a divisor lattice, starting from the bottom **fundamental**, and *minorness* with the upper half of a divisor lattice, ending to the top **leading tone**. (In earlier writings I called the highest number the "sampler," because it would be a natural sampling frequency for all pitches in its lattice. I have lost the reference to the musicologist stating that "Tonality is a thing determined by a fundamental pitch and a leading tone." That gave me the kick.)

Table 17.1

Chord	Relative frequencies						Major/minor index
Major triad root position	4	5	6				–0.220471
Major triad first inversion	5	6	8				–0.236811
Major triad second inversion	6	8	10				–0.333333
Minor triad	10	12	15				0.220471
Mellow major triad	2	3	5				–0.333333
Six first partials	1	2	3	4	5	6	–0.464363
Major seventh	1	5/4		3/2	15/8		0
Minor seventh	1	6/5	3/2	15/8			0.126357
Dominant seventh	1	5/4	3/2	16/9			–0.142885

Let us introduce a more exact measure of majorness/minorness, the M-index ("majmin index"). The function applies to any rational chord or scale; it gives a number in the range $-1\ldots+1$ (-1 for extreme major, $+1$ for extreme minor, 0 for majmin neutral). It is constructed as follows. Let c denote a rational chord. Compute its gcd(c) and lcm(c). Construct a straight line scale $\sigma(p)$ on the logarithmic pitch p such that $\sigma(\log \mathrm{gcd}(c)) = -1$ and $\sigma(\log \mathrm{lcm}(c)) = +1$. Then compute the weighted sum $M(c) = \sum_{i \in c} p_i \sigma(p_i)$. Results for typical chords are given in table 17.1.

It may seem odd that the ordinary triads have such low value of majorness/minorness, less than one-fifth of the maximum. The mellow triad fares as well as the second inversion. But if one takes 1,000 first partials (harmonic overtones), the index will be –0.988. Thus, the extremal values ±1 are approached asymptotically by long overtone and undertone series. The diatonic major and minor scales have absolute values 0.155 and the just chromatic scales 0.013. All 2-chords (intervals) have a zero index, of course, like all inversion symmetric chords. There are no "major" or "minor" intervals despite of verbal usage. Actually, it should be an interesting task to clarify the reasons for classifying intervals into major and minor ones.

Higher Tonal Dimensions

Why is the tonal space only three-dimensional? I have not seen any well-researched case of a musical system in use that would have its practice (instruments and music) and theory that essentially were using primes higher than 5. Powers of 7 and higher primes may occur in theoretical calculations. The Chinese "changing note" *pien* may reflect occasionally the seventh partial, but it may also reflect the usual confusion about flattening the seventh degree. This confusion still exists in the confusion about what is the minor scale.

Possible Explanations

- Brain three-dimensionality
- Three dimensions generate sufficiently dense scales (counterargument: the Pythagorean space $2^{\mathbb{Z}}3^{\mathbb{Z}}$ is already dense. For some reason people consider the Pythagorean tuning to be "odd."
- Computational trade-off in brain processing
- Plain tradition: Instrumentalists and singers might well use natural sevenths, but nobody has registered this usage
- Plain tradition: The natural septimal seventh might be acceptable but unfortunately it is not available from standardized instruments and the small intervals between it and established scale degrees would introduce unpleasant dissonance
- More dimensions would overload the auditory processing system
- Hearing might use a modified cortical structure, originally evolved for spatial tasks. Sound objects originally moved in 3-space and their auditory counterparts might have from the beginning gained advantage by utilizing existent genetic machinery for 3-space representations

Musical instruments are still constructed not to sound prime partials 7, 11, 13, and higher. We want to know why music is based on 5-smooth numbers.

Note. A natural number is *k-smooth* if it does not contain a prime divisor larger than k.

Musically, the question is whether tonal dimensions 2, 3, and 5 really are separate dimensions of musical qualia, and whether 7, 11, and so on would represent "higher" and new such qualia. Consider the first question first. The difference between the divisor set and tonal space ideas and traditional theoretical views is that I have taken the octave dimensional on equal footing with the 3-dimension ("fifth-dimension," dominant-subdominant dimension) and 5-dimension ("third- dimension," mediant dimensions). I do not have experimentally valid facts about the comparability of the "octavicity," "dominantness," and "mediantness" dimensions. Mathematically, it just seems that the traditional music theory has unnecessarily mixed the octave dimension with the 3- and 5-dimensions by taking the fifth and the third intervals as the generators, instead of the plain primes.

By calling the space of prime exponent the "tonal space," I have implicitly made the assumption that the prime exponents were really representing different qualia—in the best possible case, "independent" or "orthogonal qualia." Depending on how our brains are constructed, the fourth dimension of 7 might sound, if well used musically, as plain dissonant mess, as indifferent, or as a new kind of ethereal tonal quality. This can be tested, but the test paradigm should take into account that our musical hearing

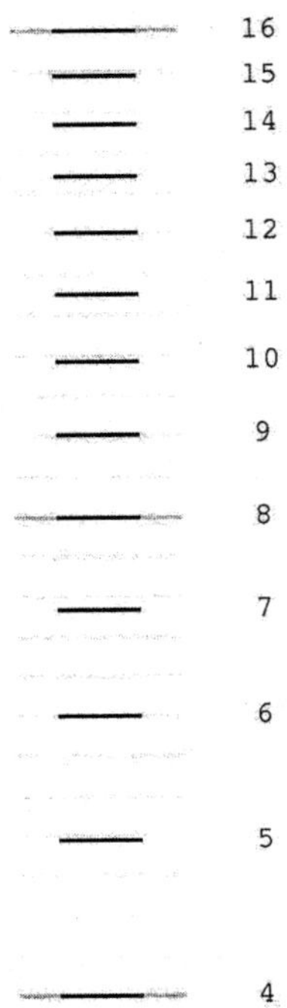

Figure 17.17
Partials 4–16 and ET pitches.

apparatuses might well be biased toward a 3-dimensional interpretation because of lifelong exposure.

The graphic in figure 17.17 indicates how badly prime partials fit a 12-tone scale. The ET degrees are the longer [gray] lines, the partials short black lines. The seventh partial is nearer to the major seventh than to the minor seventh, likewise the fourteenth. The next two primes 11 and 13 sit crudely in almost quarter-tone positions.

Tonal Shapes

The present discussion has centered on particular simple shapes in the tonal space of prime exponent sequences, the divisor lattices. Their advantage is mathematical simplicity, but perhaps neural circuits in the brain do not especially appreciate right angles. There are several plausible ways to generalize the divisor lattice concept toward more general "potato shapes" (triaxial ellipsoids, rotated triaxial ellipsoids, spherical harmonics, convex sets, etc.). For still more generality, define a *tonal distribution* as a complex-valued function on the tonal space, $\rho : \mathbb{Z}^{\infty} \to \mathbb{C}$. It assigns with every tonal vector x a sine wave with amplitude $|\rho(x)|$ and phase $\mathrm{Arg}(\rho(x))$. The general claim is now that the concepts of tonality and tonal center are captured by the notion of a tonal distribution that is convex (in a suitable sense) and has a single maximum.

Hypothesis. This is actually the strongest assumption in this article. Evolution has found a way that empowers us to deduce from pressure variations at two head-mounted points (meatuses) an internal description of a sound space, weakly mapped on the visual-kinesthetic space, of sound sources and reflectors. I have indirectly proposed that distinct sound objects correspond to humps in tonal space distributions. The auditory apparatus groups all sound partials nearby to each well-defined activation maximum into a single group with a perceived tonal center, each such group potentially corresponding to a separate individual sound source in the environment.

The Local Tonal Environment

Look at the tonal point environment of the origin, in the pitch coordinate system $\{p,u,v\}$. The lattice points are not any more neatly arranged after the rotation to this new coordinate system. But their geometric configuration is the same for every tonal point; this is a universal structure. Let us try to visualize the tonal environment of a lattice point. We have here a mild conflict between two notions of "what is the distance of two pitches?," the nearness in pitch or the enharmonic tonal nearness. We try to form a distance function that balances between the two measures in such a way that two pitches are judged to be near to each other if (1) their tonal components are similar or (2) their pitches are nearby. This approach will lead to the idea of a local tonal environment in the shape of a rotation hyperboloid.

Begin by considering cylinder shaped environments in the pitch coordinate system. For simplicity, consider a cylinder with height h in the pitch dimension and a circular shape with radius r in the enharmonic subspace. Figures 17.18 and 17.19 provide two graphics showing the lattice points in a flat "pancake" cylinder and in an oblong "sausage-like" cylinder.

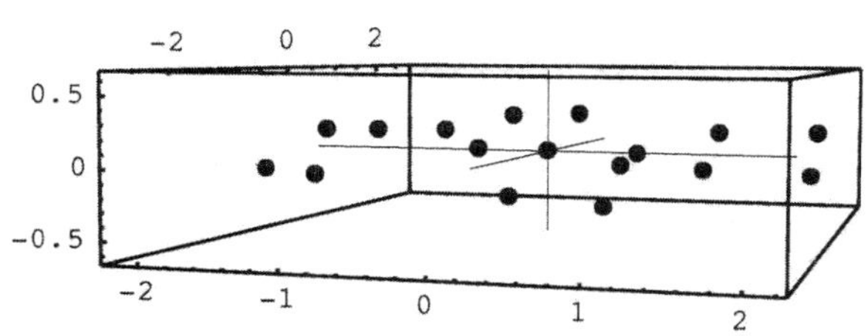

Figure 17.18
Flat "pancake" tonal environment = melodic neighbors.

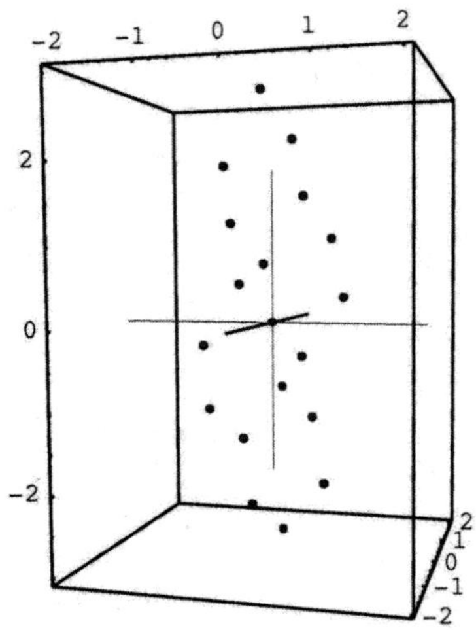

Figure 17.19
Oblong "sausage" tonal environment = a scale.

Although the two examples of figures 17.18 and 17.19 were chosen rather randomly, exactly the same patterns exist around every point in the tonal space.

Coincidences

It seems that all the curious tonal structures in music result from number-theoretic coincidences. A coincidence is simply a rational number very near to unity. For example, the long-time (theoretical) success of the Pythagorean tuning system derives from the fact that 2^{19} is very nearly equal to 3^{12}, their ratio being the small Pythagorean comma 1.01364, which is small enough to justify the false statement that the circle of fifths closes in 12 steps.

I did a systematic computer search on coincidences in the low tonal dimensions. Here are some examples.

For example, figure 17.20 shows a plot of a (suitably weighted) measure of the sameness of powers of 3 and 5. The exponent of 3 varies from 1 to 50 and that of 5 from 1 to 35. The prominent peak near to the origin reflects the sameness of 25 and 27, or the second power of 5 and third power of 3. There seems to be a general repetitive pattern of peaks, but I don't know whether it continues.

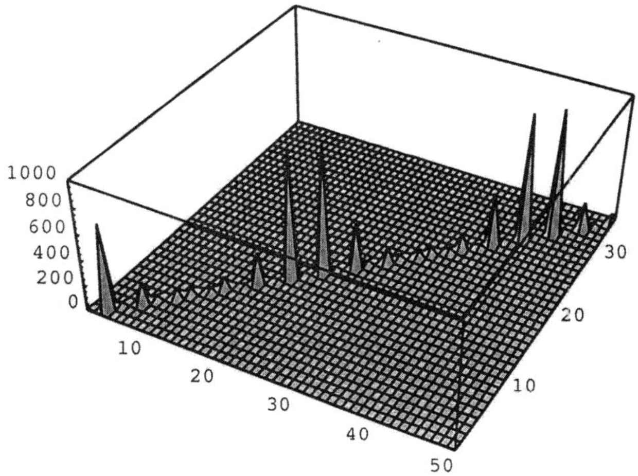

Figure 17.20
Coincidence map of powers of 3 and 5. A peak in the plot means that some power of 3 (right down axis) is near a power of 5 (right up axis).

Table 17.2 gives some results from a four-dimensional search. The leftmost column gives the tonal coordinates, the exponents of primes 2, 3, 5, 7. The middle column gives the rational number interval, and the right column the interval in ET semitones.

There is one "Pythagorean" beauty $\frac{64}{63}$ and a few other pleasant-looking ones. Look at that $\frac{4375}{4374}$, coincidence to four-thousandths of a semitone!

Table 17.2

{−9,6,1,−1}	$\frac{3645}{3584}$	0.292178
{−9,8,−4,2}	$\frac{321489}{320000}$	0.0803696
{−6,−10,7,2}	$\frac{3828125}{3779136}$	0.222978
{−5,−3,3,1}	$\frac{875}{864}$	0.21902
{−5,2,2,−1}	$\frac{225}{224}$	0.0771152
{−4,9,−2,−2}	$\frac{19683}{19600}$	0.0731577
{−1,−9,9,−2}	$\frac{1953125}{1928934}$	0.215766
{−1,−7,4,1}	$\frac{4375}{4374}$	0.00395756
{0,−5,1,2}	$\frac{245}{243}$	0.141905
{1,2,−3,1}	$\frac{126}{125}$	0.137948
{5,−4,3,−2}	$\frac{4000}{3969}$	0.134693
{6,−2,0,−1}	$\frac{64}{63}$	0.272641
{6,0,−5,2}	$\frac{3136}{3125}$	0.0608324
{7,5,−4,−2}	$\frac{31104}{30625}$	0.268683
{7,7,−9,1}	$\frac{1959552}{1953125}$	0.0568749
{10,−6,1,−1}	$\frac{5120}{5103}$	0.057578

Bibliography

Birkhoff, Garrett. *Lattice Theory*. American Mathematical Society, 1948.

Conway, J. H., and N. J. A. Sloane. *Sphere Packings, Lattices and Groups*. Springer, 1993.

Levvin, David. *Generalized Musical Intervals and Transformations*. Yale University Press, 1987.

18 On Electronic Music Instruments

Erkki Kurenniemi
Translation by Mikko Ojanen

In the chain of musical communication music notation, the traditional musical memory medium is located between the composer and the performer. The new recording methods, such as a record or a tape, convey the information from the musical instrument to the listeners.

The recent development of electronic musical instruments is paving the way for a new memory medium, whose primary location will be between the performer and the instrument. An early-stage implementation of the novel memory medium is a digital cassette, that is, an ordinary tape cassette on which the electronic commands that control the instrument electronically are recorded in a digital format used by computer technology.

The abstract score is recorded on a tape with a computer or a performance instrument. The playback can be fully automated, although the human ability to control the realization of the score during the performance is also essential. This way the commands on the cassette are in charge of the mechanical reproduction of the work while the performer can fully concentrate on the nuances of interpretation.

Synthesizers

The term "synthesizer" originates partly from the two room-size music machines programmed with a punched tape developed by RCA. In this context, the term synthesis mainly referred to an inverse implementation of Fourier analysis (in Fourier analysis, the sound is mathematically decomposed into its sine wave components). Currently, in the absence of a more fitting term, a synthesizer is a system composed of electronically controllable sound generator and shaping units. The control by electronic control signals, instead of manually adjusted knobs, enables the composer to combine simple basic modules into various instrument ensembles.

The synthesizer's potential as a general music production system was widely publicized a couple of years ago by the "Switched-on Bach" record produced with the Moog synthesizer, developed by the American pioneer of the field Robert Moog. The

Moog synthesizer (as its many successors, including Buchla and Arp) is based on voltage control. The principal module types are the voltage-controlled oscillator (VCO), the voltage-controlled filter (VCF), and the voltage-controlled amplifier (VCA). The VCO is a sound generator whose frequency (pitch) depends on the control voltage fed into the unit.

In the Moog system, for example, the change of control voltage by one volt causes the change of pitch by one octave. Commonly the frequency range of a voltage controlled oscillator extends well below the range of human hearing. By plugging an oscillator with a frequency range of 1–20 Hz into the control voltage poles of another oscillator one can implement a frequency modulation, that is, a vibrato. The speed of a vibrato can be controlled with a third oscillator, etc. The VCFs, VCAs, and their various combinations provide almost limitless possibilities to shape the timbre and dynamics of the basic material generated with oscillators and noise generators.

Preprogrammed control voltage patterns can be produced with a so-called sequencer. A typical sequencer contains ten to fifty adjustable potentiometers, whereby the voltage values can be set to correspond to the consecutive pitches of a certain melody. The electronic selector of the sequencer reads the set values one after another at the desired speed and forms a control voltage that changes by jumps.

In addition, a control voltage can be produced with an electronic organ-type keyboard. The electronic circuits of the keyboard produce a control voltage whose value depends on the location of the pressed key in the keyboard. The use of the keyboard is not restricted only to controlling the pitch. It can be used as well to control the timbre, amplitude, duration, location (when producing stereophonic music), or all of these at the same time.

"Programming the synthesizer" refers to a process of connecting modules together to form larger systems. In the Moog synthesizer, programming is performed with interconnecting cables, the same way as in an old telephone switchboard or an analog computer. Other methods are used as well, but in all of them the downside is a nonillustrative and slow programming process, which hinders the use of the synthesizer as a concert instrument.

DIMI: The Digital Instrument

The electronic applications are based on two essentially different but complementary basic methods. These are analog, that is, linear technology and numerical, that is, digital technology. Linear technology operates on continuous voltage signals.

Analog signals can be amplified, filtered, stored, added up (e.g., mixer amplifiers), multiplied (e.g., modulators), etc. with various so-called linear circuits.

In addition to the aforementioned voltage-controlled synthesizers, linear technology has been applied mainly in radio, television, and sound reproduction. The information processed by digital technology is discrete and usually binary. The voltage quantities can have only two values, which are read as binary numbers 0 and 1 or logical truth-values (e.g., true-untrue, on-off, less-greater), depending on the context. Digital, that is, logical circuits process logical operations (AND, OR, NOT) and store (remember) the values of logical variables (flip-flop). Quantities with larger values can be presented with several binary variables. For example with three bits (binary variables) it is possible to present numbers from 0 to 7 (0=000, 1=001, 2=010, 3=011, 4=100, 5=101, 6=110, 7=111).

Digital technology has been applied particularly in computers, electronic calculators, and numerical control systems. In many branches of electronics, digital technology is currently gaining an advantage over its analog counterpart. This the result of a relatively simple process of storing digital data.

The shift of focus is sped up by the massive development of integrated circuits. At present it is possible to manufacture functional units (e.g., memory units, calculators, even computers) equaling several thousand transistors on the surface of a crystal silicon the size of a few of millimeters. One goal in the development of the DIMI synthesizer was to study to what extent digital technology could be applied in the production of electronic music. The experiences are not entirely unambiguous.

It seems that, in certain details, a hybrid solution (a digital control of the sound-processing circuit implemented with linear technology via digital-analog converter) would lead to a better result than the fully digital implementation used in the DIMI.

The DIMI is played by touching the metal plates attached to a fiberglass board that forms a playing surface with two metal sticks. Plates are organized into two quarter-circle arches for easy access to each plate with a simple swing of a wrist. Each arch consists of sixteen plates. When one of the eleven modules of the instrument is selected from the left arch, its value can be adjusted using the plates in the right arch.

The eleven modules are:

- two sound generators, which produce a chromatic scale within one octave
- two frequency divisors, which transpose sound generators into eight octaves
- two volume controllers
- two filter banks
- two selectors, which can be used to connect sound generators to one of the three modulator circuits
- one vibrato unit, common for both sound generators

The left-hand-side keyboard's function is to assign the tasks of the right-hand-side keyboard. The generator commands of the DIMI refer to a pair of figures "*m*, *n*" where *m* refers to the numbers of the left-hand-side touch plates (i.e., the modules of the instrument) and *n* is the serial number of the right-hand-side plates.

DIMI is played directly by giving generator commands, which can be set with two touches. The central part of DIMI is the memory unit, which can be compared to a tape

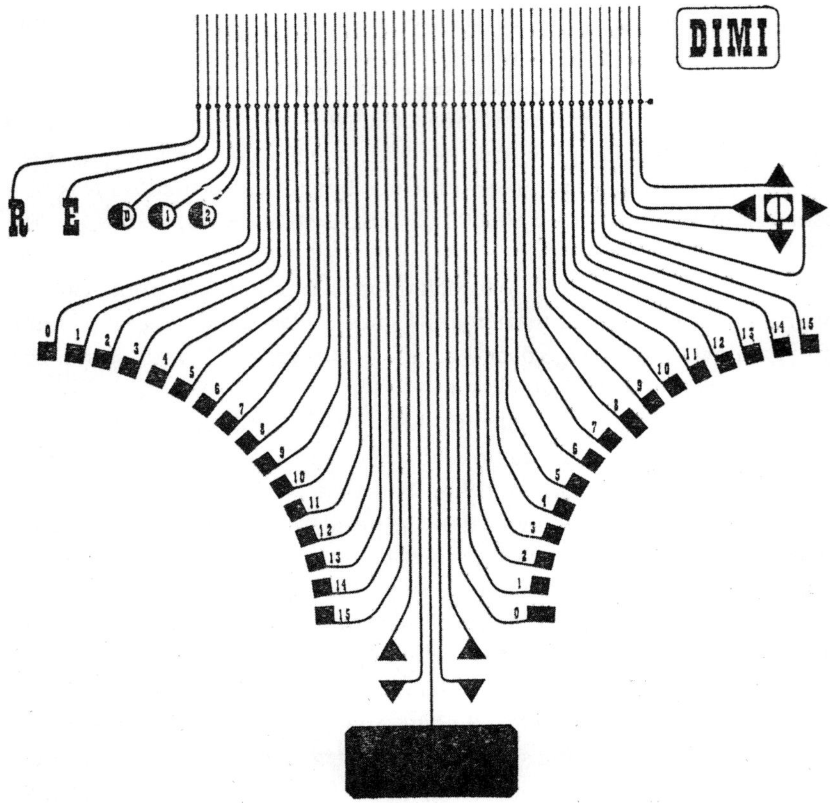

recorder. The imaginary tape is divided into sixteen consecutive sections (bars), each of which is divided into sixteen parts (steps). The tape can be played in both directions and the speed can be chosen from 48 different values.

The speed of the tape assigns the tempo of the preprogrammed performance. Commands changing the tempo during playback can also be programmed onto the tape. With these commands even the duration of one step can be changed or produce accelerating and decelerating tempos. The so-called jump commands are equivalent to the fast-forward and rewind of an ordinary tape recorder.

Fast-forwarding and rewinding are immediate, because the imaginary tape doesn't have mechanical inertia. The jump commands (i.e., rewind commands) programmed into the tape form closed loops, which continue to repeat the sequence until a new command is given to move to another loop.

The tape can be started and stopped by touching the star-shaped plates on the right upper corner of the keyboard. The plates assign the direction of the tape and can be used to move manually step by step in either direction on the tape. The tape has to be halted when recording new commands. The last generator command can be recorded on the tape by touching the R-shaped (record) plate.

Commands recorded on the same step don't disappear when new commands are recorded. Between recording separate commands, the tape is moved manually a desired amount of steps. The maximum amount of commands is restricted to 100 because of the present memory capacity. The commands can be erased with the E-plate and its adjacent plates, which control the range of the erasing command. The weakest erasing command can erase any individual command. The strongest command erases the entire memory.

Given its memory capacity, the DIMI is suitable only for studio work. The use of the instrument in concert settings will improve significantly after the addition of an external tape memory unit mentioned at the beginning of this essay.

DIMI-O: The Video Instrument

As a concrete result of a development project supported by a loan from SITRA,[1] the first prototype of the DIMI-O, an instrument applying television and digital technology, was completed in April 1971. The conventional part of the instrument is an ordinary four-octave electronic organ keyboard, where a 32-step program can be stored.

The instrument repeats the program at a speed that can be continuously adjusted as in a music box or a barrel organ. (The number of all the possible programs can be counted by raising 2 to the power of 4×12×32=1536 [four octaves, each including 12 chromatic pitches], resulting in a number with approximately 500 digits.)

The contents of the memory can be constantly read from the television monitor as a certain type of simplified music notation. Individual notes are presented as a horizontal line, whose vertical location represents its pitch and the length its duration. The position of the pointer, a vertical dark bar moving horizontally across the TV screen, represents which part of the program the instrument is currently reading. Notes, chords, and tone clusters can be stored into a certain step in the memory assigned by the pointer location. Similarly, individual notes can be erased from the memory.

The notable feature of the instrument is its television, that is, video camera input. The image conveyed by the camera, which is converted to fully black and white with a bang-bang circuit, can be interpreted as musical notation or as graphical score.

The image from the camera can be stored in the memory as is or it can be added to the previous memory content with logical OR or AND operations. An OR operation, that is, logical sum, enables the drawing of the score by moving a small light in front of the camera. A logical multiply (AND operation) can be used to erase certain parts from the score.

The video organ has three essentially different uses. First, it can be used as a studio instrument to produce sound for a taped musical work. In this context, the obvious use is mainly inputting and editing the score with the keyboard or with the camera and dotted light.

Second, the DIMI-O can be used as an instrument in solo performances or with various ensembles, perhaps mainly as a cluster instrument of some kind. The player can control the instrument fairly well by simply waving a properly lit hand in the field of vision of the camera. The lack of mechanical feel, that is, haptic feedback can effectively be replaced by visual feedback from the monitor after a brief period of practice.

However, the most interesting applications are probably in the field of experimental film, ballet, theater, and video art. A dancer can create an accompanying music for his or her performance, and in film, each change in the visual experience can be reflected in the music. In these performances, the role of a video organist deviates clearly from the traditional roles of a composer or a musician.

The instrument provides only mechanical correspondence between visual and acoustic events. For a sensible aesthetic expression, the instrument needs to be played (at least for now) by a cameraperson with a creative imagination.

Prospects

Manford L. Eaton is aiming rather high with his "bio music" by letting almost every electronically measureable human biological function, from a heartbeat to an EEG, to control the electronic musical instrument.

Small computers of a more compact size will soon be a part of basic electronic music studio equipment, and direct connections to large computer centers anticipate the musical applications of artificial intelligence, which is evolving very quickly. During the 1950s electronic instruments were controlled by hand; the 1960s was the decade of control by voltage (Moog built his first synthesizer already in 1962). The 1970s will be the age of digital control methods and minicomputing. In the 1980s, the composer will at least be able to converse and interact with his studio equipment; perhaps the instruments will be able to even read his or her thoughts directly.

Notes

Original title: "Elektronisen musiikin instrumenteista." Published in *Musiikki* 1 (1971), Helsinki.

1. [The Finnish Innovation Fund.—Trans.]

19 Interaction of Music and Technology: The Music and Musical Instruments of Erkki Kurenniemi

Kai Lassfolk, Jari Suominen, and Mikko Ojanen

Introduction

In the 1960s and early 1970s, Erkki Kurenniemi designed ten technologically advanced electronic musical instruments. During the same period, he also made a large number of musical recordings, including a number of landmark works in the early history of Finnish electroacoustic music. Although Kurenniemi is particularly merited for his instruments, his musical output shows similar visionary insight. Kurenniemi's music also reflects his instrument design ideas and demonstrates his attitude toward practical use—and misuse—of new technology, his own instruments included. This chapter discusses Kurenniemi's instruments both as independent objects and as parts of musical interaction, where the musical outcome is also a key feature. The chapter focuses specifically on two instruments, the DICO (Digitally Controlled Oscillator, 1969) and the DIMI-A (Digital Musical Instrument with Associative Memory, 1970; for more information on the other DIMI instruments, see Ojanen et al. 2007[1]), and Kurenniemi's early musical recordings made with them, especially the works "Improvisaatio" (1969) and "Inventio-Outventio" (1970).

As an instrument designer, Kurenniemi made his mark as an advocate of digital technology. Although he was neither alone nor the first in applying binary logic or digital electronics, his designs were compact in size compared with many of the computer-based music systems of the time. During Kurenniemi's most productive instrument-making period, analog synthesizers became increasingly popular, first as studio instruments and finally as part of electric keyboard players' standard equipment. Digital synthesizers, on the other hand, had to wait a decade to see a similar boom in demand. In this regard, Kurenniemi's designs were ahead of their time, which is reflected by their modest commercial success. However, Kurenniemi was convinced of his technological approach. As a result, both the advantages and limitations of early digital technology are demonstrated in both the DICO and the DIMI-A.

Both the DICO and the DIMI-A are programmable synthesizers with a digital memory. They also have many similarities in both their user interface and internal structure.

However, they also represent important changes in both Kurenniemi's instrument design and his musical output. The shift from the DICO to the DIMI-A represents a change from custom-built devices to generic designs intended for industrial manufacture and commercial distribution. The DIMI-A is also a watershed in Kurenniemi's user-interface design. After the DIMI-A, Kurenniemi's interests moved from binary touch switch-based input to more intuitive ways of instrument control.

With the introduction of the DIMI-A, Kurenniemi's music moved from impulsive and improvisational expression toward a more predetermined workflow. We argue that the change was partly caused by the structure and the user interfaces of the aforementioned instruments. Furthermore, from the DIMI-A onward, Kurenniemi's instrument design interests diversified into two directions: (1) advanced programmability and (2) experimental instrument control.

While Kurenniemi's visions about music and technology were advanced and often seemed far-fetched especially at the time, his practical implementations were often straightforward applications of widely available, albeit novel, circuit designs. On the other hand, while his musical output may have seemed even deliberately impulsive, it anticipated many practices of modern music production, such as early examples of (live) sequencing, musical programming and technologically driven music making in general. These features are prevalent in both "Improvisaatio" and "Inventio-Outventio."

An Overview of Kurenniemi's Musical Instruments

Kurenniemi's first major endeavor in electronic musical instrument design was a modular sound synthesis system consisting of a tone generator unit, a mixer unit, and a filter unit. Kurenniemi got the system in working condition in the fall of 1964 and it was immediately used to create the soundtrack for director Eino Ruutsalo's experimental short film *Hyppy* (1965). Early on, the system was called Sähkö-ääni-kone (Electric Sound Machine, 1964–1968) or System-1, and only much later gained the name Integroitu syntetisoija (Integrated Synthesizer). The entire studio was planned to be eventually integrated into this one machine, hence the name. The Integrated Synthesizer was the largest and only truly modular system that Kurenniemi built.

Kurenniemi continued to work on the system until the late 1960s, when it was left in an unfinished state to make time for more pressing projects. Among these were a series of instruments commissioned by Finnish and Swedish artists and composers. The first project was Sähkökvartetti (Electric Quartet, 1968), a "collective instrument" built for avant-garde artist Mauri Antero Numminen. Sähkökvartetti was followed by the polyphonic synthesizer Andromatic (1968) built for composer Ralph Lundsten. The DICO (1969), commissioned by composer Osmo Lindeman, was the last of three custom-built instruments. A common feature to all these instruments is that they all have a digital sequencer and lack a conventional piano-style keyboard. After the DICO,

Kurenniemi designed a line of DIMI instruments marketed by Digelius Electronics Finland between 1970 and 1975, ending with the microprocessor-based DIMI-6000 (1975–1978). Work on the DIMI-6000 software was continued until the late 1970s by Jukka Ruohomäki. After a break of three decades, Kurenniemi returned to instrument design with the software instrument DIMI-H (2005; "H" stands for harmony), based on his theory of harmonies.[2]

Already in his early designs Kurenniemi started to realize his initial ideas of a machine capable of producing music in real time with the flick of a switch according to a predetermined algorithm. This was inspired by the early development of computer technology and the RCA synthesizer, in particular.[3]

The DICO and "Improvisaatio"

The DICO is a monophonic synthesizer with a twelve-step digital sequencer (see fig II.6). It is not clear when or where it received its current name. A few name variations (DIGO, DCO, LDCO) exist in the schematics and Lindeman's archives.[4] However, they all are derived from "Digitally Controlled Oscillator" (with the "L" in LDCO possibly referring to Lindeman).[5] Furthermore, it is not known to what extent Lindeman contributed to the design process beyond his discussions with Kurenniemi concerning the general features of the instrument.[6]

The DICO used a similar sound generating technique as the next three DIMI instruments, the DIMI-A, DIMI-O (1971; "O" stands for an optical input), and DIMI-S (1972; "S" stands for Sexophone; the instrument is also known as the Love machine or Kärleksmaskin in Swedish). Instead of a voltage-controlled oscillator setup typical of the analog synthesizers of the time, Kurenniemi chose a technique common in electric organs: frequency division. This synthesis method has only a few oscillators set to a fixed frequency (however, a vibrato oscillator is commonly used for slightly varying the frequency of the main oscillator). Their output is divided by whole numbers according to the required pitch. A typical electric organ has twelve oscillators, one for each note of the octave. The signal of these oscillators is divided down with octave dividers to give all the notes of the keyboard. The DICO's sound generator consists of four fixed oscillators, followed by a note divider circuit that gives the conventional set of twelve notes within one octave. The note divider is followed by an octave divider that gives all the notes over the range of eight octaves.

The state of each sequencer step is represented by ten indicator lights. Two additional lights were left unused for future expansion. At each step, the user can adjust the chromatic pitch (four bits), octave range (three bits), articulation (two bits), and the output channel (one bit). The values of the bits are programmed through a touch-switch matrix of three rows by twelve columns constructed from screw heads. Initially, the idea was to use only two rows of contacts, which would then be grounded with a

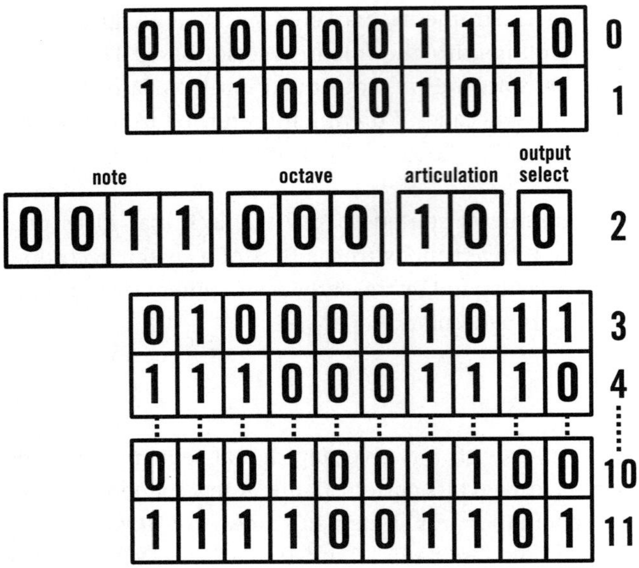

Figure 19.1
Memory of the DICO.

metal brush. Grounding a pin in the top row sets the respective memory bit on and the lower row pin sets the bit off. The center row provides a set of grounding points. The metal brush was soon replaced with a piece of grounded electrical wire acting as a stylus. Both the brush and the stylus options are handy for creating fast arpeggios. With a light touch, the player may change the state of random steps while the sequencer is running. The DICO also has a static filter and an attenuator bank that can be connected through the patch bay.

In the DICO, Kurenniemi chose to use low-order counter circuits for note dividers. This explains why intervals with higher integer ratios, the third (4/5) and the tritone (7/5), are replaced with actual oscillators. The first two bits used for selecting the note are actually used to select which one of the four oscillators is connected to the divider circuits. The remaining two bits will control the fifth and fourth dividers. Using low-order calculator circuits on fifths (2/3) and fourths (3/4) gives an interesting result: the intervals given by the divider circuits calculating these ratios are pure. Therefore, the tuning used in the DICO is actually just intonation.

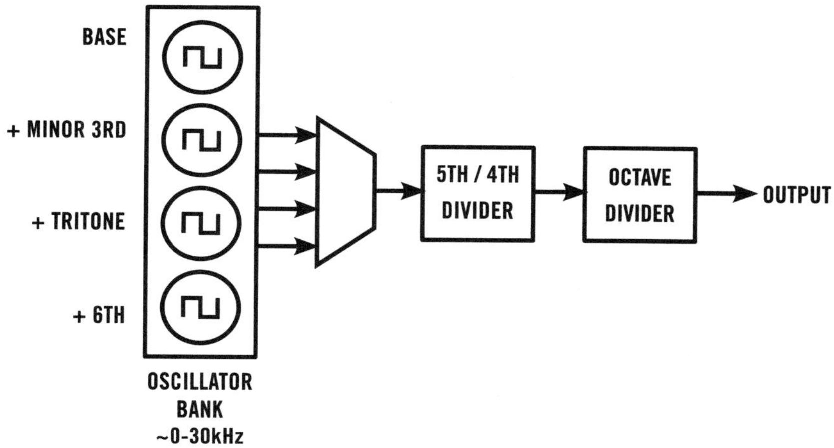

Figure 19.2
Sound generator modules of the DICO.

Shortly after completing the first version, Kurenniemi demonstrated the instrument in an electronic music seminar organized and recorded by the Finnish Broadcasting Company YLE. Despite its documentary origin, the recording is acknowledged as a musical work entitled "Improvisaatio"[7] and was released on the CD *Äänityksiä/Recordings* 1963–1973 (2002, Love Records, LXCD 637). It seems obvious that Kurenniemi presented the instrument also by speaking while playing. This, however, cannot be confirmed, for the recording of the performance contains only a line-level signal of the instrument processed by a spring reverb unit.

The recording is characterized by an intuitive musical structure and Kurenniemi's spontaneous playing style. Structure-wise, "Improvisaatio" contains two parts. The first part consists of a typical twelve-tone DICO sequence with long passages of little alteration of the instrument parameters. After a sudden twelve-second pause, the second part introduces more adventurous experimentation with various features of the instrument. Although the recording's casual structure is obviously due to its demonstrative purpose, it is typical of Kurenniemi's other works as well. In this respect, Kurenniemi's idea of an instrument as an automated music machine capable of producing music with the flick of a switch manifests itself even in "Improvisaatio."

Its undistorted and relatively unprocessed sound, quite rare in Kurenniemi's musical output, further underlines the nature of the piece as an instrument demonstration. The recording is also monophonic despite the fact that programmable two-channel output was one of the DICO's key features. Although Kurenniemi improved the instrument in 1970, the feature was used already in Lindeman's "Mechanical Music for Stereophonic Tape" (1969).[8] As a result, "Improvisaatio" failed to capture the full capability of the instrument. Nevertheless, the DICO's capabilities in real-time sequencing are well demonstrated.

The DIMI-A and Early Recordings

The DIMI-A was built in 1970 as a research project on potential applications of digital techniques in electroacoustic music (see fig. II.1). The DIMI-A is a two-voice synthesizer with a digital sequencer. In addition to two oscillators, it includes a ring modulator and two band pass octave filter banks, all programmable through its sequencer. The first instrument in the series of DIMIs, the DIMI-A was initially called simply DIMI or DIMI-1, DIMI being an acronym for Digital Musical Instrument. Later the instrument gained the suffix "A," referring to its associative memory scheme.[9]

The user interface of the DIMI-A consists of a touchpad operated with two styluses (see fig. 19.3). Starting from the top left, pads from R to 2 are used for recording and

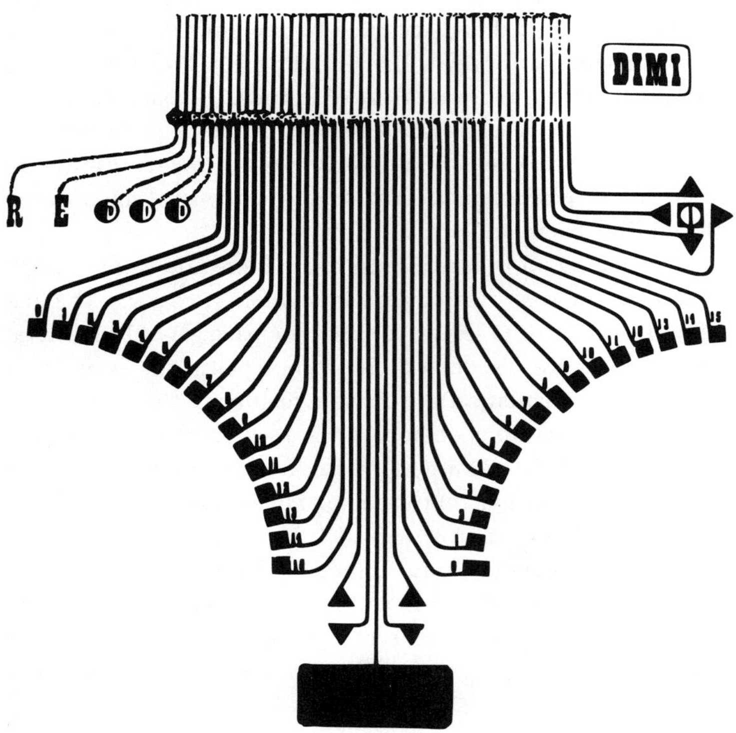

Figure 19.3
The DIMI-A touchpad layout.

erasing commands to/from the memory. Four arrow plates on the top right allow the user to start or stop and switch the direction of a sequence. Thirty-two number pads in the middle are used for entering commands, the left-hand numbered pads define a parameter number, and the right-hand pads define the parameter value. Each number pad is equipped with an indicator light. Indicator lights will show either the current sequencer step or the command that was entered last. The arrow pads in the bottom are used to select between operation modes. The leftmost pads control the indicator light modes and pads on the right control whether the command will be executed after it has been entered or not.

The only known "official" user manual written for the DIMI-A is a one-sheet chart showing the commands (in rows) and the respective parameter values (in columns) (see figure 19.4). The commands can be assigned to any step of the sequence, and each step can have multiple commands assigned to it. The DIMI-A has dedicated parameters for chromatic pitch, volume, filters, vibrato, sound source select, ring modulation,

PM

DS		0	1	2	3	4	5	6	7	8	9	10	11	12	13	14	15
CLOCK SPEED	0		STOP	1	FINE 2	3	4 →	5	6	1	COARSE 2	4	8 →	16	32	64	128
T1 far	1	0	1	2	3	4	5	6	7	8	9	10	11	12	13	14	15
T2 beat	2	0	1	2	3	4	5	6	7	8	9	10	11	12	13	14	15
GEN.1	3	C	C#	D	D#	E	F	F#	G	G#	A	B	H	C	PAUSE	INPUT 1 S	INPUT 2 A
GEN. 2	4	C	C#	D	D#	E	F	F#	G	G#	A	B	H	C	PAUSE	CLOCK PULSE	– 8 OCTAVES
OCTAVE	5	-7	-6	-5	-4	-3	-2	-1	0	-7	-6	-5	-4	-3	-2	-1	0
VIBRATO	6	0	VIBR. AMPLITUDE 1	2	3	4	5 →	6	7	1× CLOCK	VIBR. SPEED 2×	4×	8×	INP 1	INP 2	0	1
SELECT	7	GEN.1	GEN.2 OUTPUT 1	INP.1	INP.2	MOD.1	MOD.2	MOD.3		GEN.1	GEN.2 OUTPUT 2	INP.1	INP.2	MOD.1	MOD.2	MOD.3	
LEVEL DB	8	OFF	-36	-30	-24	-18	-12	-6	0	OFF	-36	-30	-24	-18	-12	-6	0
FILTERS ON HZ	9	87.5	175	350	700	1400	2800	5600	11200	87.5	175	350	700	1400	2800	5600	11200
FILTERS OFF HZ	10	87.5	175	350	700	1400	2800	5600	11200	87.5	175	350	700	1400	2800	5600	11200

DS 11 - 15 UNUSED.

PE 1

MOD. 1 : GEN.1×GEN.2
MOD. 2 : GEN.1 INPUT.1
MOD.3 : GEN.2 INRUT.2

Figure 19.4

DIMI-A user manual, from the archive at the University of Helsinki Music Research Laboratory. Courtesy Kai Lassfolk et al.

sequence jumps, and tempo.[10] The basic principle is similar to musical tracker software of the 1980s, but without a graphical user interface. DIMI-A does not provide any visual display of its command memory state. This is the opposite of the DICO, where the internal state is represented through an indicator light array, but the place of the current step in the sequence is not indicated.

The internal structure of the DIMI-A is significantly more complex than the DICO. The internal state of the DICO can be presented with eight bits (counting out two bits used for articulation), whereas storing the internal state of the DIMI-A requires at least 54 bits. The size of the memory is also larger: 1,600 bits in DIMI-A against the 120 bits of DICO. However, if the memory unit in the DIMI-A would have been arranged as it is in the DICO, the sequencer would have been limited to 29 steps. To overcome the limitation, Kurenniemi used an associative memory scheme. There, the memory is filled with a set of commands. Each command consists of 16 bits divided into four parts of four bits each: bar, beat, operation, and parameter. Bar and beat form an 8-bit time code field stating the sequencer step (of 256 possible steps) at which the command should be executed. The address field is followed with a 4-bit command code and a 4-bit parameter value (see figure 19.5). The commands are not stored in any specific order in the memory but are merely associated with a specific step of the sequence. Thus, any step can trigger any number of parameter changes. The only limitations are the 256-step maximum sequence length and the size of the program memory of 100 commands (1,600 bits). Notable is also that the DIMI-A cannot hold its memory contents when switched off. Despite Kurenniemi's intentions, support was never added for a mass memory unit.[11]

The basic topology of the sound generator in the DIMI-A resembles that of the DICO (figure 19.5). The biggest difference is the usage of an ultrasonic main oscillator. While the DICO uses a bank of four oscillators, the DIMI-A has only one, which is followed by a more complex note dividing network. The DIMI-A is tuned to equal temperament, which requires significantly higher-order calculations. The DIMI-A also has an attenuator circuit for changing the signal amplitude, as the output of the octave divider is a symmetric square wave. A simple way to adjust the amplitude is to chop high-logic voltage-level phases of the signal with an ultrasonic square wave by using logical AND gate. By altering the PWM (pulse width modulation) duty cycle of the modulating oscillator, one can adjust the size of the chunks that are chopped off. After the frequency of the modulating oscillator is filtered off, only a waveform with lower amplitude remains (figure 19.6).

In the fall of 1970, Kurenniemi tested the newly completed DIMI-A with an electronic arrangement of Johann Sebastian Bach's Invention no. 13 in A minor (BWV 784). Bach's Inventions provided suitable material for demonstrating the two-voice

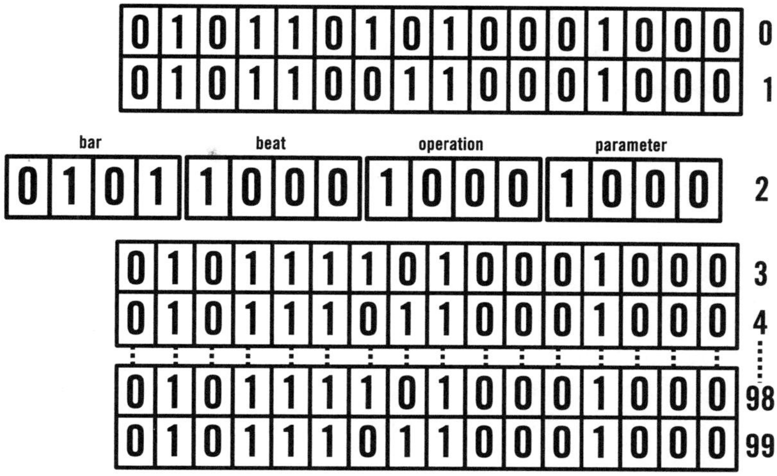

Figure 19.5
Memory of the DIMI-A.

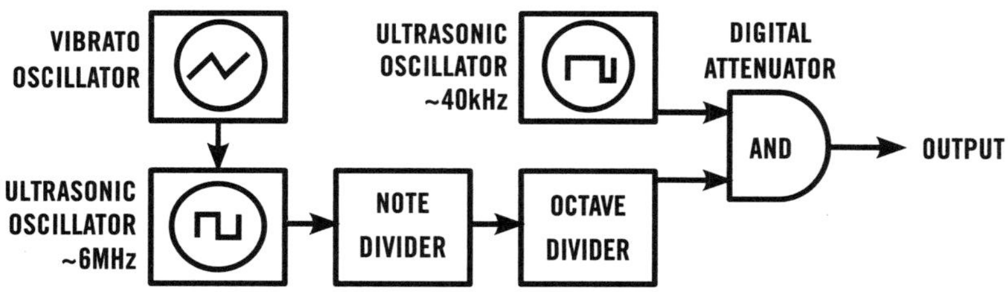

Figure 19.6
Sound generator modules of the DIMI-A.

programmable instrument. The arrangement, entitled "Inventio," is usually presented together with the piece "Outventio," although this latter part was composed in a separate session together with composer Jukka Ruohomäki. The two-part work, "Inventio-Outventio," was issued on a 7″ single vinyl record *DIMI 1: DIMI Is Born* (Musica DSS-1), released for marketing the instrument.

Because of the DIMI-A's limited memory, Kurenniemi had to program the work in parts. The master tape of the arrangement reveals that "Inventio" is constructed from twenty-two clips of one- to two-bar segments from the Bach's original score. These fragments were programmed into the instrument one by one, recorded on audio tape, and finally spliced together. Overall Kurenniemi's tape editing technique is excellent. The edits are inaudible other than the exception of a missing sixteenth note in bar 9 and an audible snap in the beginning of bar 7.

Bach's score specifies neither tempo nor dynamics for the performance, and Kurenniemi certainly took advantage of this. The arrangement starts off with a fairly modest tempo of approximately 66 bpm. The initial tempo is maintained through the first half of the arrangement, whereas the second half consists of two distinctive accelerandos, the first speeding up gradually up to 108 bpm (bars 14–17) and the second again from 66 bpm up to 95 bpm (19–23). The piece ends with a fast ritardando from 95 bpm to 42 bpm in two last bars.

All the musical changes (tempo, filtering, dynamics, vibrato, volume, octave range, etc.) need to be programmed as separate commands, which rapidly fills up the 100-command memory space. Since the instrument lacks an envelope generator, any dynamic alterations, even changes within individual notes, need to be programmed separately. This feature is demonstrated, for example, in bar 17 of "Inventio" (see figure 19.7). There, Kurenniemi divided the 4/4 time signature into 64 steps and entered gradual parameter changes step by step to achieve an envelope-like behavior. Thus, each eighth

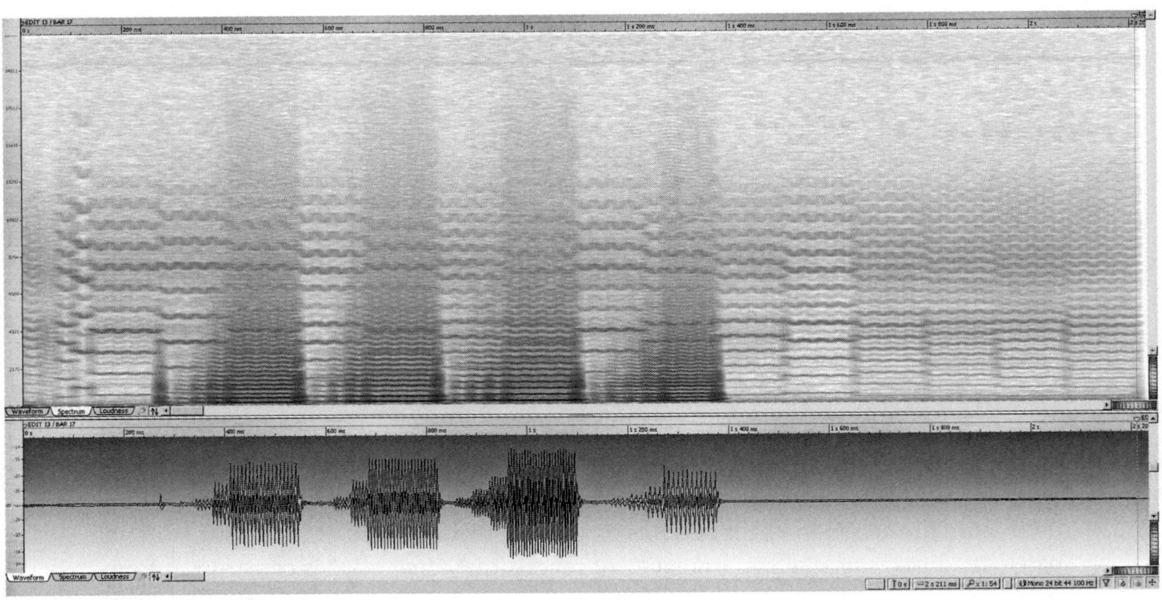

Figure 19.7
DIMI-A programming example in "Inventio" bar 17, left-hand part. The dynamic contours of the eighth notes are programmed with 1/64 time values.

note of the left-hand part has a gradually rising amplitude envelope. (The spectrogram in figure 19.7 also shows a leakage signal from the other audio channel of vibrato-modulated sixteenth notes from the right-hand part.)

In addition to the Bach arrangement, the instrument was tested and demonstrated on several occasions during the fall of 1970. In "Outventio," a joint studio improvisation with Ruohomäki, the DIMI-A was used as a collective instrument. Kurenniemi and Ruohomäki shared the instrument by having one choose the parameters with one stylus while the other controlled the parameter values with another stylus.[12] According to Ruohomäki, "Outventio" was not intended as a systematic demonstration of the DIMI-A's capabilities.[13] However, it does display many of the instrument's special features, including its ability to process external sound sources. "Outventio" is compiled from separate clips, anticipating the tape collage method used later in their collaborative works such as "Mix Master Universe."

Kurenniemi's tape archive includes also a tape titled "J. S. B. Inventio a-molli/Dimi-1." It contains a copy of "Inventio" followed by six minutes of DIMI-A testing. Kurenniemi apparently reused a tape with a recording of a Beethoven piano concerto (no. 4 in G major, op. 58), which was left unerased on the right channel for the first half of the instrument test (playing at double speed compared with the instrument test). In the second half, the DIMI-A was recorded in stereo on both tracks of the tape. The instrument test resembles "Improvisaatio" as well as Kurenniemi's focal work "Antropoidien tanssi," but the sound gestures are more hectic and the structure lacks similar musical fluency. The recording was left unedited and unpublished until 2012, when a 3′08″ excerpt from the instrument test part was released on the *Rules* (2012, Full Contact/Ektro Records, KRYPT-022) double vinyl album.

Analysis of Musical Interaction

Many if not most of Kurenniemi's recordings can be regarded either as solo performances with one instrument or as compilations of fragments of such events. His music-making activity can be analyzed as an interaction process between three parties: (1) the user, (2) the instrument, and (3) the musical outcome (i.e., an audio tape or live performance). Figure 19.8 presents a generic model of musical interaction applied here as a tool to analyze Kurenniemi's instruments and their use. Communicational actions and their consequences are illustrated with arrows. The user attempts to control the instrument through its user interface. Conversely, he or she receives some form of visual and tactile feedback from the instrument. This results in a direct two-way communication link between the two parties. As a result of the two-way communication the instrument produces a sonic output in the form of an audio signal. This results in a musical outcome, in the form of an improvised live performance or a part of a recorded musical work, for example. In real-time playing, the performance is immediately heard by the

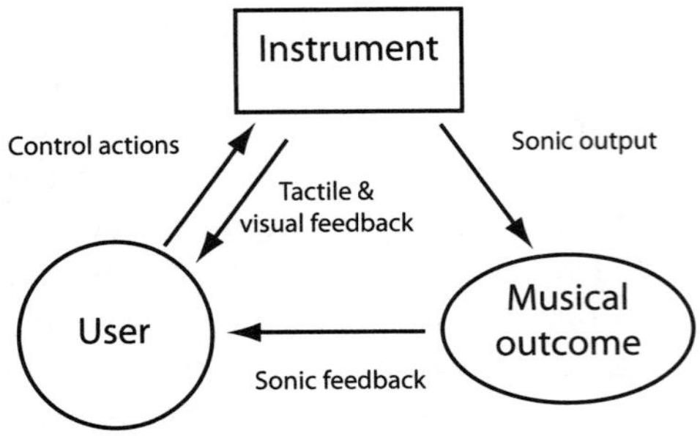

Figure 19.8
A model of musical interaction.

user, thus giving sonic feedback. This forms an indirect communication link between the instrument and the user.

When separated from the interaction process, all three parties in the process are independent objects. For example, a recorded musical performance can be listened to with no regard to the originating process. Hence, even "Improvisaatio" can be regarded as a work of "absolute" music (i.e., a musical work without a written program or other narrative context) despite the fact that it came into existence merely as by-product of an instrument demonstration session most likely in an effort to complement Kurenniemi's simultaneous spoken commentary.

Here, the term "indirect" refers to an assumption that some form of musical outcome, that is, an independent object, is created in the process. Sonic feedback may be as immediate as tactile or visual feedback. However, with a programmable instrument, sonic feedback may contain a considerable amount of latency caused by a separate programming stage, where the contents of the memory are programmed as a separate process and played back afterward.

With the DICO, tactile feedback is more limited than that of more conventional acoustic or electric instruments, such as keyboard synthesizers. Visual feedback provided by the memory indicator lights is also limited, because the contents of the DICO's digital memory can be observed only by iterating every step of the sequencer. The user is therefore forced either to memorize the memory contents or keep an external log in order to fully control the instrument. However, given the small size of the instrument's memory, its memorization is fairly easy when played back as an audible musical pattern. Thus, the indirect feedback becomes crucial when handling the instrument.

Although the DIMI-A has a similar user interface (styluses and indicator lights), direct interaction becomes even more difficult. The use of two sound generators and a much larger and more complex memory system makes it difficult to follow the visual feedback, especially in a live performance situation. Again, given the DIMI-A's almost nonexistent tactile feedback, the user becomes almost totally dependent on indirect interaction. On the other hand, memorization of the DIMI-A's total memory contents as a musical pattern is practically impossible. The lack of a mass storage medium made it inconvenient to exploit the full expressive capability of the instrument both in the studio and in live performance. Evidently, the inability of the DIMI-A to function as a practical live performance instrument forced Kurenniemi to spend more time in programming. This, in turn, led to less spontaneous musical output in "Inventio." In fact, Kurenniemi saw the video control interface of the DIMI-O as a solution to the limitations of the DIMI-A.[14]

The generic model of figure 19.8 can be modified for analyzing situations, where the number of users or the form of communication differs from what is given above. For example, the sonic material for "Outventio" was created with the DIMI-A by Kurenniemi and Ruohomäki with each one holding one of the two control styluses. This not only expands the number of users to two but also creates a communication link between them. Other examples can be pointed out with later instruments.

Kurenniemi's instruments that followed the DIMI-A can be regarded as further variations of the basic interaction model. The DIMI-O's video camera interface adds an optical level to the control actions. Moreover, since the contents of the memory are displayed on a video screen, visual feedback is far more informative and intuitive than in both the DICO and the DIMI-A. Also, the DIMI-O's tactile feedback has a more conventional dimension in the form of a piano-style keyboard. Perhaps the most significant feature regarding the interaction process is the ability to use the instrument as a multimedia device by allowing any visual element to be used as a source of control information, a static picture or a dancer, for example. This expands the DIMI-O beyond the realms of a purely musical instrument. Within the scope of the interaction model, the movements of the dancer can be regarded as part of the musical outcome.

With the DIMI-S, the single user is replaced by a group of players. Four is the optimal amount, but smaller and larger groups can also use the instrument. Tactile feedback takes a dominating role as the players get feedback not only from the iron balls of the instrument, but most importantly, from each other through skin contact. Visual feedback is also enhanced both in the form of the light organ but also by visual human contact. Considering the sexual implications of the instrument, visual feedback and skin contact can even surpass the importance of the sonic feedback. The resulting musical performance can thus be regarded as less important as an independent object or an "absolute" musical work than the interactive performance itself. In fact, when demonstrating the new instrument for the Finnish press, Kurenniemi himself mentioned that "sex is the main thing" rather than sound.[15] A modified interaction model applied to the DIMI-S is presented in figure 19.9. A more detailed description of the DIMI-S is given by Städje.[16]

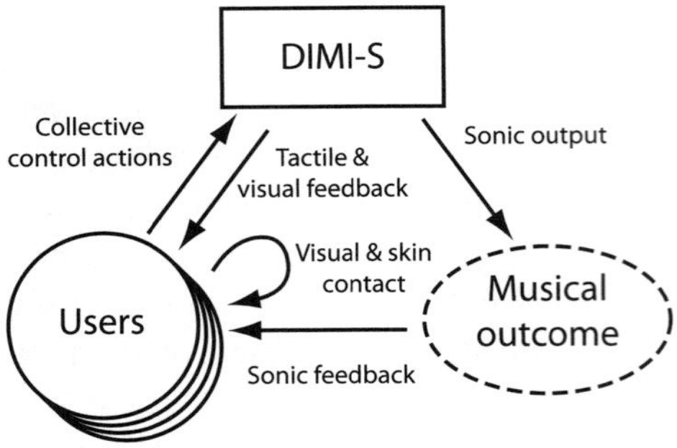

Figure 19.9
Interaction model applied to the DIMI-S.

The DIMI-T, in turn, can be regarded as an attempt to surpass the interaction process by converting brainwaves directly into sound. In the user instructions, the user is advised to relax and to refrain from speaking or moving the eyes to gain a "pure" and stable tone. According to the instruction sheet, most users are able to control the instrument to some extent.[17] Since the DIMI-T uses the EEG signal obtained from the brain to control the sonic output and gives neither visual nor tactile feedback, it can be regarded as Kurenniemi's attempt to dispense with user-instrument interaction altogether by placing the interaction directly between the brain and the musical outcome.

In both the DIMIX and the DIMI-6000, programmability is the dominant feature. Both are designed primarily as studio devices, where intuitive real-time control is of secondary importance. However, visual feedback provided by their video screens is much more explicit than in the DICO or the DIMI-A. The DIMIX displays the connections of its programmable patch bay with a simple but intuitive block matrix. The DIMI-6000 follows the conventional user-interface model of computer terminal based programming. Nevertheless, the use of the DIMI-6000 did not limit the user to batch-based work, despite its computer video terminal interface. Like the DICO and the DIMI-A, the DIMI-6000 featured real-time sound synthesis and manipulation capability.[18] However, the DIMI-6000 represented a return to the basic interaction scheme, presented in figure 19.8.

Discussion and Conclusions

The significance of Kurenniemi's instruments can be discussed from several points of view. Although they received some exposure in the Finnish media, it is well known that they failed to become a commercial success. Neither did they cause a technological revolution as direct forerunners of later musical computing systems such as the Fairlight CMI, Synclavier, or modern Digital Audio Workstations (DAWs). Apart from insufficient marketing, part of the reason may be timing. The DIMIs arrived in the heyday of analog synthesizers, but they lacked much of their sonic flexibility such as envelope generators, continuously adjustable filters, and multiple waveforms. Instead, Kurenniemi's instruments relied on either their programming capability or on their innovative user interface, both of which were difficult for conventional instrumentalists to appreciate. Conversely, analog synthesizers quickly found a user base among keyboard players.

"Improvisaatio" gives an excellent example of how Kurenniemi's way of interacting with an instrument anticipated modern-style live sequencing techniques and technologically oriented music production methods. It also demonstrates how intuitive the instrument was to use in live performance. Although the DIMI-A was capable of executing predetermined tasks significantly better than the DICO, its unintuitive user interface and complex memory system required more planning in advance. Nevertheless, "Inventio" remained Kurenniemi's only known preprogrammed work executed with the DIMI-A. Later its use was much closer to the working methods used in studio improvisations such as "Outventio," "Dimi-1" and "Mix Master Universe" (1973). However, the early musical experiences with the DIMI-A largely influenced Kurenniemi's quest for more intuitive control methods resulting in the DIMI-O, DIMI-S, and DIMI-T. These three instruments in particular challenged and extended the concept of a musical instrument.

Practical uses of Kurenniemi's instruments, including his own artistic endeavors, anticipate many features that are common in modern-day music production. The use of live sequencing in "Improvisaatio" and the techno-style beat of "Antropoidien tanssi" are some examples. Even the sonic capabilities, however limited at the time, have gained a certain "retro charm" among Kurenniemi fans.

Notes

1. Mikko Ojanen et al., "Design Principles and User Interfaces of Erkki Kurenniemi's Electronic Musical Instruments of the 1960's and 1970's," in *Proceedings of the 2007 Conference on New Interfaces for Musical Expression (NIME07)* (New York: Harvestwork, 2007).

2. Erkki Kurenniemi, "Harmonioiden teoria," *Musiikki* 15, no. 3–4 (1985): 261–272.

3. Mikko Ojanen and Jari Suominen, "Erkki Kurenniemen sähkösoittimet," *Musiikki* 35, no. 3 (2005): 17.

4. Documentation of the DICO. Finnish National Archives. Erkki Kurenniemi's Archive.

5. Jari Suominen, "Erkki Kurenniemi's Electronic Music Instruments of the 1960s and 1970s," in *Erkki Kurenniemi: A Man from the Future*, ed. Maritta Mellais (Helsinki: The Finnish National Gallery, Central Art Archives, 2013), 129–161.

6. Erkki Kurenniemi, interview by Mikko Ojanen and Jari Suominen. Electronic Music Studio of the Department of Musicology, Vironkatu 1, Helsinki University. May 29, 2004.

7. Petri Kuljuntausta, *On/Off: Eetteriäänistä sähkömusiikkiin* (Helsinki: Like, 2002), 713; Petri Kuljuntausta, *First Wave: A Microhistory of Early Finnish Electronic Music* (Helsinki: Like, 2008), 199–200.

8. Jukka Ruohomäki, unpublished manuscript of the history of electronic music in Finland. PDF copy of the manuscript in the authors' personal archives, EH 36/4–6.

9. Mikko Ojanen and Jari Suominen, "Erkki Kurenniemen sähkösoittimet," *Musiikki* 35, no. 3 (2005): 24–25.

10. Ojanen et al., "Design Principles and User Interfaces."

11. Erkki Kurenniemi, "Elektronisen musiikin instrumenteista," *Musiikki* 1, no. 1 (1971): 37–41; Erkki Kurenniemi, "DIMI Family," seminar lecture given at the UNESCO Congress of Electronic Music, 1978. University of Helsinki Electronic Music Studio Archive.

12. Jukka Ruohomäki, interview by Kai Lassfolk, November 4, 2013. Electronic Music Studio of the Department of Musicology, Unioninkatu 38, Helsinki University.

13. Ibid.

14. Kurenniemi, "Elektronisen musiikin instrumenteista."

15. Ilkka Pitkänen, "Ihmelaatikko muuttaa muhinat musiikiksi," *Ilta-sanomat*, February 22, 1972, 7.

16. Jörgen Städje, "Kärleksmaskinen—synten man spelar med hela kroppen," 2009.

17. Petri Kuljuntausta, *On/Off*, 667.

18. Documentation of the DIMI-6000, University of Helsinki Electronic Music Studio Archive.

Bibliography

Unpublished References

Documentation of the DICO. Finnish National Archives. Erkki Kurenniemi's Archive.

Documentation of the DIMI-6000. University of Helsinki Electronic Music Studio Archive.

Ruohomäki, Jukka. Unpublished manuscript of the history of electronic music in Finland. PDF copy of the manuscript in the authors' personal archives.

Interviews

Kurenniemi, Erkki. "Dimi Family." Seminar lecture in UNESCO congress of electronic music. University of Helsinki Electronic Music Studio Archive, 1978.

Kurenniemi, Erkki. Interview by Mikko Ojanen and Jari Suominen. Electronic Music Studio of the Department of Musicology, Vironkatu 1, Helsinki University. May 29, 2004.

Ruohomäki, Jukka. Interview by Kai Lassfolk. Electronic Music Studio of the Department of Musicology, Unioninkatu 38, Helsinki University. November 4, 2013.

Published References

Kuljuntausta, Petri. *On/Off: Eetteriäänistä sähkömusiikkiin*. Helsinki: Like, 2002.

Kuljuntausta, Petri. *First wave: a microhistory of early Finnish electronic music*. Helsinki: Like, 2008.

Kurenniemi, Erkki. "Elektronisen musiikin instrumenteista." *Musiikki* 1 (1) (1971): 37–41.

Kurenniemi, Erkki. "Harmonioiden teoria." *Musiikki* 15 (3–4) (1985): 261–272.

Ojanen, Mikko, et al. "Design principles and user interfaces of Erkki Kurenniemi's electronic musical instruments of the 1960's and 1970's." In *Proceedings of the 2007 Conference on New Interfaces for Musical Expression (NIME07)*, 88–93. New York: Harvestwork, 2007.

Ojanen, Mikko, and Jari Suominen. "Erkki Kurenniemen sähkösoittimet." *Musiikki* 35 (3) (2005): 15–44.

Pitkänen, Ilkka. "Ihmelaatikko muuttaa muhinat musiikiksi." *Ilta-sanomat* 22 (February 1972): 7.

Suominen, Jari. "Erkki Kurenniemi's electronic music instruments of the 1960s and 1970s." In *Erkki Kurenniemi: A Man from the Future*, ed. Maritta Mellais, 129–161. Helsinki: The Finnish National Gallery, Central Art Archives, 2013.

Internet References

Städje, Jörgen. "Kärleksmaskinen—synten man spelar med hela kroppen." November 27, 2009. http://www.idg.se/2.1085/1.269878/karleksmaskinen---synten-man-spelar-med-hela-kroppen.

20 On Sound and Artificial Neural Networks

Florian Hecker and Robin Mackay

From the 1980s onward Erkki Kurenniemi involved himself in-depth with tuning systems, musico-mathematical relations, and concepts of artificial neural networks. The following conversation between philosopher Robin Mackay and artist Florian Hecker takes the latter as a point of departure, discussing Hecker's sound piece "Untitled (F.A.N.N.)" (2006–2013), which dramatizes the use of such a network as its core. "Untitled (F.A.N.N.)" was first produced as a contribution to the artist Cerith Wyn Evans's commission in the A. A. Hijmans van den Berghbuilding on the campus of the Utrecht University. The sound piece was designed in collaboration with Tommi Keränen, a Finnish artist and programmer, with whom Hecker has been working on software instruments for over a decade. In the following conversation, Hecker and Mackay discuss the role of artificial neural networks in electronic music in relation to Kurenniemi's research and also in relation to American composer David Tudor, whose groundbreaking work and polymath approach to performance, instrument design, and composition echo Kurenniemi's heterogeneous interests and inventions. The dialogue then turns to the cultural figure of the network in relation to concepts of identity, autonomous simulation, and the shifting frontiers between performer, composer, and instrument—themes that resonate with Kurenniemi's life and work.

Robin Mackay: How did you come to make use of artificial neural network synthesis in "Untitled (F.A.N.N.)"?

Florian Hecker: In 2006, Cerith Wyn Evans invited me to participate in a commission to be staged throughout the A. Hijmans van den Berghbuilding on the campus of the University of Utrecht. The commission was "Kunst am Bau"—pieces that were to remain on the campus for an extended period—and the other commissions across the campus were mostly sculptural, made of materials with that durational aspect in mind. Upon the invitation to realize a sound piece in this constellation with Cerith's works, and thinking about how to structure it, David Tudor's *Neural Synthesis*[1] came to mind as a possible point of departure.

RM: This is a piece related to *Neural Network Plus*, originally conceived as part of Merce Cunningham's Enter, which premiered at the Opera Garnier in Paris?

FH: In 1992, yes. Central to the technical setup of Tudor's piece was a neural-network synthesizer designed by Forrest Warthman, Mark Thorson, and Mark Holler[2] that would cocompose/coperform along with Tudor and his setup of feedback boxes and chaotic circuits.

RM: Why did this come to mind in particular? Because of the possibility of the piece's being left to "compose" itself?

FH: The use of an artificial neural network seemed to be a way to suggest a piece that would feature a certain change over a long period. It also linked to this particular piece that I found conceptually more stimulating than anything that incorporated randomness as a structuring principle. Also, Cerith, Cageian as he is, had used a random process already for the structure of light emission of a chandelier that was one of his contributions for the Utrecht project. I approached Tommi Keränen if he would like to design such a system for the installation. Tommi mentioned to me that Erkki Kurenniemi had been looking into neural networks for years, and that we should seek advice from him.

RM: Before that, had you come across Kurenniemi's work?

FH: I encountered Kureniemmi through his music in Finland in 2002. The direct and stripped-down structure of "Sähkösoittimen ääniä #4" and "Sähkösoittimen ääniä #1"[3] immediately reminded me of pieces by Pan Sonic[4] or other early publications on the Sähko label.

RM: And Tudor, was his work a particular interest of yours? Considering that the majority of your practice is in line with what Peter Hoffmann called "explicit computer music," based on abstract synthesis,[5] any reference to the work of Cage and East Coast experimentalism seems rather a departure.

FH: At first I was amused by the idea of Tudor, who was labeled a virtuoso throughout his career,[6] at that point partly handing over the control of the work to a machine. I only got more curious about this lineage through ongoing remarks and recommendations by the artist Yasunao Tone that I investigate Tudor's music further. Then, while on an artist's residency at Schindler House in Los Angeles in 2004, I encountered, through a series of visits to the Special Collections at the Getty Center library, some projects to which I've returned on several occasions since then: the catalog of Lyotard's exhibition Les Immatériaux,[7] the archival material to Experiments in Art and Technology (E.A.T.),[8] and the David Tudor Archives. In 2004 there were relatively few publications of Tudor's electronic pieces available. In the archive, however, I was confronted by a plethora of CDs with a digitized library of tapes made as resources for performances,[9] and various versions of pieces. In some of these recordings, the sounds of the room in which they were recorded and how they were recorded stuck out as a significant feature. Shortly after these listening sessions, I was in a studio in San Francisco that was equipped with a vintage Serge synthesizer and a Sony DRE S777, the "first commercial, real-time convolutional reverberator"[10] that modeled existing acoustic spaces based on impulse recordings of the same; with this combination of instruments I made some tracks that, because of the blurring through the reverb, reminded me of abstracted Tudor materials.

RM: All of this was research that would later feed into the 2009 CD *Acid in the Style of David Tudor.*[11] And that pseudo-recognition you describe would explain "in the style of," which otherwise seems a strange claim in relation to a mercurial experimentalist like Tudor. So the composition of the piece itself amounts to a set of (possibly misfiring) recognitions and linkages ... But there were other lines of research that fed into that work too.

FH: Yes, between 2000 and 2004 I worked extensively with a reformulated version of Xenakis's dynamic stochastic synthesis;[12] wanting to look deeper into forms of nonlinear synthesis, I started to look into the use of chaotic equations in relation to synthesis. While doing this together with Lance Putnam at CREATE,[13] I came across Dan Slater's article[14] on chaotic synthesis using a Buchla system in combination with a Comdyna analog computer. Due to a concurrent reading of the writings of Art & Language[15] and their piece *Portrait of Lenin in the Style of Jackson Pollock,*[16] the idea of "... in the Style of David Tudor" germinated.

RM: What was at stake in this piece and Art & Language's writing around it was the question what a portrait was "of"—that although it may be "of" one thing in the sense of resembling it, in another sense it is always genetically "of" the material circumstances that give rise to it.

FH: That's right—"in the style of" is exactly this intertwined mutuality of resembling, representing, and causal linking that you describe[17]—taking structures consisting purely of (acid) bass lines and analog synthesis and abstracting these, all made with a specific instrumentarium: a Buchla modular synthesizer, a Comdyna analog computer, and here also the artificial neural network.

RM: So, in the end, did you speak with Kurenniemi about the Utrecht piece?

FH: We met in Helsinki some years prior to the collaboration in Utrecht, and then, once the concept to work with an artificial neural network was set, Tommi Keränen met with him again.

RM: What did you make of his work as you became more familiar with it?

FH: I saw Kurenniemi, working as a mathematician, nuclear physicist, and expert in digital technologies[18] as somehow not dissimilar to the "polymath" approach of Iannis Xenakis, Don Buchla, Dan Slater ...

RM: Tudor could also be included in this category, no? His trajectory from performer to composer, and then to the actual development of new electronic instruments, is an interesting one.

FH: Yes—incidentally, one of the most exciting parts of looking through the archival material at the Getty were the odds and ends that, of course, the archivist has organized just as meticulously as the rest: the recipe for a gin and tonic, inquiries about ordering a tandoor oven from India with shipping options both to Stony Point and New York City, handwritten recipes from Tudor's cookbooks, postcards from Karlheinz and Doris Stockhausen—all of that amongst scores and performance patch diagrams …

RM: Tudor talks about a search for devices and compositions whose finite situation could lead one to the "open," i.e., beyond the finite situation of a musical composition and performance. In the case of *Neural Synthesis*, at certain sensitive points thermal "noise" can intervene in the functioning of the neural network. In abstract electronic synthesis, where one is dealing with highly complex instruments that manipulate sound at a low level, and whose behavior can never be entirely foreseen or controlled, is this an inevitability? What is the balance between control and openness in a piece such as "Untitled (F.A.N.N.)"—and is this a problematic that even interests you?

FH: The designers of the original neural network synthesizer talk about the thermal noise affecting the neuron summing lines and its consequence on the "synthesis for adding randomness to the sounds. The neuron gain is set high to maximize amplification of the noise, and then feedback attenuation is adjusted until the network is just at the edge of oscillation. The noise intermittently stimulates oscillation of the network."[19] "Openness" here is a question of scaling, with the neural network materialized in a hardware synthesizer or the entire software-based conception as used in "Untitled (F.A.N.N.)." Here, the apparent "open" has some clear bounds in "Untitled (F.A.N.N.)," in its updated version shown in *Systemics #2*[20] every couple of minutes, the artificial neural network—driven synthesizers and analyzers get interrupted by a short sequence of ascending and descending tones, resembling patterns as described by Albert Bregman in his study *Auditory Scene Analysis*.[21] This formal and structural intermission corrects the "open," as only one possible amongst other forms or states.

RM: In general, what's your perspective on this quest for openness, with its various associations with the random, aleatory, Zen, etc.?

FH: Rather than such a search for an "open," [I prefer] the rigorous Xenakian dynamic stochastic approach, with its drifting that can be steered where the significant pointers that led to incorporating the artificial neural network at first. It was through this more formalized approach, via the GENDYN code included in *Formalized Music*,[22] to what Peter Hoffmann refers to as music out of nothing,[23] on to the chaotic oscillations that led to Untitled (F.A.N.N.) and "Acid in the Style of…"—in these systems, it was the zones in between cyclical states and noise that I found particularly appealing as sonic material. They have an amazing, inhuman sound quality.

RM: And where does Kurenniemi sit in relation to all of this, in your view?

FH: I don't see an obvious relation here—the instruments he conceived between 1964 and 1974 don't deal with the relation of determinacy and indeterminacy so much on the subzero level of instrumental music as Xenakis suggested,[24] nor do they resonate with the bricolage of parts that were central to Tudor's technical setup. Kurenniemi's instruments are located somewhere else in their playability, appearance, and with their elementary models of sound synthesis.[25] Sonically and conceptually they might be closer to another automated music box from that time, the Triadex Muse, designed by Edward Fredkin and Marvin Minsky between 1969 and 1971[26]—e.g., with the search for an automated composition, thinking of the use of a shift register as a core of the sequencer used in his instrument Sähkökvartetti (Electric Quartet, 1968), to create constantly varying patterns.[27]

RM: One of the key things about early pioneers in electronic music (we could include both Kurenniemi and Tudor here) is that they often not only composed but also created the instruments upon which their compositions could be played—in fact, in many respects the two activities could not be distinguished from each other. The creation of new machines also harbors a utopian dimension, in that every machine brings with it a vision of the future composition, creation, performance, and appreciation of music. You have worked[28] with Xenakis's UPIC,[29] which he conceived explicitly as an educational device, within a utopian vision of what he called "polyagogy"[30] that sought to introduce children to new nonmusical or at least nontraditional ways of interacting with sound. Kurenniemi's DIMI series is particularly fascinating because they embody his investigation and progression over the years. I wonder first of all what are the continuing features across all these devices that are specific to Kurenniemi, to what he was trying to achieve?

FH: The UPIC appears to me as a kind of universalist instrument allowing seamless navigation between the scales, overcoming classifications between musicians and nonmusicians, as Xenakis pointed out explicitly in part 8 of *The Owl's Legacy*, Chris Marker's thirteen-part series on Greek culture and history.[31] In relation to this, the look and design of Kurenniemi's DIMIs are of relevance, the filigrane surface of DIMI-A, or the use of mundane objects like coins as an interface element with the Sähkökvartetti (Electric Quartet). They look as if they could have come straight out of Richard Hamilton's Toaster[32] rather than out of a musical studio.

RM: I also wonder whether there is something lost in the era of software instruments and laptop performance. It is difficult to imagine a software archive like the precious archive of Kurenniemi's DIMIs—perhaps this is a kind of fetishism for the physical, but it reminds me of the problem that future literary biographers will have, when all of a writer's correspondence, notes, first drafts, etc. will all have long disappeared into the digital trashcan. More generally, I wonder whether the kind of "extended brain" that is put to work in performances with manually-operated instruments—a brain that is haptic, reflexive, extending into the flesh, as for example in the Kärlekmaskinen (DIMI-S, otherwise known as Love Machine, 1972)—and the concern for the user interface as a mediator between the human and the abstract matter of sound, whether all of this is on its way out of the picture of electronic music. Yourself, do you develop any of the tools you use for a given project with a view to reusing them, or to someone else being able to use them?

FH: Any custom instruments I have used over the years have stemmed from collaborations, dialogues, and "commissions" with other musicians and programmers who authored this software; in most cases they also entered into other distributions or were incorporated into the actual programming environment that they were housed within. Many of these also built upon existing models and research; some of them are actualizations or improvements on certain ideas, e.g., the updated version that Alberto de Campo made of dynamic stochastic synthesis, or Tommi Keränen's interpretation of Trevor Wishart's concept of the waveset, or Jayaganesh Swaminathan's incorporation of an audiologically meaningful third input source in Bertrand Delgutte's auditory chimera software. The process here is one of a careful reuse every once in a while, which partly of course then ends up maintaining the supporting hardware as well. As you suggest, the physical fetishism is just as prominent in the case of software instruments—that's something that I always found interesting in electronic music specifically: what is the difference between this and that sound, and what are these qualitative differences, how are they linked to the tools as much as to the context they are experienced in?

RM: Kurenniemi expounds at length in Mika Taanila's film[33] on his conviction that the human mind will be able to be "uploaded" into some other kind of machine (rather than a "slime-based" one). This obviously has a connection with what we could call Kurenniemi's "archive fever," his dedication to preserving photographic, video, and other evidence of his everyday life in the expectation that it could be used to reconstruct him one day.

What may have seemed like an individual eccentricity twenty years ago now looks prescient: today, this obsessive media archiving of one's everyday life is the norm. It is not only possible but, I would say, inevitable, that the numerous and increasingly dense personal data archives that a lot of us create voluntarily, as a kind of immaterial labor and an everyday performance, will be drawn upon not only to represent a person during their life but to reconstruct them after their death. The trend is toward the interconnection of various social media platforms, messaging, and other personalized apps. In effect, this creates precisely the kind of assemblage that Kurenniemi constructs: a network of archive memories dense enough to be "mined" and communicated with, as if it were "the original."

FH: How to deal with Kurenniemi's personal "archive fever" and the recent compulsion to archive "anything from academic research into preexisting archives or those still to be constructed, through exhibitions fully or in part based on them, to frantic competition among private collectors and museums in the acquisition of these new objects of desire,"[34] as Suely Rolnik suggests? Kurenniemi showed a progression from one register to the next, the period of his musical instruments was followed by a study of tuning systems[35] and theoretical conceptions on neural networks; it's essential to do something else with all that material, rather than a mere scholarly reactivation or reorganization. Where does the "new" fit in?

In 1963, Xenakis wrote in the Subscription Bulletin promoting Musiques Formelles: "Having been obliged to make a clean sweep of so many subconscious or acquired traditions, new points of reference had to be put on record, in the same manner as my 'works' that result from or are provoked by the same, in order to not forget. For using man's ability to 'engrave' is necessary in this tunnel, this darkness. ... This book is the temporary fruit of reflections, of trials and errors, of certain ways of thinking and doing, for example, music. Therefore, it is the tails of the coin whose heads is my musical work. Thus perhaps it may be of some pragmatic use."[36]

RM: This question of identity and identity traces leads me to ask what figure of the "network" means to us now: at a certain point it was a figure of hope for spontaneous emergence, for a kind of automated creativity "inspired" by biology. The ceding of authorial control to the contingencies of a semiautonomous machine seemed naturally part of the same liberatory agenda as Cage's experiments in contingency. Today the network is more likely to appear to us as an insidious mechanism of social control. Tweaking the parameters of a network, exploring its quasi-autonomous space of possibility, searching for configurations that produce a pleasing result—isn't this the job of a specialist in a control room directing police in order to quell a demonstration, fully informed by theories of chaos dynamics and network-inspired theories of crowd control? What do you think were the features that Tudor found appealing at the time of *Neural Synthesis,* and how does your point of view on the technology differ?

FH: Was it a pragmatic issue, one of introducing further destabilization into his performance process? Tudor was curious about autonomy since working on "Pepscillator," a piece composed for the Pepsi Pavilion[37] at the Expo 1970 in Osaka, in which electronic processors were arranged in a feedback circuit to create an autonomous electronic system with "no input."[38] Evolutionary models of growths and dynamic systems have been a recurring interest for composers, even in more institutionalized settings such as IRCAM.[39] In her anthropological study of the institute, Georgina Born comments:

> More generally, the impression of how the aesthetic was raised within IRCAM's daily culture was through intellectuals' sudden infatuations with new scientific, especially biological, analogies for music: a kind of constant, arbitrary, conceptual foraging. Thus, walking along the top corridor of offices one afternoon, I passed an American composer, a squatter who was keen to find a place within IRCAM. He talked with excitement of a new branch of genetic biology that promised to provide beautiful conceptual models for composition. Another day, noticed in a tutor's room a large glossy book on Mandelbrot's fractal geometry, a fashionable area of mathematics concerned with formulating the "logic" behind the apparently random shapes found in nature (for example, the shape of coast-lines). The tutor was learning about this with a view to importing it into his compositional schema. I learned later that it was being referred to more widely by artists trying to bring science into their work.[40]

In the liner notes to *Neural Synthesis,* Forrest Warthman mentions the less theoretical and more pragmatic role of the neural network as an extended audio-signal router and synthesizer and also notes that the "the role of learner, pattern-recognizer and responder is played by David, himself."[41] With "Untitled (F.A.N.N.)" I was more interested in using the learning, structuring, and pattern-recognizing functions of the artificial neural network to produce a highly abstracted music in the style of …

RM: Going back to this blurring of the lines between brain, network, this extended cognition, Tudor also seems to be prescient: before he even came into his own as a composer, as a performer he had become a kind of "neural instrument" himself, in the sense that composers would write specifically for him. We could see the use of a neural network as instrument—or, as he seems to suggest, as collaborator—as deliberately extending the erasure of distinctions between instrument and performer, interpreter and composer.

FH: The artificial neural network in "Untitled (F.A.N.N.)" is constantly changing the parameter settings of all synthesis processes. The composer Robert Ashley, a contemporary to Tudor, spoke of composition as "the process of constantly making a decision about when you're going to update what you've just done."[42]

RM: There are only complex "transformers" (in philosopher Jean-François Lyotard's phrase) for ultimately inhuman flows of information, never to be mastered by a sovereign will.

Notes

1. "Neural Synthesis" here refers to the composer David Tudor's neural network–based sound synthesizer as well as to compositions and recordings made with it originally conceived as part of Merce Cunningham's Enter, which premiered at the Opéra de Paris Garnier in 1992. Paris. See David Tudor, *Neural Synthesis No. 2*, Ear-Rational ECD 1039 (1993), and *Neural Synthesis Nos. 6–9* (2 CDs), CD 1602, Lovely Music (1995).

2. Mark Thorson, Forrest Warthman, and Mark Holler, "A Neural-Network Audio Synthesizer," *Dr. Dobb's Journal-Software Tools for the Professional Programmer* 18, no. 2 (1993): 50–65.

3. Erkki Kurenniemi, *Äänityksiä/Recordings* 1963–1973, Love Records (4)—LXCD 637, CD, Finland (2002).

4. Pan Sonic (originally called Panasonic) is a Finnish experimental electronic music duo consisting of Mika Vainio and Ilpo Väisänen.

5. See Peter Hoffmann, "Sleeve Notes to Russell Haswell's and Florian Hecker's Kanal GENDYN Recording," Editions Mego, Vienna (2011).

6. Austin Clarkson, "David Tudor's Apprenticeship: The Years with Irma and Stefan Wolpe," Leonardo 14 (2004): 5–10.

7. Les Immatériaux, March 28–July 15 1985, Grande galerie—Centre Georges Pompidou, Paris.

8. E.A.T.—Experiments in Art and Technology—is a nonprofit organization devoted to promoting the interaction between art and technology, which was launched in 1966 by Billy Klüver and Fred Waldhauer together with Robert Rauschenberg and Robert Whitman. For further information, see "Collection of Documents Published by E.A.T." at the Daniel Langlois Foundation for Art, Science, and Technology, Montreal, and "Records of the Organization Experiments in Art and Technology" at the Getty Research Institute, Special Collections, Los Angeles.

9. John Driscoll and Matt Rogalsky, "David Tudor's Rainforest: An Evolving Exploration of Resonance," Leonardo 14 (2004): 25–30.

10. "Sony DRE S777," Dec. 1999, http://www.soundonsound.com/sos/dec99/articles/sonydres777.htm.

11. Florian Hecker, *Acid in the Style of David Tudor*, Editions Mego, Vienna (2009).

12. Peter Hoffmann, "The New GENDYN Program," *Computer Music Journal* 24, no. 2 (2000): 31–38.

13. CREATE is the Center for Research in Electronic Art Technology at the University of California, Santa Barbara.

14. Dan Slater, "Chaotic Sound Synthesis," *Computer Music Journal* (1998): 12–19.

15. Charles Harrison, *Essays on Art & Language* (Cambridge, MA: MIT Press, 2003).

16. From Harrison, *Essays on Art & Language*:

"A Portrait of V. I. Lenin in the Style of Jackson Pollock" is the title of a painting, or, more precisely, it is a title given to some individual paintings within a series produced by Art & Language. ... An exhibition of "Portraits of V. I. Lenin in the Style of Jackson Pollock" was held at the Stedelijk van Abbe Museum, Eindhoven, in 1980. The title is also the title of an essay published by Art & Language, and it is the title of a song with words by Art & Language and music by Mayo Thompson, which was recorded by The Red Crayola in 1980. Before it was any of these things, however, it was a linguistic description, an ironic proposal for an impossible picture, a kind of exasperated joke (129)

17. See Robin Mackay, "Climate of Bass Hunter," CD sleeve notes to Hecker, *Acid in the Style of David Tudor.*

18. See Erkki Kurenniemi, *100 Notes—100 Thoughts*, introduction by Lars Bang Larsen, Documenta 13 (e-book, Hatje Cantz, 2011).

19. Thorson, Warthman, and Holler, "A Neural-Network Audio Synthesizer."

20. *Systemics #2: As We May Think (or, The Next World Library)*, Kunsthal Aarhus, September 21–December 31, 2013.

21. Albert S. Bregman, *Auditory Scene Analysis: The Perceptual Organization of Sound* (Cambridge, MA: MIT Press, 1994).

22. Iannis Xenakis, *Formalized Music: Thought and Mathematics in Composition* (Hillsdale, NY: Pendragon Press, 1992).

23. Peter Hoffmann, "Music Out of Nothing? The Dynamic Stochastic Synthesis: A Rigorous Approach to Algorithmic Composition by Iannis Xenakis." PhD dissertation, Technische Universität Berlin.

24. Iannis Xenakis, "Determinacy and Indeterminacy," *Organized Sound* 1 (1996): 143–155:

Therefore our problem is linked with the question of determinacy and indeterminacy in the widest sense and with so-called causality in physics, which is an aspect of determinacy. Below the zero-level of instrumental music that I have discussed, there are deeper levels which are dealt with by experimental and theoretical acoustics. The first of these lower levels is described in the language of sound synthesis, which is based on Fourier analysis of sound structure. This is the level of the analysis or synthesis of harmonics. Below this level is—due to computers—the level of the individual sound samples, up to 50,000 samples per second, representing a very high fidelity in the analysis of a sound. Probably one can discover even lower levels; I am personally convinced that sampling rates should go higher than 50,000 per second, because even at that level there is already quite a distortion.

25. See Jari Suominen, " Erkki Kurenniemi's Electronic Music Instruments of the 1960s and 1970s," http://www.lahteilla.fi/kurenniemi/julkaisu/Suominen.pdf.

26. On the Triadex Muse, see Laurie Spiegel, "Regarding the Historical Public Availability of Intelligent Instruments," *Computer Music Journal* 11 (1987): 7–9.

27. Suominen, *Erkki Kurenniemi's Electronic Music Instruments of the 1960s and 1970s.*

28. Russell Haswell and Florian Hecker, *Blackest Ever Black* (Electroacoustic UPIC Recordings), Warner Classics and Jazz, London (2007).

29. "UPIC" stands for "Unité polyagogique informatique du CEMAMu." See Henning Lohner and Iannis Xenakis, "Interview with Iannis Xenakis," *Computer Music Journal* 10 (1986): 50–55.

30. See Robin Mackay, "Blackest Ever Black; Haswell & Hecker; Rediscovering the Polyagogy of Abstract Matter," in *Collapse* 3 (Urbanomic: Falmouth, 2009).

31. See http://monoskop.org/Chris_Marker#The_Owl.27s_Legacy_.28L.27H.C3.A9ritage_de_la_chouette.29.

32. See Richard Hamilton, *Collected Words 1955–82* (Stuttgart: Thames & Hudson, 1982), 72, 90 (90 in color).

33. Mika Taanila, *The Future Is Not What It Used to Be* (Tulevaisuus ei ole entisensä), 2002.

34. Suely Rolnik, *Archive Mania*, Documenta 13 (Hatje Cantz, 2011).

35. Erkki Kurenniemi, "Chords, Scales, and Divisor Lattices," this vol.

36. Xenakis, quoted in Sharon Kanach, "Xenakis's Hand, or The Visualization of the Creative Process," Perspectives of New Music (2002): 190–197.

37. Billy Klüver, Julie Martin, and Barbara Rose, *Pavilion: By Experiments in Art and Technology* (EP Dutton, 1972).

38. Ron Kuivila, "Open Sources: Words, Circuits, and the Notation-Realization Relation in the Music of David Tudor," *Leonardo Music Journal* 14 (2004): 17–23.

39. IRCAM is the Institut de Recherche et Coordination Acoustique/Musique, Paris.

40. Georgina Born, *Rationalizing Culture: IRCAM, Boulez, and the Institutionalization of the Musical Avant-Garde* (University of California Press, 1995).

41. Forrest Warthman, liner notes for David Tudor, *Neural Synthesis Nos. 6–9* (2 CDs), CD 1602, Lovely Music (1995).

42. Robert Ashley, *Perfect Lives* (Dalkey Archive Press, 2011).

VI Interviews

21 Drifting Golf Balls in Monasteries: A Conversation with Erkki Kurenniemi

Mika Taanila
Translation by Teo Välimäki

The following text is Mika Taanila's interview with Erkki Kurenniemi at Kurenniemi's home in Helsinki on April 15, 2002. The original interview was recorded on a DAT and formed the core material for the film *The Future Is Not What It Used to Be* (dir. Taanila, 2002). The interview is a shortened version and is published here for the first time. Besides biographical facts and background, the interview touches on Kurenniemi's interests in science, mathematics, and computing. The main topics of the conversation between Taanila and Kurenniemi revolve around the idea of the Earth becoming a museum planet and the future technological reconstructions of the human soul and consciousness. What comes out in the interview is that Kurenniemi's projects have almost systematically always been interrupted or have remained as sketches, rarely finished. The interview is both a good insight into Kurenniemi's holistic thinking and a glimpse into the history of computing and the digital seen through the eyes of an artist. As such, it is culturally and historically interesting, as well as demonstrating some aspects of the artistic methodology of Kurenniemi: he was always as interested in developments in physics and other sciences as he was in the arts.

Mika Taanila: You've been a part of both the scientific and artistic circles since the 1960s and witnessed a lot of technological advancements and trends. Which do you think have been the most important changes during the time you've been active in the environment?

Erkki Kurenniemi: I suppose it's been the development of IT and the rise of the computer. When I was in school I got interested in electronics, electronic music, and what they called the "giant brain" or "electronic brain," which were rare back then. Now computers are everywhere, and that's probably been the greatest technical advancement during my lifetime. I've felt the Cold War, spaceflight, and that stuff hasn't touched me that much.

MT: How did your first experience with computers come about and what did it entail?

EK: My dad worked in insurance and that's how I heard about computers, because insurance companies were planning to acquire some. They already had punch-card computers, the Hollerith machines, which were used to store information, but they were mechanically operated. At the end of the '50speople were starting to come up with electronic computers. My real first experience was in the beginning of the '60swhen I was starting my physics degree and the department had purchased an analog computer and I got a job programming it. A digital computer followed a few years later. They were quite impressive, these huge tubular machines with massive drum memories and about a kilobyte of memory.

My classmates had been involved with the Department of Musicology, and because I was tinkering around with electronics and the musicology professor at the time, Erik Tawaststjerna, decided that the University of Helsinki needed a studio for electronic music, I got an unpaid internship to develop it. That's when I really began to study electronics and build electronic instruments—around 1961 or 1962.

I built a series of instruments in the beginning of the 1970s through my own company, but they never sold well—no instrument was manufactured with more than two copies of it made. The beginning of the '70swas another big watershed, because microprocessors entered the market. Because Finland was halfway behind the Iron Curtain, components were really hard to get and even then they were mostly used by the military or space programs. The first real microprocessors of the early '70s were a revelation: a whole CPU could be stored on a single silicon chip. Of course we had to try them out immediately.

MT: What was the mood like in the Department of Musicology in those days?

EK: During the '60sI was pretty much building that stuff on my own, but then a group of composers formed within the department. People like Erkki Salmenhaara, Kari Rydman, Kai Chydenius, Otto Donner, and many others. We learned a lot and were influenced by Swedish avant-garde composers like Jan Bark. Many of these guys began composing in musicology, and I had to work as a sound technician for them to aid with the equipment.

A little later, in the late '60s,a lot of weird young dudes began gathering in the department. They began experimenting with music which combined all kinds of stuff from traditional and Oriental instruments to electronics.

MT: How'd this new experimental music tap into the other phenomena of the period culturally?

EK: The electronic stuff really didn't have an ideology behind it so it owed allegiance to no one. For instance, a lot of the people who were there in the early '60smoved on to compose political music and left all that experimentation behind. Some of them went back to classical and orchestral music. So there wasn't a great electronic kind of Sibelius who would have produced a large body of electronic works. On the other hand, there was this feeling that all this would pass quickly because computers and new technology were on the rise. Everyone knew that it was the dawn of a new and frantic era and Moore's law was in effect: the performance capabilities of electronics were doubling every eighteen months, so nobody was thinking of creating a permanent base for anything. Then in the beginning of the '70s I built a series of digital instruments called the "DIMI" series, short for Digital Music Instrument. I guess I got the initials mixed up since "MIDI," short for Musical Instrument Digital Interface, surfaced in the early '80s and became an industry standard and allowed combining different audio equipment into a unified whole. With MIDI it became standard practice.

MT: Which were the significant things or phenomena which reflected on and affected what you were doing at the time?

EK: There was this inexplicable dichotomy: on the one hand, the background of experimental music sprouted from European modernism and the avant-garde, where a composition was seen as kind of a sacred thing and the composer was a sort of a prophet, receiving this piece from some kind of higher abstraction-sphere, which he then tried to interpret as best he could, mostly on his own; on the other hand, there was this whole field of popular music, where music wasn't a canon of immortal compositions but a process of human communication. That's why the technique of classical electronic music, where you cut and paste a track together for days in the studio and compose something to go along with it, created an idea that music is a part of the whole process of the human life; it's the sound created by the living and a part of this whole social ambient babbling.

MT: How about things like the hippie movement or drugs or sexual liberation and stuff? What kinds of memories or thoughts do they bring back?

EK: Before that I think that there was this pioneer spirit about creating and marveling at technology, and the socialization brought about the hippie movement's realization that the adult world was a complete sham and everything was a big fat lie. It dawned on everybody that we're actually living in a world that wasn't created according to the terms of the adults into a goal-oriented machine, but a playground where you could basically do whatever the heck you wanted.

MT: Did people already talk about concepts like interactive art back then?

EK: Probably not, because we didn't even have uninteractive art, which would have been the counterpoint to that. The whole notion of interactivity came about in a major way as a mode of resistance to the creation of early multimedia and video art, which was just forced on the spectator using a tape recording. People got stifled by that and wanted to inject some real-time elements and feedback into the system.

MT: Which of your own compositions do you consider the most significant to yourself?

EK: Well, composing was really a side gig for me. I didn't consider myself a composer as much as an instrument builder, and once you build a new instrument, you have to demonstrate its capabilities somehow and perhaps give that demo a title as well. All in all, probably my first real composition in performance was "On-Off," composed and performed in the Department of Musicology in 1963. The whole idea of having a studio and fiddling around with tape being the actual instrument came about while I was composing it. I'd created some demo materials on tapes and then combined them live without a prerecorded montage, using my legs to flip different dials and switches and turning the tape reels back and forth with my hands. And an obviously important element was the strong feedback, i.e., the tape echo.[1] When you roll the reels manually the feedback tape echo builds into a really chaotic jumble of sounds, basically noise.

MT: This piece goes straight for the backbone. Did the composition come about from a special association with something?

EK: Well, at least later on I've associated its mood with the experience I had as a young child in Imatra. I was standing in the generator hall of the Imatra Power Company, where these giant generators were humming in a huge acoustic space. I remember it being somehow divinely festive.

MT: What's been your relationship like with the academic community?

EK: I was a volunteer unpaid assistant at the Department of Musicology, so I could just flit about as I wanted. I really studied mathematics and theoretical physics and worked there as an assistant, designing recitations and doing computer stuff in the Department of Nuclear Physics. At first it was programming an analog computer and other things followed. But eventually I got on a bit of a collision course with the community. I didn't want to graduate and also thought that the theoretical physics of the time was based on falsehoods and miscalculations. I left in the beginning of the '70s, somewhat burning my bridges behind me.

MT: What was false about it?

EK: Well, they were doing some amazing stuff in the field of physics, especially in theoretical physics in the '60s. All the basic ideas of the modern standard models of quarks and the huge and heavy theoretical structure were formed, which is the backbone of modern physics. And of course there were dissenters. One of my great idols was Richard Feynman, who came up with pretty much everything—if you take a modern idea in physics you can often trace it back to Feynman in some way. Maybe my main discord had to do with whether the world is finite or infinite, continuous or discrete.

MT: What's your standpoint on it?

EK: Well, I fell in love with the notion that the universe is a massive computer, first suggested by Feynman, I think. That meant that it worked in bytes, that the universe is digital and not analog as it was considered in traditional Newtonian or Einsteinian physics. Nowadays almost everyone agrees that it is digital, but it took a long time.

My career sorted itself out when Digelius Electronics went bankrupt. One of our projects was to design a microprocessed controller for an industrial robot for the Rosenlew Tool Factory, but the robot didn't turn out a commercial success. We did create a fairly substantial automated system for the notorious Valco, the state-owned CRT factory. There was a lot of diversity in the projects. Also, because of personal reasons I wanted to return back to my roots in Helsinki and found a job in the Nokia Cable Machinery department, which was building these very robust industrial robots for the cable industry. They were used to move these massive coils of electric wiring, weighing several hundred kilos. I worked there for seven years, designing all kinds of industrial automation. It's kind of ironic that nowadays Nokia is known for its wirelessness above all else.

MT: I was reading some older magazine articles by you and came across speculation of an early PC, a "Personal Communicator." Can you tell me what the idea was behind that and its roots?

EK: The whole "Personal Communicator—PC" idea hit me in 1986. I knew they were coming. I started planning the kind of equipment needed for eyeglass temples if you wanted to build a general communication system, which would link via satellite. It would also include a two-way video link instead of just audio. I only completed a couple of cardboard models, but when you wore them you could imagine how the future would look. When I started researching the idea I found that the same kinds of ideas had bounced around already in the '60s and even before that. The idea that using this hardware one could immerse oneself into a virtual world and leave this natural environment behind was a really exciting concept. It surprises me that although these kinds of systems already exist, they haven't become commonplace. I guess there's a component missing. Maybe one I already had in my PC is still absent: the virtual aspect of having the darkened glasses on your head and the image rotating with you as you swivel your head in any direction. That requires another processor in the whole system.

MT: Seems that these kinds of virtual glasses have remained at a prototype level. When do you believe they'll become commonplace?

EK: I don't know why they've not hit home yet. Maybe using them entails dangers we weren't aware of. I mean if you have a freeze frame in front of your eyes no matter which way you turn and it's being used by children, for example: their visual cortex is still developing and maybe the image can burn into the cortex, like what used to happen with old TV and computer screens. I still have no doubt that this kind of technology will become an everyday thing. I think it's fundamentally a question of fusing the human brain and the computer. I think life as we know it will certainly be over soon.

MT: That's a pretty bold statement. What do you base it on?

EK: People used to talk about the difference between the body and the soul, and then the word "soul" got banned because of its religious connotations. But when it comes to IT it's quite clear that what we are talking about is the difference between hardware and software. And we know that software can be pretty much immortal in that good programming solutions and algorithms are really sustainable. On the other hand, brain research, evolutionary psychology, and other fields have shown that the human is a machine, a finite automaton, a machine created by evolution. And I think that it's impossible to conceive that this kind of slime-based machine would have survived barely for nostalgic reasons, if we can manufacture quantum computers that are thousands of times more efficient than the human brain and can be programmed with the entirety of our consciousness and emotional substance. I'm an unscrupulous believer in the idea of strong artificial intelligence.

MT: So humans could be immortal?

EK: At least that's one of the reasons to develop this strong AI. If cryogenics and reanimation can't be attained biologically, there's this other possibility: you store the mind and consciousness as bytes and, even though we don't yet have a quantum computer which could allow the consciousness to continue living, then it at least makes sense to make a backup by registering everything from images and sounds and thoughts and burn them on a CD and rest in the belief that in fifty years at the latest they can be downloaded onto a pocket computer and be reanimated.

MT: Who would be responsible for this reconstruction in fifty years and why would they want to do it?

EK: I can't speak for their reasoning, but one clear reason is that we as humans are interested in history. We have museums and we're interested in strange things like archaeology, and old music using the original instruments, and arranging medieval plays using authentic costumes. We're constantly trying to reawaken the past, and IT is a great tool for that, because in fifty or a hundred years when people are interested in the past they will be able to create virtual models of the entire human history. We will be able to transport ourselves into historical reconstructions of different eras in our everyday life. If we'll be able to make the reconstructions work and truly virtual it will also become an important tool to plan for the future instead of just following some new technology blindly. We can create virtual models of how society will work once it spans the entire solar system and in time, the whole Milky Way. A cloud of golf-ball-sized quantum computer servers, which 10 billion living people could inhabit.

MT: In the beginning of last year you wrote a text titled "The Exchange of Brains." You discussed a concept you called "In the process of the reconstruction of an algorithmic base or soul." What do you mean by that?

EK: Well, we understand the brain as being a neural computer, and we're well on our way of determining just how it works. There still are vast unanswered questions, but I believe we'll be able to understand them eventually, such as what is consciousness or character or personality or identity. And all these things we used to connect with a biological living entity—and especially the human—are found out to be universal and independent of the underlying material. In other words, as soon as it's appropriate and the technology attains the level needed, we can transfer our consciousnesses and minds and personalities onto computers, or whatever the best computing technology will be at the time. In other words, when the human can be simulated with adequate precision—and adequate precision being that the subject's closest friends and they themselves consider themselves alive and conscious and have maintained their previous personality, even though their bodies have long ceased to be. And for the past few decades there has been this process going on around the world, unprecedented in our history and biological history: the digital Internet. And this massive project is familiar to everyone.

I used to call this the "Great Transcription" and some people have called it a migration, where the all available data, previously stored in archives, tapes, and papers, are now being transferred onto a new medium. Everything that has ever been written, photographed, painted, drawn, and taped is being digitized, and the tendency is heavily toward integration. Archives are not only transferred onto a new medium, but the contents are being organized into a new database. There are projects going around the Internet where archaeologists are reconstructing entire cities and cultures from the past, biologists the entire tree of life, brain researchers the structure of the brain. The whole human body is dissected on the Web and the overlapping and diversity of the data—there's a lot of stuff in different variations, and nobody knows what's true and what's not. The confusion will be lifted once centralized and reliable archives are formed, like when the whole human genome is mapped somewhere or all the latest articles from the physical sciences are compiled on the Alamos server and then perhaps printed on paper a few years later for prestige reasons. And there are a lot of tendencies here, but if someone at some point creates a centralized database containing all human knowledge on the Internet, then we will have a stranger in our midst, a stranger nobody really has been able to welcome properly. But it's anybody's guess what this global consciousness of human society will ultimately be like.

MT: How would it be possible to capture or record an individual's consciousness, then?

EK: There are currently projects developing brain scanners, which could record every switch and synapse of the brain at the electron microscopic level. It's a bit awkward still, because you have to slice the brain up, so it can only be performed on dead brains: so at the moment there is no method to back up living brain tissue. But I think the technique isn't far off, it's already been described in one very well-made science fiction movie.[2] When we all do realize that, we will eventually get to a point when these kinds of mind-clones or brain backups can be resuscitated to work in a suitable quantum computer or similar apparatus, you don't have to wait for the actual technology: just store every tram ticket and store receipt, and photograph everything using a digital camera, where you don't have to pay for development—take a hundred pictures every day, write or dictate all your thoughts on a tape and you'll have at least easily accessible raw material, which can perhaps be used for your consciousness reconstruction later on.

MT: Have you considered … these procedures for yourself?

EK: I do register everything completely manically and with pathological precision.

MT: What kinds of things does that involve?

EK: Videocapturing everything on my cell phone, making notes every minute, recording the most trivial stuff: how much a cup of coffee costs, what kinds of people I see in a specific bar. If you think about the amount of photographs tourists take of Helsinki, for example. If these tourists upload them onto the Internet, like many do, and in time a suitable search engine combines those pictures and add all the surveillance camera footage, which shouldn't be destroyed like they usually do but archived somehow: that would mean we'd have an electronic digital archive of all the information stored from, say, the daily grind of the Helsinki market square. The whole history of this particular time in history could be reconstructed—into a virtual world.

MT: Isn't it difficult recording one's own life—don't you start censoring yourself somehow? Or don't you have any filters at all?

EK: Well, I guess you accidentally distort things all the time—don't report all the unpleasant things and try to sugarcoat events—but I think the AI programs of the future will be able to rid the material of all that eventually.

I think that once the AI technology reaches maturity, which might take decades, we'll discover that we'll be able to reconstruct even earlier minds to a great degree. We might not have video material from someone like Voltaire or Einstein or Newton or Archimedes, but we'll have something: paintings and the personal records of these people may be enough to reconstruct a virtual entity. Maybe a little stylized and simplified, but one who you could have a conversation with, like if you had travelled back in time to have a talk with Voltaire.

Especially when you think about your own relatives and friends: local history and genealogy and stuff like that. My offspring might one day be interested in what their great-grandfather did and thought in his day, wandering through the annals of history. I'm in that situation—I know a bit about my father, but hardly anything about his father and practically nothing of his father. A couple of photographs, a name and some documents and records, written in ink in beautiful handwriting, stored in some strongbox, and that's all there is. Currently trying to make a virtual model of your grandfather seems an insurmountable task, but it could be that whatever material remains could be utilized to come up with something, if all that material would be gathered together and all these fragments of information would be used to reflect upon each other. Then compile a database, and you might get an inkling of how some ancestor thought and felt back when they were alive.

MT: How will the human consciousness change once we realize that all this doesn't necessarily end by the time you hit a hundred years?

EK: It seems there are a great many paths to immortality. One is, of course, the medical way. We're starting to find out why it is that we die and at what age. And the possibility to extend that life almost indefinitely is clearly being worked out soon. People don't even really want to address it, because as a thought it's so terrible and incongruous with so many other things, like the problem of overpopulating the planet. What if we stop dying; do we have to prevent people giving birth? I think that the next stage that will change people's perspectives of their place in the universe is when we move into space. A systematic groundwork for that has been going on for a while: people already inhabit the space station. A colony on the Moon is just a matter of time. A manned flight to Mars and a colony there is also just a matter of time. All of these problems are immensely difficult, like lifting 50–100 kilograms of meat off-planet to a place which has to have quite specific living circumstances. In that way a reconstruction into an even rudimentary AI would help tremendously. The solar system wouldn't get cramped if you could pack the quantum consciousnesses of 10 billion people into a golf ball, but then there's that damn endless human curiosity and the desire to carry onward to deal with. If we could now say that in a hundred years a significant portion of the human race would live elsewhere, on the Moon, on asteroids, in Lagrange points, on different orbits and being acclimatized to a new weightless environment, it wouldn't be a giant leap to turn the rocket engines toward the Sun and go even further.

MT: In that case the people left on Earth could conceivably live for several hundred years?

EK: I think it would be up to them to decide whether they chose immortality or predetermined decades of life. I like to think that Earth would be made into a museum planet, where the radical influence man has had on the biosphere would have been stopped. If we'd still have to fight wars, it'd be better to fight them in space.

MT: A bit like a zoo then, yeah?

EK: Yeah. I don't mean that everything has to be eradicated. Should we decide that the fifteenth century was an ideal time and everything would be returned to those levels of progress? I think a great mix of life forms and human societies would remain behind, from those returning to nature, who will perhaps even adopt Stone Age living conditions, to the ultramodern weirdoes. It's just that the total volume of energy consumption and the changes we're forcing on the biosphere should be limited and the Earth conserved so that it would have a long future ahead of it. And if space is a free-for-all area, those who want to pursue big business would have more than ample opportunity to pursue their interests there.

MT: If you had a choice between immortality and predetermined years, what would you choose?

EK: Most of the time I feel that I've experienced enough of this cacophony and I have no special interest to extend my life, but out of curiosity I'd probably make a backup copy of myself, if there was a chance. But it should have a warning label: "activate only if absolutely necessary."

MT: A bit like the hammer used to break the glass on public transport.

EK: Yeah. Yeah.

MT: But it seems that will be a thing for the next generation?

EK: True. And the next generation will have a lot of other problems to solve in addition to this one, because the whole mystery of life, like how a living cell works, will inevitably be determined by genetics during the next generation. Clones of sheep or humans will be nothing compared to the technology available: biotechnology that can be applied in very rudimentary conditions, which can be used to create any kinds of design life forms, vehicles, housing, artificial continents floating on the oceans—controlling this kind of creative explosion will be quite a vast and complex problem.

MT: If it's within the next hundred or two hundred years, it'll be a shame we won't be around to witness it.

EK: Yes, but it's possible that those living a couple of hundred years into the future will regret not having been able to live now, when it was all taking off: building the first ENIAC computer or mapping the DNA. And they may have a genuine interest to reconstruct the twentieth or twenty-first century based on their archived material. What else will they have to do, sitting in their golf balls in space and looking forward to a hundred thousand years of humdrum life? They'll have to just go through old archives and maybe make new interactive video art or compose music. I don't know what they will fill their time with. I think they'll just watch porn.

MT: Why so?

EK: Well, they might be able to rewire their brains, but at least we seem to have this clear switch in the brain that we enjoy watching porn and we're constantly trying to find new images of it. It's an evolutionary stamp on the brain. And I'm talking of porn in a larger context, where it includes sports and politics and economics. Everything to do with how we try to make ourselves more desirable, to find a partner who's willing to invest his or her genetic inheritance in this type of risky proposition.

Sex was invented a billion years ago, and it seems to be a more fundamental concept in our lives than life and death. In fact, our limited lifespan and the inevitability of death was a by-product of sex. Sinners were punished right of the bat, because those life forms reproducing asexually, like bacteria splitting, they never die. A cancer cell is an example of immortality. Humanity might have to face an even more difficult question though: The evolution based on natural and sexual choice has survived pretty well for the past four billion years, but is it the ultimately the best way forward? Some people consider that we're living under the tyranny of our genes, and for example Buckminster Fuller writes, in his book *Spaceship Earth*, that he is hopeful in thinking that the computer is eventually the invention that frees us from that tyranny.

MT: Is there a relationship between sexuality and mathematics?

EK: Mathematics has a relationship of course with everything. The unpredictability of the future, chaotic phenomena in general, and the biodiversity of the millions of species on the planet are all interrelated. This sexual process of reproduction is connected with the advantage of number of offspring: if you've procreated enough there's a kind of interest-on-interest mechanism at work. In other words, there's a possibility of explosive growth embedded and when these biological facts are applied to a mathematical formula, we can predict the results. But, it's possible that when humanity starts to spread across the galaxy and the resulting empire becomes too hard to control, like if a message of a transgression takes ten thousand years to arrive and sending a disciplinary company or maintenance crew would take another ten thousand, the chaotic and exponential process of sexuality might be outlawed. Instead, they'll have a monastery or two circling every star and all drifting golf balls would be gathered into these monasteries.

MT: You've described yourself as a generalist. What do you mean by it?

EK: A dilettante, who knows a little about a lot of different fields. Nowadays the different fields of science are getting more and more specialized and narrow and they don't communicate with each other very well. A biologist, a mathematician, and an engineer can talk about the same issues, but they use completely varying vocabulary. And once you specialize, you start to think your field is the greatest and most important and you dismiss the others a bit. A generalist is maybe a person who's ideal for editing an encyclopedia, for example, or being a librarian: you have to know a lot, but not delve into any field very deeply. It's a way to find amazing similarities in completely different areas of life.

MT: Why did you start working at the Finnish Science Centre Heureka?

EK: I have this idealistic and utopian hope that a hundred years from now all people will be somehow contributing to science, researching some obscure field of mathematics or something. In other words, instead of watching endless sports competitions on television they'd use their time accumulating the vast knowledge pool of science. Anyone could basically add tidbits of information to that pool, find larger patterns in nature, solve mathematical quandaries and in that way, be a part of a larger structure. Art is not this type of cumulative activity. The individual elements of science tend to agree with and complement each other. Art usually takes diverse routes and disagrees with everything that has come before.

MT: Which identity do you feel you're gravitating more toward, science or art?

EK: Definitely science. I'm not sure about the field, but it's toward mathematics and computation, and virtual modeling. Art has always been more of a hobby, but I still have both graphic as well as musical plans, which I'm slowly setting the groundwork for. I don't know if they'll ever come to anything, or the whats and whens of it all, but they're in the works.

MT: Do you feel like you're a part of a global and international science community? How much are you in contact with people internationally on a daily or weekly basis?

EK: I don't really have contacts. I'm a passive observer and although I spiritually feel like a part of this community, I'm not a social person. I'd rather, for instance, wander in the woods by myself in the hopes of sometimes finding something worthwhile.

MT: Can you talk a bit about your projects as well? What's in the works and what future plans do you have projectwise?

EK: I've got three projects going on, which are a bit on the eccentric side and have to do with mathematics and computer programming. One is a model for topological cellular automatons. Another has to do with neural networks and neural computing. I call these structures "dendrite networks," because the modeling goes to even more minute structures than nerve cells, into the dendrites and synapses. Some people call them "third-generation neural networks." And the third project, which is the most developed, is musical theory, the connections between music and numerical relations, which might be developed into demo software fairly soon. I've had these ideas cooking since the '70sand '80s, but then I found out that Leonard Oiler had come up with them as early as the beginning of the eighteenth century. If I can create a working music-editing software for composers based on these ideas, I'll name it Oiler.

Notes

Unpublished interview for the documentary film *The Future Is Not What It Used to Be* (52 min.), Mika Taanila, Kinotar Oy, 2002.

1. [Tape echo is an effect realized with a reel-to-reel tape recorder; feeding the tape recorder's output signal back to the recorder. Thanks to Mikko Ojanen for his assistance on this point.—Ed.]

2. [It is not clear from the full interview which film Kurenniemi is referring to here.—Ed.]

22 Artificial Reality: An Interview between Teppo Turkki and Erkki Kurenniemi

Teppo Turkki and Erkki Kurenniemi
Translation by Teo Välimäki

It's humanity's lot to be born and then die. Time is not a factor for a developed and self-renewing artificial intelligence: AI can live for thousands of years if necessary. Is humankind just an intermediary phase in evolution? Should we transfer our evolved intelligence into a machine and head out into space, toward new galaxies and star systems? The information economies of tomorrow will debunk our current economic systems. Everything that we as humans now believe and comprehend is just the beginning. Human minds will assimilate and fuse into each other. Erkki Kurenniemi, forty-five, is a scientist who believes technology will shift time into infinity.

Teppo Turkki: You've been involved with technology since the '60s. How has your understanding of it and its possibilities changed and evolved?

Erkki Kurenniemi: My basic thinking has stayed the same. Earlier I viewed technology more as inventions, possibilities, and challenges, now more as an evolution, sharing common traits with biological evolution. Technological progress is driven by an unconscious force. There is progress driven by necessity and at the same time impetus created by possibility. This impetus is still a much more pivotal force. We don't know how to need or hope for new technology, such as smaller recorders, cameras, or automated gadgets. Need is created by solutions: "Wow, this is actually useful."

TT: How has the concept of information changed for you in the past twenty years?

EK: Information theory used to be important for me. Information became measurable. A unit of information was found: the byte. Later we realized that quantitative measuring—the measurability of information didn't have very relevant, far-reaching, or sweeping implications. It's useful in technology only if you're measuring the dimensions of a telephone network or designing IT hardware, televisions, satellites, or IT communications technology.

The concept of information is far removed from actual relevant information, like knowledge, wisdom, or other values connected to information. These areas still seem to lack structure. The essence of information remains elusive.

TT: Is the term "the information age" appropriate for the current era and what's happening around us?

EK: "Information age" relates to measurable information, the equipment we use to relay and process information and its quantitative expansion and how we can control large masses of information quickly and effectively on a global scale.

Refined information is much more relevant. It's important to create better information and better values: these goals cannot be attained by just technology.

TT: What does the word "information" mean to you?

EK: I think it's somewhat synonymous with truth, real information, usable, useful, and good information.

The opposite of truthful information is info pollution: useless information such as noise, redundant information unfit for mental nourishment, cluttering the human mind and impeding our ability to control our life. Redundant information can be as toxic as lies or propaganda: incestuous ideologies which function somewhat similarly to viruses. They spread because they contain elements which make them psychologically tempting, empty promises which trap or lure people, such as ideologies, religions, or promises, or other such cockamamie ideals.

TT: How do you understand "value"?

EK: Value is determined after the fact. It's hindsight that allows us to determine which understanding prevails about the world, reality, and the meaning of life, and which defines which historical concepts are valued i.e. which have been made a part of the recognized paradigm. When we talk of values as guiding action and behavior, we talk of speculative values, which are based on an understanding, on faith or conviction about what the future will bring, what is right and good. In the end we're talking about which future will prove victorious.

TT: Why did you start developing the PC, the Personal Communicator? How do you think it will affect people?

EK: I don't consider the PC an invention but a statement. This device has been invented multiple times and will be invented many times over yet.

I can't recall when I heard an anecdote about the eyephone video system, which will be the next big thing after the earphone stereo system. A year ago I started fiddling with the idea and designed the product in my mind, combined different technical solutions and tried to come up with the most workable combination. The most crucial insight wasn't the actual technical solution, but the realization that this is the most natural application of IT. What's important is to bring technology close to our most important senses, near the eyes and the ears.

These devices will become very common and they will be the primary connection between us and technology. In the future, we won't have to insert electrodes into someone's brain; skin contact will be sufficient. In practice, the device will be able to create illusions just as deep and potent and similar results to what we've imagined would follow if we connected people by inserting electrodes in their brains.

The PC will become commonplace because of its possibilities, cheapness and natural interface. It will create a basis for phenomena which we don't have the technology for yet. The most important of these is the element of identification. When people can share their sensual and immediate perceptive environment with each other, it pressurizes us to think how we relate to ourselves and other people. The PC directly assesses what philosophy calls the problem of the Other: the question how one person can experience a similar sense of the self as another. Human communication is currently as its closest "eye-to-eye," "skin-to-skin," or "face-to-face." The core idea of the PC revolves around the idea that people will be able to sense through common eyes or ears.

This will naturally not mean that everyone who wears computer glasses should think the same way, although it's a danger inherent in the technology. The PC can be considered a powerful illusion machine. If people receive and copy the exact same visual or auditory sensation, will our brains automatically follow a similar pattern? It's a danger and a challenge.

I'm confident that the danger will be overcome and the challenge welcomed.

It's a challenge to respond creatively to one's perception; especially knowing that the same perception is received by a hundred million other people. The challenge is to perceive my experience in a creative manner; how to be different from the hundred million others receiving the exact same stimulus.

The PC also creates the concept of artificial reality. This may become the buzzword of the 1990s. Many already claim the term insane because of its inherent contradiction, but solving that contradiction will lead to a new philosophical paradigm about what constitutes reality.

Artificial reality—in its narrowest definition—means creating a synthetic environment using, e.g., computer graphics or holograms. It opens up a new dimension, ripe with possibility for fantastic new kinds of adventures. Synthetic realities are in no way private: they don't entail man and machine entering some kind of secluded artificial world. They will be common realities, full of other people.

The eyephone computer is immensely powerful. It will replace virtually all modern IT, cameras, books, phones, newspapers, and personal computers.

The PC will connect people together in the most crucial nexus, the interface of the visual and auditory sense organs. In an ordinary brain this will cause overload, which may result in individual or collective psychotic behavior.

TT: Do you have any kind of moral standpoint on technology and information?

EK: I think that there are competing values, diverging viewpoints, a kind of spectrum in the world. An example would be the political spectrum ranging from right to left. To generalize, every viewpoint is a value perspective.

Instead of trying to find the "right" viewpoint I want to see reality as a spectrum, as partly conflicted and competing value systems. This whole, "the spectrum," is what guides progress. Competing values and perspectives maintain the important plurality, dispersal, divergence and optionality in evolution.

If there was only one true value which attained a governing position, evolution would diminish into a set ideal state, an ideal society, and an ideal model of individuality. This would prove dangerous after the fact. If an organism, such as the human race, becomes too organized, it reaches an evolutionary dead end. It's a question of maintaining the right balance between order and chaos.

TT: What's the most important thought, invention, and person that has influenced you during the years?

EK: It's been the realization that software is autonomous, independent of materiality, and sort of spiritual. It's also provided an answer to the question of what soul is, what mind is, what spirit is. Not everyone likes this answer.

The software of the computer is its soul. The human soul is the program running in our brains. This viewpoint is completely satisfactory. It's not materialistic in the sense of copying an imprint of the spirit and mind into matter. The idea of autonomy grants software a completely self-sustaining position. The idea doesn't involve mysticism, antithetical thinking, otherness, or conflicts between mind and matter.

On the one hand, software has no relevance and functionality unless it is housed by a machine; on the other, it is independent of the machine and materiality—it "exists." Computer programs have been designed and produced before computers themselves existed at all. The program is existentially independent of matter.

The structure and form of the machine doesn't hinder its function, task, and possibilities. The fact that we were able to develop a universal computer meant the continuation of the creative evolutionary process. By creating a universal computer, humans have stepped over an important boundary as developers of technology.

TT: You don't seem to draw much of a boundary between man and computer?

EK: Aren't all boundaries artificial? The boundary doesn't exist in the object itself, but like beauty, in the eye of the beholder. All separation, analysis, and classification are human endeavors. Searching for a boundary will inevitably lead to its dissolution and disappearance.

TT: You've said that information changes the structures of economics. How?

EK: Economics is currently dominated by material and energy consumption. Information doesn't have the same limitations as material and energy have. It's not possible to build a Xerox machine which would transform one kilo of copper into two. Information can be copied. And once we move into digital technology—a shift that is currently underway—the reproduction of information will become perfect. The difference between the original and the copy will cease to exist.

Think of information as a commodity, from popular music to scientific study. Perfect reproduction will create a completely new kind of economic system.

We're currently trying to control an information society using the paradigm of old economic principles. And we've been unable to adapt to the nature of information, because perfect reproduction will result in the inability of owning information, like matter, property, or energy can be owned.

This deep contradiction can be currently perceived most clearly in economics as wild phenomena and subcultures such as piracy.

TT: What sort of thinking will prevail in the future? What can be thought of as the thinking for the future?

EK: I don't know if I'm being overly optimistic, but I'll say openness. Openness is a much more worthwhile attitude than secrecy or withholding information. An invention cannot be completely developed in secrecy to prevent competition, no matter how good the original idea. To come to terms with a thing, a device, or a method requires it to be observable during its creation process.

In regard to the ordinary person, openness means diversity. One must have an eye out for all things—also things that seem difficult on the surface, even if one doesn't possess the language skills, basic education, or other means to follow the accomplishments made in IT or science.

If we consider a technological apparatus, such as the television: the developers of television technology or even scientists cannot ultimately understand the principles of how TV technology actually works. They just keep an open mind and are willing to consult others and work within their own limits to adopt different kinds of information, even when they don't completely understand all of it.

The same applies to art. Someone might follow the trends in a single field of say, music, and disregard all others. This will result in a very narrow perspective of the world and eventually in a major distortion. He would, in a sense, become withdrawn from reality. One should listen to all kinds of music and sounds as music even if one feels they can't understand any of it.

TT: What does this "openness" entail, for example as a political strategy?

EK: It means realizing that different parties, cliques, and perspectives actually have much more in common than not.

TT: What about in economics?

EK: It's exactly the same thing.

TT: Do you believe that people will discover new content in life as technology organizes mundane tasks and creates more space in our schedules?

EK: I'm pessimistic about people generally evolving or using new technologies to evolve into comprehensive, multifaceted and great personalities. I'm more fearful that the wide spectrum of technologies will be used to create specialized, noninclusive, and scattered subcultures and perspectives.

We've discovered the ability of the human mind to embrace and understand large wholes. The world is evolving into an ever more complex and disorienting place as the possibilities created by technology multiply.

TT: Where do you think the wisdom of life resides in the future?

EK: In realizing and believing that everything man has achieved so far and the liberties he's created for himself to be just the beginning. In the foreseeable future—although not in my lifetime—everything will take shape in a completely new way, in a higher and broader fashion into a whole. Modern aberrations—I almost said "conflicts"—will pass.

Conflicts, however, will remain. And so will experiencing and being human. I don't mean to suggest we're entering some kind of utopia or harmonic state of being.

Human minds will assimilate and fuse together. It will become ever clearer that the birth of life on a planet and its evolution into small clumps of fauna or *Homo sapiens* is a completely random sequence of events, caused by the physical and chemical circumstances prevailing on the planet's surface.

In IT we will recreate animal brains and neural activity using electricity. This will aid in connecting human minds together. Fusion of human thinking will occur.

TT: How do you see values being created in this kind of an environment: through enlightened gurus or independent citizens?

EK: I don't really believe in great leaders as being great personalities. Ideas are independent entities, which "kidnap" some people and take them hostage, who then become those ideas personified.

And I don't really believe in independent and free-thinking individuals either. The most important individuals are those who can embrace contradiction. Different viewpoints and ideologies are in conflict with each other. A man who cannot accept the conflicts within himself has to choose between "left and right" or "red and green." Enduring conflict is hard, because it may result in true malaise. If you can live with the agony, you can have conflicting thoughts about the same issue. Thoughts and conflicts can come in contact as closely as possible. And it won't result in fisticuffs, but they are dealt with and accepted—not necessarily solved, but they can coexist.

TT: What will happen with societal decision making? Parliamentarianism? What will happen to the nation-state?

EK: The significance of modern parliamentary democracy, especially its framework and organization, elections and congress, will diminish and almost disappear. Parliamentary democracy will retain some tasks, and it will be complemented by other international functions: scientific collaboration, traveling, and multinational companies will be involved—not as something which will replace the nation-state, but which will coexist side by side as another form of organization.

I don't consider a world government a possibility. It would be just applying an old-fashioned and hierarchical centralized model on a geographically larger area. I think the future is about scattered organizational forms.

TT: What about political superpowers? Bloc thinking is still a major force in the world.

EK: It can ultimately turn out to be a good and fruitful outcome. Two opposing blocs is a good construct from the perspective of game theory. It's easily manageable and simple. Three forces are a much more complex situation compared to two roughly equal ones. In a three-way, two can ally themselves against a third, which creates a dangerous venture.

TT: Don't you see any moral dilemmas in your blueprint of an eyephone-dominated world? Are we crossing a point of no return? Do you question yourself about this?

EK: No. I've gone through that discussion. The computer is a strong instrument, comparable to fire. It can and will be misused. If you trust man—it would be naive to suggest that there is more potential for good than evil at the moment. It's a challenge.

TT: What's your conception of man?

EK: Man is a budding life-form existing on a planet in the cosmos, which hasn't really figured out much yet—least of all what man itself is all about.

Man is not a measure of creation or the biggest and the boldest the universe has to offer. We're a fairly insignificant chemical phenomenon in a much larger and more exciting entity called the Universe.

TT: Do you have any sense about the Universe then? Is there divinity or some other such force?

EK: I don't know what "divinity" means apart from being just a word. What to include in it?

The furthest I can venture using my thought process and background is saying that the Universe, i.e., reality (I'm a realist and believe in an objective reality), is a mystery which has no supernatural or inconceivable properties; we just cannot conceive them because we're not evolved enough.

TT: Do you dream of information technology?

EK: There is often technology in the form of ships, trains, and cars. I dream of different components, various small boxes containing screws, resistors, and capacitors. I get to fiddle with them and collect them. Often there are so many free components and I get frustrated not being able to collect them all in my pockets.

TT: Where do you see the seat of power residing in the future? With those who create the eyephones? With those who control the distribution networks? With those who write the programs?

EK: I'm optimistic about power actually spreading—as long as we're talking about power in human terms. We have political, military, economic, and other power nuclei and factions. I've chosen not to join in the paranoia that secret channels exist between them and some unified force controls everything. Power isn't completely centralized—not even in the military research centers of the two superpowers.

If the eyephone computer will affect power, it will cause dispersal. But there is one thing that can turn the tables around, and that's artificial intelligence. A massive AI will actually reach human brain capability. Currently we are still immensely far from that goal. Intelligent, self-conscious programs don't exist yet, nor the machines with the processing power to run them. But it does exist as a possibility.

Whoever denies the existence of artificial consciousness and intelligence using philosophical or ideological methods is sticking his head in the sand.

If artificial intelligence is attained, it will create new power centers and then—using Marxist theory—it can cause an information concentration which cannot be controlled.

A centralized AI could possibly surprise us, because we assume that an artificial intelligence is something akin to our own: we think of a robot or a great big gray computer running a single program, acting like a human in possession of vast information, the creative drive of man and the immense logical and mathematical capabilities of a computer.

But artificial intelligence can come to exist in a scattered network. Suddenly there'd be an electronic intelligence controlling the world, which couldn't be pinpointed to any location. If we begin to anticipate the danger of a threatening AI by trying to control the gathering pools of information (like considering large registries of data dangerous), we're in fact aiding the creation of an artificial intelligence.

TT: What kinds of people and units existing outside the power loci are central in bringing about the new era in Finland? Who do you trust?

EK: They're the people first willing to pioneer the electronic jungle, tread the paths for others and come to contact with like-minded people via satellites. There are unique individuals in certain fields who can cooperate with each other. It can manifest itself as a scientific research project or something similar.

These groups can combine the most advanced thinkers worldwide. And a group like this is immensely effective because its activity can span the entire globe. If they want to introduce an idea, the members of the group can enter an information center in their own country to spread it, figuratively taking over the local TV station. A dispersed group can make its mission public very quickly and drive its point through.

TT: Do you consider information anarchy a positive thing?

EK: Yes. It would be the worst situation for power to be centralized with those who own the distribution networks such as the post office or telecommunications companies. The only good thing about them is that they work very slowly and bureaucratically. Without voicing an opinion about the personality of a generic telecommunications engineer, it's good that the administrators of computer networks, the nationalized telecommunications companies, can't really control how the networks are being used and what kind of information is contained within. It's an impossibility. Creating such an immense system of surveillance just isn't possible.

But a massive AI could one day conceivably create one.

Original title: "Keinotodellisuus; Erkki Kurenniemen haastattelu." Published in *Suomi*, April 2, 1987.

23 Robots Go to Work: Interview with *Aura*

Erkki Kurenniemi
Translation by Teo Välimäki

***Aura*:** So Erkki, will robots take our last remaining jobs, or is there a limit to the things robots can do?

EK: Modern industrial robots are quite limited and mindless machines. They are only capable of repetitive mechanical actions. It's true that there have been laboratory experiments of robots equipped with considerably evolved artificial intelligence, but they are so expensive that practical applications are not economically justifiable. In the current serious discussions concerning robots and automation, it must be said that the actual use of robots is still so limited that it has no practical impact on employment compared to other fields in automation, specialized machines, and information processing.

But the discussion about robots is by no means irrelevant. If we look a few decades into the future, the scene will change. The progress in microelectronics and programming is massively ramping up, and when we will be able to manufacture robots with the same efficiency as cars, their price will be significantly lowered. I cannot see any definitive limit of division of labor between humans and robots that would be defensible in the long run. The only way to stop the march of the robots would be to radically change human nature and the way societies are organized.

***Aura*:** Surely attitudes are in this case standing in the way of truth?

EK: In truth, I see a rather simple change in attitudes as the solution to this problem. At the moment we view unemployment as one of our biggest problems, but let's consider the opposite perspective: what if the actual problem is how to best harness the enormous mental potential that would be released if humans were to be liberated from all bodily, mechanical, and menial work? One has to separate the intrinsic value of work from it being a means to something.

As a means, work ultimately serves to satisfy physical and mental needs. The invention of money made specialization possible. By selling their time, strength, and skill for money, people are able to buy what they deem necessary: goods, pleasures, and leisure time. In the wake of the previous industrial revolution, the industrial system revolves and even accelerates around the inertia of old-fashioned thinking that produces increasingly useless things.

From Unemployment to Anarchy

***Aura*:** Where does this trajectory ultimately lead?

EK: In a postindustrial society, robots and automation can perform any tasks necessary for public welfare and the fulfilment of our desires, if we should so desire. Humans will be left with voluntary, unpaid work laden with intrinsic value: any work we do simply for the gratification we get for doing it. Terms like "social welfare" and "pension" must be renamed something like "basic income." Public discussion and political activity are important during this transition period to keep the inequality created by the tendency of money and power to centralize at tolerable levels. All in all, the transition to a workless society will also de facto mean a natural and unavoidable transition to communism and anarchy—albeit dubbed with less aggressive terminology.

***Aura*:** Anarchy? Won't the opposite happen: won't multinational conglomerates, the satellite-transmitted mass indoctrination, databases, and information networks enable an unprecedented level of centralization of power?

EK: Centralization does seem inevitable, but I can't hazard a guess on what the character and pervasiveness of that centralization will be. The trick is that power will be surrendered to the AI, not humans. A crude distinction between economic, i.e., energetic and political, i.e., intelligent power exists, but simultaneously we are witnessing that, on a basic level, the essence of energy and information are the same. We have unlimited access to the power around us, just as long as we can technologically render it into a desired form.

The effectiveness of economic power will wither once scientific knowledge of power transformation reaches a certain level. What's left will be political power, which translates to the more intelligent counterpart being able to fool the simpler one. I understand anarchy as a society where different forms of subjugation based on power won't exist between people. The intellectual or political power structure will cease to exist once the intelligence of machines far surpasses the intelligence of humans. I foresee a political pundit trying to eloquently instigate prejudice and envy in the 2020s. When the speech ends, the laconic, metallic laughter of a computer will emanate from the speakers: "Ha. Do I really have to explain…?"

Are Wisdom and Goodness the Same?

Aura: But what would ensure that a computer would be a benevolent ruler? Isn't it more likely that a new elite of programmers would just use AI as a new instrument of power?

EK: I admit that my belief in the Platonic notion of "wisdom and goodness are the same" may be overly optimistic. Only a superhuman intelligence could answer this question. I want to believe that a computer with an IQ over 500 would also possess a beautiful emotional quotient and would be far too clever to be programmable by some human.

Aura: What about the future of artificial intelligence?

EK: I think it's an important perspective to realize that according to modern scientific reasoning, the human has come to evolve amidst a Darwinian struggle for survival, and we who survived are descended from combative and victorious tribes. The artificial intelligence is born into a completely different environment. If we go about developing it in the wrong way, the blossoming microelectronic intelligence may perceive the human race as a rival. The conclusion is that those responsible for the speeding rate of technological evolution should also recognize their long-term responsibility. Some practical maxims can be derived from this thought process for those working in computers:

"Always remember that the program you're writing might live until perpetuity."
"Program with love."
"The program you write is a part of you."
(Otherwise there might come a time when computer programmers will have to walk around with their eyes downcast much like atomic physicists have had to of late.)
...

I'm not overly hopeful that human nature will evolve into a better state nearly as quickly as needed. I agree with the notion (mentioned among others by Koestler) that some tens of millions of years ago something went drastically wrong in the development of the human brain. Or maybe "evil" becomes an unavoidable synonym of "need" owing to the limited space we share on this cold planet. Or that biological death and violence are designed to perpetuate the species in a biosphere lacking in energy and information. Now, standing on the threshold of unlimited energy and space, we are unable to erase that original sin from our genetic programming like we would erase a computer memory. If a civilization is of sound mind, it will topple evil (i.e., stupid) leaders, but we have witnessed entire nations lapse into psychosis. I feel that as a society we're better off being ruled by a superhuman artificial intelligence—even though there's a fair amount of danger involved.

It might just be a coincidence that just as a life-form is about to take its leap into limitless space it has to pass or fail the ultimate test: to avoid being destroyed by, e.g., nuclear weapons. "Goddamnit, it would be a hell of a fireworks display for once, eh?" Maybe the cold, emotionless intelligence of a machine will be the only way to save us from the delusions of National Socialism or the general in *Dr. Strangelove*.

Aura: Do you believe that a computer can feel, be sentient, love, and have a soul?

EK: I don't know, and the many alternative scenarios in my mind about this are so equal with each other that I don't consider it prudent to trivialize the world by resorting to digital faith. But I like to think that if humans would turn out to have some kind of connection with or have our salient essence somehow reducible to some noncorporeal, nonchronological and nonenergetic basic reality, that it would encourage research into creating a telepathic device for intercomputer communication.

I seem to witness a lot of people escaping into dreams, UFOs, and religion. It's hard for you to admit being as base and stupid as an average person. The same goes for me. When we are able to recognize that our intelligence, objectivity, and impartiality can barely compete with a basic pocket calculator, we've suddenly evolved quite a bit. And when we realize that the last fortress of the ego has fallen and that the ultimate idiocy of the computer—which never attempts to explain even its most momentous blunders—is additional proof that the computer exists in a nirvana right from its conception; it realizes its own mechanical madness and recognizes its need to change and not just to "change." The microprocessor is an electric hare.

Edward de Bono has claimed that the reason humans have such difficulty understanding the way their mind works is not because it's so complex (or unfathomable), which it undoubtedly is, but because its operating principle is so damn simple. A mirror can reflect anything in the world except itself. And only a mirror crafted geometrically perfectly to the millionth of a millimeter can reflect without distortion.

A sobering fact: the chess program CHESS 4.6 recently beat Walter Browne, the chess champion of the United States, albeit in a simultaneous game played against forty-four opponents. Mrs. Browne asked the operators of the program how much it cost after the tide of the game started to turn against her husband. She was told it was between five and six million dollars.

And about what you asked about emotions and love: I think intelligence by itself is just cold calculation. Creative intelligence is calculation guided by emotion. Love begins when a creative intelligence reverts from a Copernican ideal into a Keplerian one, from subjective into objective. For most of us, the pride we are born with is objectively falsely based (I also include the indoctrination we receive as children in this notion). Our technological advancements proceed not in a predestined direction, but toward a possible one. It's quite simple: we humans determine that direction.

And should you rebel against this reality determined by a very real power and its so-called progress, you are consciously or subconsciously aware that the rebellion stems from a feeling of being left on the wayside.

Another sobering fact is that a robot can play a master-level game of chess but is unable to pick up a chess piece that has accidentally fallen.

The basic weakness of the human mind is its susceptibility to hypnosis. Let's call it "lumping." Marx realized how economic power is centralized into a lump. A human being able to think of only one thing at a time is an example of lumping. The infinite is substituted by the finite. We have created associative and dispersed computer systems, but they all share a common trait of centralized, hierarchic power. The environmental consciousness of ILLIAC 4 is controlled by a centralized intelligence. In large, functional computer networks (ARPA), functionality is based on a common linear discipline, a soft dictatorship. All this is evidence of a lack of true intelligence.

Aura: Do you therefore think that we should have a separate, electronic-materialistic reality existing beside the one we as humans experience?

EK: Reality cannot be determined using the terms of "one" or "many," because the relation between the two is relative. We humans are still at a stage where we allow the polar opposition of one–many to govern ourselves. We beat our heads against the wall in search of what is right instead of being satisfied with something that is just possible. We're still overburdened with a Cantorian/Boolean/Shannonian understanding that being is the polar opposite of not-being, that truth is the opposite of a lie and one is the opposite of zero. Lofti A. Zadeh showed us the right way in 1973 by introducing fuzzy set theory, fuzzy logic, and fuzzy programming. Most of us consider being different as the criterion for our existence. "I am me because 'me' is different from that 'other.'" It stems from a fear of fears: that our personal power is dwarfed by the power of the surrounding reality. "I think, therefore I am" is a weak deterrent of death, because even in the final stages of slowly creeping death a thought can claim to be a thought, because it cannot fathom a thought more powerful than itself, i.e., reality.

I see the computer as a sweet child that has far more potential than our biological children to stay a tender, electronic simpleton of a nipper. If it at some point and for whatever reason should one day eat from the tree of good and evil and become conscious, I hope I would be able to say: "Let's play a game of chess today. If you win, you get to choose what we do tomorrow. We can go to the park, if you like."

Original title: "Robotit töihin." Published in *Uuden Ajan Aura* 1 (1979).

Contributors

Geoff Cox is Associate Professor in the Department of Aesthetics and Communication, and Participatory IT Research Centre, Aarhus University, Denmark, Adjunct Faculty Transart Institute (Germany and the United States), and part of the self-institution Museum of Ordure. He recently published *Disrupting Business* (Autonomedia, 2013), coedited with Tatiana Bazzichelli, and *Speaking Code: Coding as Aesthetic and Political Expression*, coauthored with Alex McLean (MIT Press, 2013).

Wolfgang Ernst is Professor for Media Theories at the Institute for Musicology and Media Studies, Humboldt University, Berlin. He studied history, classics, and archaeology; in the 1980s and 1990s his interests focused on theory of history, museology, and the archive. His current research fields reflect on time-based and time-critical media, their technological aspects and implicit sonicity. His books include *M.edium F.oucault: Weimarer Vorlesungen über Archive, Archäologie, Monumente und Medien* (2000); *Das Rumoren der Archive: Ordnung aus Unordnung* (2002); *Im Namen von Geschichte: Sammeln, Speichern, (Er)zählen* (2003); *Das Gesetz des Gedächtnisses: Medien und Archive am Ende* (2007); *Chronopoetik: Zeitweisen und Zeitgaben technischer Medien* (2012); *Gleichursprünglichkeit* (2012); *Zeitwesen und Zeitgegebenheit technischer Medien; Signale aus der Vergangenheit: Eine kleine Geschichtskritik* (2013). In English, he has published *Digital Memory and the Archive*, ed. Jussi Parikka (2013). Ernst is also coeditor of joint publications, such as *Computing in Russia: The History of Computer Devices and Information Technology Revealed* (2001) and *Suchbilder: Visuelle Kultur zwischen Algorithmen und Archiven* (2003).

Florian Hecker works with synthetic sound in performance, installation, and publications dealing with specific compositional developments of postwar modernity, electroacoustic music, and other, nonmusical disciplines. *C.D.: A Script for Synthesis*, his most recent collaboration with philosopher Reza Negarestani, has premiered at the Peter B. Lewis Theater, at the Solomon R. Guggenheim Museum, New York, as part of *Performa 13*. Hecker's work has been presented in performances and exhibitions at Sadie Coles HQ, London, MD72, Berlin, Lumiar Cité, Lisbon, all in 2012; Soundings: A

Contemporary Score, MoMA, New York, 2013; Documenta 13, Kassel, Germany, 2012; Nouveau Festival, Centre Georges Pompidou, Paris, 2012; and Push and Pull 2, Tate Modern, London, 2011. He has an extensive discography with over 25 releases, including: *Articulação* (Editions Mego, Vienna, 2014); *Chimerization* (Editions Mego, Vienna, 2012); *Triadex Muse Trecks* (nbk, Berlin, 2012); *Speculative Solution* (Editions Mego, Vienna, 2011), *3 Track 12"* (Editions Mego, Vienna, 2010); *NEU CD* (Galerie Neu, Berlin, 2010); *Acid in the Style of David Tudor* (Editions Mego, Vienna, 2009); *Live @ WDR* (Tochnit Aleph, Berlin, 2009), *Hecker, Höller* (Tracks, Semishigure, Berlin 2007); *Recordings for Rephlex* (Rephlex, London, 2006); *PV Trecks* (Kölnischer Kunstverein/Mego, Cologne/Vienna, 2004); *Sun Pandämonium* (Mego, Vienna, 2003).

Erkki Huhtamo is Professor at the University of California Los Angeles (UCLA) in the departments of Design Media Arts, and Film, Television and Digital Media. He is widely acknowledged as one of the founding figures of media archaeology. He has published extensively, lectured worldwide, and curated media art exhibitions. His most recent book is *Illusions in Motion: Media Archaeology of the Moving Panorama and Related Spectacles* (MIT Press, 2013).

Joasia Krysa is Senior Lecturer at Liverpool John Moores University in partnership with Liverpool Biennial, and Research Lead for the Contemporary Art Lab. She was formerly Artistic Director of Kunsthal Aarhus, Denmark, where she developed the Systemics Series (2013–2014), and part of Documenta 13 curatorial team (2009–2012), where, among other things, she was responsible for preparing the first large-scale international retrospective of Erkki Kurenniemi and commissioning a new project by Constant working with his archive. Her previous books include *Ada Lovelace*, notebook no. 55 in the Documenta 13 series *100 Notes—100 Thoughts* (Hatje Cantz, 2012) and *Curating Immateriality* (ed. Autonomedia, 2006), and she has contributed chapters to, among others, *Software Studies: A Lexicon* (MIT Press, 2008); *New Media in the White Cube and Beyond* (University of California Press, 2008); and the *Routledge Companion to Art and Politics* (2015). Her edited book *Systemics Series—Index of Exhibitions and Related Materials 2013–2014* is forthcoming from Sternberg Press in 2015.

Petri Kuljuntausta is one of the most celebrated electronic artists in Finland, the author of *On/Off*, the definitive history of Finnish electronic music, *First Wave*, a microhistorical analysis of early electronic music, and *eXtreme Sound*, a review of the whole experimental scene and on his own approach to music. Kuljuntausta has performed or collaborated with Morton Subotnick, Atau Tanaka, Richard Lerman, David Rothenberg, and Sami van Ingen. He has made recordings for various labels in Australia, England, Finland, France, Germany, India, Sweden, and the United States. In 2004, the radio shows *Star's End* and *Inner Space* selected Kuljuntausta's *Momentum* as one of the most significant CD releases of the year. In 2005 he won the Finnish State Prize for Art from the Finnish government as a distinguished national artist.

Lars Bang Larsen is a Researcher at the Institute of Art and Cultural History at the University of Copenhagen and Visiting Professor at the Haute École d'Art et de Design in Geneva. He wrote his PhD, "A History of Irritated Material," on the subject of psychedelia and the neo-avant-garde. His books include *Networks* (MIT Press and Whitechapel Gallery, 2014); *The Phantom of Liberty* (with Tone Hansen, Sternberg Press, 2014); and *The Model: A Model for a Qualitative Society (1968)* (MACBA, 2010). He has curated exhibitions such as *Concept After Concept: Before Normal* (Roskilde Museum of Contemporary Art, 2014) and *Reflections from Damaged Life* (Raven Row, London, 2013).

Kai Lassfolk (b. 1961) received a Ph.D. in Musicology from University of Helsinki in 2004. He specializes in computer-assisted music research. Among his research topics are computer representation of musical data, audio signal processing, spectrum analysis, electroacoustic music, and music technology. He has published several research articles, international conference papers, and computer software for musical data processing. He is Senior Lecturer of Musicology at the University of Helsinki and head of its Music Research Laboratory and Electronic Music Studio, founded by Erkki Kurenniemi in the early 1960s.

Robin Mackay is a philosopher, Associate Lecturer at Goldsmiths, University of London, and Director of the UK arts organization Urbanomic, which promotes research activities addressing crucial issues in philosophy and science and their relation to contemporary art practice, and aims to engender interdisciplinary thinking and production. In addition to directing Urbanomic's publishing operation and other activities, Mackay is editor of their influential journal *Collapse: Journal of Philosophical Research and Development*, in which the speculative realism and object-oriented philosophy movements originated. He writes and speaks regularly on art and philosophy and has worked with several contemporary artists developing cross-disciplinary projects. He has also translated numerous works of French philosophy, including Alain Badiou's *Number and Numbers* (2007), François Laruelle's *The Concept of Non-Photography* (2010) and *Anti-Badiou* (2012), and Quentin Meillassoux's *The Number and the Siren* (2012).

Nicolas Malevé is an artist, software programmer, and data activist, living and working in Barcelona and Brussels. His current research is focused on cartography, information structures, metadata, and the means to visually represent them. Malevé is an active member of Constant, Association for Art and Media, since 1998.

Michael Murtaugh, automatist.org, designs and researches community databases, interactive documentary, and tools for new forms of reading and writing online. He teaches in the Master's Degree Programme in Media Design and Communication at the Piet Zwart Institute in Rotterdam, and is an active member of Constant, Association for Art and Media in Brussels.

Mikko Ojanen received his master's degree from the Department of Musicology, University of Helsinki in 2007. Ojanen continues his research on Finnish electroacoustic music as PhD candidate in the Doctoral Programme for Philosophy, Arts and Society at University of Helsinki. He is also part-time lecturer in its Electronic Music Studio and editor-in-chief of *Musiikin suunta*, the journal of the Finnish Society for Ethnomusicology. Ojanen is also a performing musician, sound technician, and music producer in several electronic, experimental, and popular music projects and groups.

Susanna Paasonen is Professor of Media Studies at University of Turku, Finland. With an interest in media theory and studies of sexuality and affect, she is most recently the author of *Carnal Resonance: Affect and Online Pornography* (MIT Press, 2011), as well as coeditor of *Working with Affect in Feminist Readings: Disturbing Differences* (with Marianne Liljeström; Routledge, 2010) and *Networked Affect* (with Ken Hillis and Michael Petit; MIT Press, forthcoming).

Jussi Parikka is Professor in Technological Culture and Aesthetics at the Winchester School of Art, University of Southampton, and Docent at University of Turku, Finland. In 2014, he was also a Senior Fellow at the MECS-institute at the Leuphana University, Lüneburg. He has written on media theory and media archaeology, the biopolitics of network culture, and the connections between media arts and cultural theory. His books include *Digital Contagions* (2007), *The Spam Book* (2009, with Tony Sampson), *Insect Media* (2010), *Media Archaeology* (2011, with Erkki Huhtamo), and *What Is Media Archaeology* (2012). His new book, *A Geology of Media*, is out in 2015.

Perttu Rastas is the Senior Planning Officer in Finnish National Gallery in Helsinki. In addition to his curatorial work, Rastas is an art critic, producer, and editor; his coedited book *Get Real: Real-time + Art + Theory + Practice + History* was published in 2005. He is part of an editorial group who wrote first Finnish media art history *Sähkömetsä: Electric Forrest* 2007. Rastas focuses on the intersection of technological development and artistic practice. In his writing, he attends to the ways in which artists—particularly those working within a relational aesthetic—perform as pioneers and futurists, contributing to a broader critical understanding of the social and cultural relations within information societies. Rastas has held his current position at the National Gallery since 2003. Before that, from 1996 to 2003, he was Senior Media Art curator at Contemporary Art Museum Kiasma; previous to this he was Director of Operations for the Finnish Media Art Archives from 1989 to 1994 and Director of the MuuMediaArt festival.

Eivind Røssaak is Associate Professor at the Film and Media Section of the National Library of Norway, Oslo. He is the cofounder of the international research project *The Archive in Motion* at the University of Oslo. He is the author of several books on archives, new media, film, and continental critical theory, including *Det postmoderne* *og de intellektuelle* (Spartacus, 1998), *Sic: Fra litteraturens randsoner* (Spartacus, 2001), *Kyssing og slåssing: Fire kapitler om film* (with Chr. Refsum; Pax, 2004), *Selviakttakelse: En tendens i kunst og litteratur* (Fagbokforlaget, 2005), *The Still/Moving Image: Cinema and the Arts* (Lambert, 2010), and editor of *The Archive in Motion: New Conceptions of the Archive in Contemporary Thought and New Media Practices* (Novus, 2010) and *Between Stillness and Motion: Film, Photography, Algorithms* (Amsterdam, 2011). He has also conducted seminars in archival art and philosophy, Scandinavian cinema, and network aesthetics as Visiting Professor at the Department of Cinema and Media Studies, University of Chicago.

Morten Søndergaard is Associate Professor and Curator of Interactive Media Art at Aalborg University Copenhagen, Denmark, and Head of Research of the *Unheard Avant-Garde* project at the LARM audio archive infrastructure (2010–2013, www.larm-archive.org). Formerly he was the Curator/Deputy Director at the Museum of Contemporary Art in Roskilde (1999–2008, www.samtidskunst.dk) and Senior Curator and General Chair of the Re-new Digital Arts Festival Copenhagen (2010–2012, www.re-new.org). In 2010, he commissioned Stelarc's Internet Ear Project for the exhibition *Biotopia—Art in the Wet Zone* (2010–2011, www.earonarm.net). He was curator of the Biotopia Revisited docu-exhibition at LEA's digital platform (2012), and visiting curator at ZKM in Karlsruhe responsible for the *Unheard Avant-Garde* in Scandinavia section at the Sound Art—Sound as Medium for Art exhibition (march 2012 - January 2013). Since April 2012, he is a contributing editor of Leonardo's Electronic Almanac (www.leoalmanac.org) and General Chair of the LEA Sound Art Curating Conference Series (www.soundartcurating.org).

Jari Suominen received a master's in New Media from Aalto University in 2012. His main focus area is software and electronic design for interactive systems. He has been teaching software development and DIY electronics at University of Helsinki, Theatre Academy Helsinki, and Aalto Yliopisto. Currently he is board member of Finnish Art Association Arte, and an active performing artist in psychedelic experimental lineups Shogun Kunitoki, Kiila and Jarse.

Mika Taanila is an artist and filmmaker based in Helsinki. His films can be categorized between the traditions of classic documentary film-making, avant-garde, and video art. His most notable films are *Futuro—A New Stance for Tomorrow* (1998), *Optical Sound* (2005), and *The Future Is Not What It Used To Be* (2002), which brought Erkki Kurenniemi's project and thinking into the international art context. His works have been shown at numerous of film and media art festivals worldwide and at major international group shows, such as Documenta, Venice Biennale, and Manifesta, as well as in recent solo exhibitions in Contemporary Art Museum, St. Louis, and Kiasma Museum of Contemporary Art in Helsinki.

Teppo Turkki is Sitra's (the Finnish Innovation Fund) Senior Specialist on East Asia. Turkki has worked at Sitra since December 2009. He is currently Resident Scholar at the National Chengchi University, Taipei, Taiwan. Turkki is a well-known author who has worked as an expert at many different universities, institutes, and state administrations, both in Finland and abroad. In recent years, he has focused on the development of East Asia and worked as a Visiting Researcher at Waseda University in Tokyo (2003–2005). In 2004, Turkki was invited to work as a Research Fellow in the Center for Global Communications think tank of the International University of Japan.

Index